LEGITIMACY AND DRONES

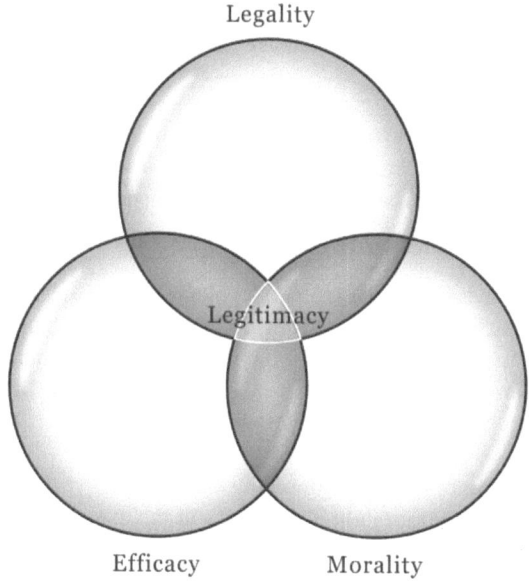

Legality

Legitimacy

Efficacy   Morality

Emerging Technologies, Ethics and International Affairs

Series editors:

Jai C. Galliott, The University of New South Wales, Australia
Avery Plaw, University of Massachusetts, USA
Katina Michael, University of Wollongong, Australia

This series examines the crucial ethical, legal and public policy questions arising from or exacerbated by the design, development and eventual adoption of new technologies across all related fields, from education and engineering to medicine and military affairs.

The books revolve around two key themes:

- Moral issues in research, engineering and design.
- Ethical, legal and political/policy issues in the use and regulation of technology.

This series encourages submission of cutting-edge research monographs and edited collections with a particular focus on forward-looking ideas concerning innovative or as yet undeveloped technologies. Whilst there is an expectation that authors will be well grounded in philosophy, law or political science, consideration will be given to future-orientated works that cross these disciplinary boundaries. The interdisciplinary nature of the series editorial team offers the best possible examination of works that address the 'ethical, legal and social' implications of emerging technologies.

Forthcoming titles:

Super Soldiers
The Ethical, Legal and Social Implications
Edited by Jai Galliott and Mianna Lotz

Social Robots
Boundaries, Potential, Challenges
Edited by Marco Nørskov

Legitimacy and Drones
Investigating the Legality, Morality and Efficacy of UCAVs

EDITED BY

STEVEN J. BARELA
University of Geneva, Switzerland

Routledge
Taylor & Francis Group

LONDON AND NEW YORK

First published 2015 by Ashgate Publishing

Published 2016 by Routledge
2 Park Square, Milton Park, Abingdon, Oxon OX14 4RN
711 Third Avenue, New York, NY 10017, USA

First issued in paperback 2017

Routledge is an imprint of the Taylor & Francis Group, an informa business

British Library Cataloguing in Publication Data
A catalogue record for this book is available from the British Library

The Library of Congress has cataloged the printed edition as follows:
Legitimacy and drones : investigating the legality, morality and efficacy of UCAVs / by Steven J. Barela.
 pages cm. – (Emerging technologies, ethics and international affairs)
 Includes bibliographical references and index.
 ISBN 978-1-4724-4687-9 (hardback) – ISBN 978-1-4724-4688-6 (ebook) – ISBN 978-1-4724-4689-3 (epub) 1. Uninhabited combat aerial vehicles (International law) I. Barela, Steven J., 1971- editor.
 KZ6687.L44 2015
 341.6'3--dc23
                                                                                2015019282

ISBN 13: 978-1-138-08693-7 (pbk)
ISBN 13: 978-1-4724-4687-9 (hbk)

To Mary

Contents

List of Figures and Tables

Figures

Tables

Acknowledgments

This volume has undoubtedly been made possible by the enthusiastic and dedicated work of each and every chapter author. There is a story behind the quest to commission, and the pleasure of working with, all of them. However, this is certainly not the place to detail the 16 different tales of encounter and gratifying labor that have brought the project to fruition. Suffice to say that all of the quality to be found in this book is directly related to their diligent participation and merits all of the gratitude I can muster.

Additionally, special thanks go to Kirstin Howgate of Ashgate Publishing. Although I had already been engaging in research on armed drones when I met her at a conference in Warsaw, it was at her suggestion and prodding that an edited volume could indeed capture the wide purview I had been envisioning as necessary on such a fast moving topic. Indeed, Kirstin stepped in at the pivotal moment to encourage me to pursue this book project for an upcoming series to be launched on emerging technologies. This brings me to the series editors. Not only am I grateful for the trust in me to put together a worthwhile publication in their series, but they also gave critical advice to expand the chapter outline to not miss key aspects and cover the subject more thoroughly. Additionally, genuine appreciation is expressed to my graphic designer (and webmaster) Troy Kendall who has created the imagery from cover to cover that captures the organization of this book and a view on the military material itself.

Finally, I wish to acknowledge the person to whom this book is dedicated: my mother. Not only has she been categorically supportive throughout my life, she has laboriously read over every word I have written during the course of my entire school career from penmanship practice of my name to an autobiography in middle school of a short life she had witnessed firsthand to monotonous typing assignments to the most basic book reports and boring essays to a doctoral dissertation. However, this time I have held back from presenting this work to her until it is in its final form—in other words, she has been freed from the duty of editing. Thus I would like to sincerely thank her for getting me to the place where I find myself today.

Steven J. Barela
Geneva, 2015

About the Contributors

Steven J. Barela is Assistant Professor at the University of Geneva in the Global Studies Institute and is a member of the Law Faculty. Along with a PhD in law, he holds an LLM and two MA degrees. His monograph exploring the theory of "legitimacy as a target" is entitled, *International Law, New Diplomacy and Counterterrorism: An Interdisciplinary Study of Legitimacy* (Routledge 2014).

Frédéric Bernard is Senior Lecturer at the University of Geneva, and has been admitted to the Geneva Bar. He has published a monograph on the relationship between the rule of law and the fight against terrorism, *L'Etat de droit face au terrorisme* (Schulthess 2010), along with several articles on the subject.

Jeffrey S. Brand is Dean Emeritus and Professor of Law at the University of San Francisco, School of Law. He also serves as Chairman of the Center for Law and Global Justice. His scholarly and professional work has focused on human rights, civil liberties, global justice, constitutional law, labor law, and various civil procedure and evidence issues.

Carlos R. Colon is a co-founder and analyst at the Center for the Study of Targeted Killing and is currently completing a book focusing on the ethics, legality and strategic dynamics of drone warfare.

James G. Devaney holds a PhD from the European University Institute in Florence and is currently a research assistant at the University of Glasgow. His doctoral research focused on fact-finding before international courts and tribunals, and in particular before the International Court of Justice.

Tom Farer is University Professor at the Josef Korbel School of International Studies of Denver University, and was the school's Dean from 1996 to 2010. His most recent monograph is *Confronting Global Terrorism and American Neo-Conservatism: The Framework of a Liberal Grand Strategy* (Oxford University Press 2008).

Gloria Gaggioli is Assistant Professor and Grant Holder of Excellence at the University of Geneva. She holds a PhD in international law, an LLM and MA degree. She has notably published a monograph on the right to life in armed conflicts, *L'influence mutuelle entre les droits de l'homme et le droit international humanitaire à la lumière du droit à la vie* (Editions A. Pedone 2013) and edited a book on the interplay between international humanitarian law and human rights law with Robert Kolb, *Research Handbook on Human Rights and Humanitarian Law* (Edward Elgar 2013). Additionally, she was formerly legal adviser at the International Committee of the Red Cross (Geneva).

Patrycja Grzebyk is Assistant Professor at the University of Warsaw and holds a doctoral degree in humanities. She also holds degrees in law (MA) and international relations (MA). She is the author of the monograph *Criminal Responsibility for the Crime of Aggression* (Routledge 2013), and the Polish version of this book was awarded the Manfred Lachs Prize.

Amos N. Guiora is Professor of Law and Co-Director of the Center for Global Justice at the S.J. Quinney College of Law at the University of Utah. He served for 20 years in the Israel Defense Forces as Lieutenant Colonel (retired) and held senior command positions where he took part in targeted killing decisions in his capacity as a JAG officer.

Delphine Hayim is a post-doctoral fellow at the Law Faculty of the University of Geneva. She holds a LLM from the Geneva Academy and a PhD in international law from the Graduate Institute (Geneva). Her current work focuses on the effects of the new generation of warfare devices on the nature of contemporary armed conflicts.

Jenna Jordan is Assistant Professor in the Sam Nunn School of International Affairs at the Georgia Institute of Technology. Her research interests include terrorism and international security, and she is currently working on a monograph focusing on the leadership decapitation of terrorist organizations.

Alexis Keller is Professor of Legal History at the University of Geneva and Visiting Professor at Sciences-Po (Paris). He is author of many books and articles on the history of legal and political thought, the history of international law and legal theory.

Robert Kolb is Professor of public international law at the University of Geneva (since 2007) and also teaches at the Geneva Academy of International Humanitarian Law and Human Rights. He has worked as a legal advisor for the ICRC and acted as counsel for the German government at the International Court of Justice.

Marek Madej is Assistant Professor at the University of Warsaw. His main areas of expertise include NATO affairs, terrorism and other asymmetric threats posed by non-state actors to state security, including technological aspects and the question of the privatization of war. He is the author of two monographs and numerous articles on these issues.

Avery Plaw is Associate Professor at the University of Massachusetts, Dartmouth. He specializes in political theory and international relations, with a particular focus on strategic studies. In 2008, he published *Targeting Terrorists: A License to Kill?* in Ashgate's Ethics and Global Politics series.

Katja Schöberl is a legal and dissemination advisor for the German Red Cross, Berlin. She previously worked for the ICRC, the Geneva Academy, and the Institute for International Law of Peace and Armed Conflict, Bochum. She is currently working on her PhD thesis on "The Geographical Scope of Application of International Humanitarian Law" and has published on the subject in English and German.

Christian J. Tams is Professor of International Law at the University of Glasgow. He is a member of the ILA Committee on the Use of Force and sits on the scientific advisory board of the European Journal of International Law. In addition to his academic work, he has advised states in proceedings before the International Court of Justice and the International Tribunal for the Law of the Sea.

List of Abbreviations

ACHR	American Convention on Human Rights
ACLU	American Civil Liberties Union
API	First Additional Protocol to the Geneva Conventions
APII	Second Additional Protocol to the Geneva Conventions
AQAM	Al-Qaeda & Associated Movements
AQAP	Al-Qaeda in the Arabian Peninsula
AUMF	Authorization for the Use of Military Force
BIJ	Bureau of Investigative Journalism
CCF	continuous combat function
CIA	Central Intelligence Agency
DoD	Department of Defense
DoJ	Department of Justice
DPH	direct participation in hostilities
DRC	Democratic Republic of Congo
ECHR	European Convention of Human Rights
ETA	Euskadi Ta Askatasuna (Basque Homeland and Freedom)
FARC	Fuerzas Armadas Revolucionarias de Colombia
FATA	Federally Administered Tribal Areas
FISA	Foreign Intelligence Surveillance Act
FISC	Foreign Intelligence Surveillance Court
FISCR	Foreign Intelligence Surveillance Court of Review
FRE	Federal Rules of Evidence
FYROM	Former Yugoslav Republic of Macedonia
GC	Geneva Conventions
GWOT	global war on terror
HR	human rights
HRL	human rights law
HuJI	Harakat-ul-Jihad al-Islami
HVT	high-value target
IAC	international armed conflicts
ICC	International Criminal Court
ICCPR	International Covenant on Civil and Political Rights
ICJ	International Court of Justice
ICRC	International Committee of the Red Cross
ICTY	International Criminal Tribunal for the former Yugoslavia
IED	improvised explosive device
IHL	international humanitarian law

ILA	International Law Association
ILC	International Law Commission (UN)
IRA	Irish Republican Army
ISAF	International Security Assistance Force
ISI	Inter-Services Intelligence (Pakistan)
ISIL	Islamic State of Iraq and the Levant
ISIS	Islamic State of Iraq and Syria
ISR	intelligence, surveillance and reconnaissance
JWT	just war theory
LLM	low-level militants
LOAC	laws of armed conflict
LTTE	Liberation Tigers of Tamil Eelam
MALE	medium altitude, long endurance (category of UAV)
MAV	miniature aerial vehicles
NAF	New America Foundation
NATO	North Atlantic Treaty Organization
NGO	non-governmental organizations
NIAC	non-international armed conflict
NSA	National Security Agency
NYC	New York City
NYU	New York University
OAS	Organization of American States
OLC	Office of Legal Counsel
OSC	Operational Security Court
PKK	Partiya Karkerên Kurdistan (Kurdistan Workers' Party)
PLO	Palestine Liberation Organization
PR	public relations
PRISM	passive radar identification system
RMA	revolution in military affairs
SFRY	Socialist Federal Republic of Yugoslavia
SOF	special operation forces
SSMS	surface-to-surface missile system
SWAT	special weapons and tactics
TSP	Terrorist Surveillance Program
UAS	unmanned aerial systems
UAV	unmanned aerial vehicle
UCAV	unmanned combat air vehicle
UN	United Nations
UNAMA	United Nations Assistance Mission in Afghanistan
UNSC	United Nations Security Council
USAF	United States Air Force
USS	United States Ship

Introduction

Legitimacy as a Target

Steven J. Barela

[W]hether it is drone strikes or training partners ... when we cannot explain our efforts clearly and publicly, we face terrorist propaganda and international suspicion, we *erode legitimacy* with our partners and our people and we reduce accountability in our own government.

President Barack Obama, May 2014[1]

Before aviators risked climbing aboard vehicles to direct their airborne flight from inside a cockpit, unmanned flying objects were our first encounter with the skies. From Chinese kites to the first hot air balloon, the human experience with flight started with the pilot's feet planted firmly on the ground.[2] Thus it is appropriate to begin this work with recognition that unmanned aerial systems (UAS) are not new, but in fact the precursor to the aircraft that we know today: those that have come to dominate our understanding flight itself. Hence, a part of the paradigm shift we are experiencing at the beginning of the twenty-first century can be considered as a reacquaintance with our past, built upon all that we have learned about effective flight control from the onboard experience, rather than as an unprecedented march into the future.

This volume will address one specific military application of the new technology: unmanned combat aerial vehicles (UCAVs, commonly referred to as drones) used for counterterrorism operations carried out across international borders. The fact that these aerial vehicles are armed with precision guided missiles and infrared cameras producing full motion video is a significant change; this enables the crew to target an individual from thousands of miles away. As a result there has been a reduction of practical constraints on exercising lethal force across borders and

---

1 President Barack Obama, "Commencement address at West Point" (May 28, 2014) (my emphasis) available at <http://www.washingtonpost.com/politics/full-text-of-president-obamascommencement-address-at-west-point/2014/05/28/cfbcdcaa-e670–11e3-afc6-a1dd9407abcf_story.html> accessed Sept 2014.

2 Charles Jarnot, "History" in Richard K Barnhart and Stephen B Hottman, et al (eds), *Introduction to Unmanned Aerial Systems* (CRC Press 2012) at 1–2. This author and editor would like to extend special thanks to Stephen Hottman, director of the Physical Science Laboratory at New Mexico State University, for sitting down and discussing the development, obstacles, and future of UAS, along with offering a tour of their UAV hangar outside of Las Cruces with an experienced pilot. This institute shares airspace with Holloman Air Force Base where military personnel are trained to fly UCAVs.

across the globe in a time when violent individuals and groups are seen by many as the most substantial threat posed to a nation. While drones have been rightly identified as just another weapons platform,[3] they have also directly altered the calculations for exercising force within the territory of other states.

Throughout this work we will see that it is this significantly increased facility for exercising deadly force in far-flung regions across the globe that is putting great stress on many of our established norms. Our existing legal and moral frameworks, not to mention methods for judging efficacy, are often insufficient, or not quite applicable, to the novel use of killer drones. Consequently, our book aims to explore this tension and consider a policy proposal meant to ease this strain.

It is also worth noting that as this technology advances there are countless new applications of the unarmed version: for example, the United Nations builds up its unmanned aircraft capacities for peacekeeping surveillance;[4] UAS are deployed to fight illegal fishing off the coast of Belize and California;[5] flying surveillance vehicles now patrol nearly half of the US–Mexico border;[6] there are inexpensive autonomous 3D robotic quadcopters that follow and film their owners engaging in action sports;[7] and private users fly unmanned aircraft into the middle of firework displays, illegally and dangerously, to capture breathtaking footage.[8] Whether one looks at armed or unarmed UAS, one overarching difficulty can be found in the rapid growth. As is always the case with new technologies, their uses are developing faster than rules, regulation, and enforcement can be instituted. Both

---

3   See eg Philip Alston, *Study on Targeted Killings, Addendum to Report of the UN Special Rapporteur on Extrajudicial, Summary or Arbitrary Executions* (UN Doc A/HRC/14/24/Add.6, May 28, 2010): "a missile fired from a drone is no different from any other commonly used weapon, including a gun fired by a soldier or a helicopter or gunship that fires missiles" at para 79.

4   Samuel Oakford, "Drones, Drones, Everywhere: UN ramping up peacekeeper surveillance flights" *Aljazzera America* (Aug 27, 2014) available at <http://america.aljazeera.com/articles/2014/8/27/united-nations-drones.html> accessed Nov 2014.

5   Brian Clark Howard, "Can Drones Fight Illegal 'Pirate' Fishing?: Conservationists test unmanned aerial vehicles in Belize and California" *National Geographic* (July 18, 2014) available at <http://news.nationalgeographic.com/news/2014/07/140718-drones-illegal-fishing-pirate-belize-ocean/> accessed Nov 2014.

6   Associated Press in Sierra Vista, Arizona, "Half of US–Mexico Border Now Patrolled Only by Drone" *The Guardian* (Nov 13, 2014) available at <http://www.theguardian.com/world/2014/nov/13/half-us-mexico-border-patrolled-drone> accessed Nov 2014.

7   Adam Clark Estes, "New Autonomous 3D Robotics Drone Follows You Wherever You Go" *Gizmodo* (Sept 8, 2014) available at <http://gizmodo.com/new-autonomous-3d-robotics-drone-follows-you-wherever-y-1631694870> accessed Nov 2014.

8   Gregory S McNeal, "Flying A Drone Through Fireworks May Land You In Prison" *Forbes* (July 4, 2014) available at <http://www.forbes.com/sites/gregorymcneal/2014/07/04/video-shows-drone-flying-through-fireworks/> accessed Nov 2014.

politics and general understanding of the shifting considerations impede adequate management of the technology.[9]

Nevertheless, parameters for better understanding do indeed exist and in the epigraph to this Introduction we find an idea that has guided the studies in this book. The President of the United States, the very same official who drastically expanded the government's use of UCAVs for cross-border counterterrorism, exhibits in a 2014 speech that he recognizes that such operations must be broadly believed to be *legitimate*, and that this legitimacy is subject to erosion. Considering Max Weber's widely accepted definition of the modern state—"a human community that (successfully) claims the *monopoly of the legitimate use of physical force* within a given territory"[10]—one can quickly grasp that lethal force repeatedly exercised in the territory of another state raises many questions of legitimacy, both at home and abroad.

One Sri Lankan senior official, Lakshman Kadirgamar, worked tirelessly to put this concept of legitimacy at the center of his nation's struggle against an armed group employing terrorist tactics. As its foreign minister from 1994 to 2001 and from 2004 to 2005, Kadirgamar helped lead the effort against the separatist Tamil Tigers—the movement that ultimately took his life. Through an effective countering of their propaganda campaigns, he skillfully achieved their diplomatic isolation, which laid the groundwork for their eventual, yet controversial, military defeat in 2009. Although this is certainly not the place to delve into an analysis of that conflict, nor Kadirgamar's philosophy for dealing with it, a recent publication doing just that has arrived at a conclusion that is directly pertinent to our study:

> [i]f there is one central message from Lakshman's speeches and actions, it is that the problem of terrorism has to be tackled within a legal framework; and the struggle between terrorists and their adversaries is to a significant extent a struggle for *legitimacy*.[11]

---

9   Dee Ann Divis, "RTCA Standards Committee Grapples with UAS Collision Avoidance Rules" *Inside Global Satellite Navigation Systems* (Aug 29, 2014) available at <http://www.insidegnss.com/node/4166> accessed Nov 2014; Conor Dougherty, "Drone Developers Consider Obstacles That Cannot Be Flown Around" *New York Times* (Sept 1, 2014) available at <http://www.nytimes.com/2014/09/01/technology/as-drone-technology-advances-practical-obstacles-remain.html?_r=3> accessed Nov 2014; Michael Cooney, "Report: Significant technical trials remain before drones can safely access national airspace" *Network World* (Dec 10, 2014) available at <http://www.networkworld.com/article/2858098/security0/report-significant-technical-trials-remain-before-drones-can-safely-access-national-airspace.html> accessed Dec 2014; and CBS News, "Pilots Report Increasing Close Calls with Drones" (Nov 27, 2014) available at <http://www.cbsnews.com/news/federal-aviation-administration-surge-in-reports-of-drones-interfering-with-aircraft/> accessed Nov 2014.

10   Max Weber, "Politics as a Vocation" in H H Gerth, and C Wright Mills, (eds and trans) *From Max Weber: Essays on Sociology* (OUP 1946) at 78 (original emphasis).

11   Sir Adam Roberts, *Democracy, Sovereignty and Terror: Lakshman Kadirgamar on the Foundations of International Order* (IB Tauris 2012) at 25 (my emphasis).

While more on the relevance of the legal framework of international law will follow,[12] here it is sufficient to underscore the fact that Kadirgamar recognized legitimacy as a principal element in such a conflict.

For additional support of that recognition one can turn to the 2006 Army Field Manual 3–24, *Counterinsurgency*, which eventually came to be employed as the guiding military doctrine for the US engagement in Iraq and Afghanistan. Although neither of these conflicts is by any means a completed operation, this document drafted under the supervision of General David Petraeus (a work that helped catapult him to the highest levels of military and civilian authority),[13] puts the concept of legitimacy at the center of the armed conflict. The Army Field Manual does not address the validity of the invasions themselves (a question for the civilian leadership), but rather provides a strategic analysis of insurgency and counterinsurgency in which legitimacy is illuminated in the very definition of the struggle:

> [A]n insurgency is an organized, protracted politico-military struggle designed to weaken the control and *legitimacy* of an established government, occupying power, or other political authority while increasing insurgent control.[14]

In fact, even a cursory reading of the Field Manual reveals a consistent return to the legitimacy theme: "[p]olitical power is the central issue in insurgencies and counterinsurgencies; each side aims to get the people to accept its governance or authority as *legitimate*";[15] "the long-term objective for all sides remains acceptance of the *legitimacy* of one side's claim to political power by the people of the state or region";[16] "[v]ictory is achieved when the populace consents to the government's *legitimacy* and stops actively and passively supporting the insurgency";[17] and in the unmistakable sub-title "*Legitimacy* Is the Main Objective."[18] What we are

---

12   Kadirgamar believed that this legal framework particularly included international law: "Lakshman, when foreign minister, always insisted that the relevant rules of the laws of war and human rights must be observed for reasons that were both moral and practical," ibid at 33; see specifically Kadirgamar's speech "Human Rights and Armed Conflict," Kotelawala Defence Academy (March 19, 1996) at 109–118.

13   See Paula Broadwell, *All In: The Education of General David Petraeus* (The Penguin Press 2012) 301pp.

14   Army Field Manual (FM 3–24), *Counterinsurgency*, US Department of Defense (December 15, 2006) at 1–1 (my emphasis).

15   Ibid (my emphasis).

16   Ibid at 1–2 (my emphasis).

17   Ibid at 1–3 (my emphasis). While these citations suggest that both sides aim to be considered legitimate, it is arguable that terrorist groups are primarily focused on destroying the legitimacy of a government and have not pivoted to building their own legitimacy claim, see Steven J Barela, *International Law, New Diplomacy and Counterterrorism: An Interdisciplinary Study of Legitimacy* (Routledge 2014) at 56–59.

18   Ibid at 1–21 (my emphasis).

seeing is that practitioners and scholars alike are increasingly acknowledging this concept as pivotal.[19]

However, what parameters actually construct legitimacy is by no means an easy or settled question. What actually creates an *uncoerced pull toward compliance* with an authority or a policy?[20] Indeed, this query is one that has confounded and inspired political and legal philosophers for centuries. Machiavelli, Grotius, Hobbes, Rousseau, and Montesquieu are some of the most well known, to name but a few. Yet this book does not set out to settle such a complex and essential matter. Rather, in the simplest sense, the chapters of this volume have been organized around some of the most accepted and straightforward principles of what constitute legitimacy. To be precise, to explore the pertinent queries regarding the legitimacy of drones the accomplished scholars of this volume analyze the principal points at issue by peering through the conceptual lenses of *legality*, *morality*, and *efficacy*. Put simply, this book begins with the uncomplicated premise that, at a minimum, sound policy should be legal, moral, and effective.[21]

---

19 See eg Deborah Cook, "Legitimacy and Political Violence: A Habermasian Perspective," (2003) 30,3 Social Justice 108; Audrey Kurth Cronin, *How Terrorism Ends: Understanding the Decline and Demise of Terrorist Campaigns* (Princeton University Press 2009); Martha Crenshaw (ed) *Terrorism, Legitimacy, and Power: The Consequences of Political Violence* (Wesleyan University Press 1983); and Michael Ignatieff, *The Lesser Evil: Political Ethics in an Age of Terror* (Princeton University Press 2004); for an excellent work on how international law fits into this question see Jutta Brunnée and Stephen Toope, *Legitimacy and Legality in International Law: an Interactional Account* (Cambridge University Press 2010); for additional works that speak to the issue of legitimacy as a part of armed conflicts involving terrorist acts, even if the issue is not explicit and placed as predominant, see Tom Farer, *Confronting Global Terrorism and American Neo-Conservatism: The Framework of a Liberal Grand Strategy* (OUP 2008); Louise Richardson, *What Terrorists Want* (Random House 2006); Brian Jenkins, "International Terrorism: A New Mode of Conflict," in Carlton D and Schaerf C (eds), *International Terrorism and World Security* (Croom Helm 1975); Walter Laqueur, *Terrorism* (Weidenfeld and Nicolson 1977); Maurice Tugwell, "Terrorism and Propaganda: Problem and Response" and Ronald Crelinsten, "Power and Meaning: Terrorism as a Struggle Over Access to the Communication Structure" in P Wilkinson and A Stewart (eds) *Contemporary Research on Terrorism* (Aberdeen University Press 1987) at 409–418 and 419–450 respectively; see generally Philip Bobbitt, *Terror and Consent* (Alfred A Knopf 2008); David Cole and Jules Lobel, *Less Safe, Less Free* (The New Press 2007); Robert Pape, *Dying to Win* (Random House 2005); Colonel Thomas X Hammes, *The Sling and the Stone: on War in the 21st Century* (Zenith Press 2006).

20 This apt description of legitimacy has been developed from the work of Thomas Franck, *The Power of Legitimacy among Nations* (OUP 1990) at 10–34.

21 There are indeed other scholars who have put forward such an organization for the analysis of this tactic. For example see Amos Guiora, *Legitimate Target: A Criteria-Based Approach to Targeted Killing* (OUP 2013) at xi: "[t]he state's decision to kill a human being, in the context of operational counterterrorism, must be predicated on an objective determination that the 'target' is, indeed, a legitimate target. Otherwise, state action is illegal, immoral, and ultimately ineffective"; see also Gabriella Blum and Phillip Heymann

No doubt the Obama administration has been keenly aware of these parameters. In a landmark speech in 2013—one that was meant to be momentous for pulling back a portion of the opaque veil of secrecy on the drone program—President Obama finally explained to the citizenry that "the United States has taken lethal, targeted action against Al-Qaeda and its associated forces, including with remotely piloted aircraft commonly referred to as drones."[22] In this formal acknowledgment the president went on to strike the precise three chords that will organize this book. He first enumerated:

> To begin with, our actions are *effective*. ... Dozens of highly skilled Al-Qaeda commanders, trainers, bomb makers, and operatives have been taken off the battlefield. Plots have been disrupted that would have targeted international aviation, U.S. transit systems, European cities and our troops in Afghanistan. Simply put, these strikes have saved lives.[23]

President Obama then continued making the case with his next key assertion, referencing both domestic and international law:

> Moreover, America's actions are *legal*. We were attacked on 9/11. Within a week, Congress overwhelmingly authorized the use of force. Under domestic law, and international law, the United States is at war with Al-Qaeda, the Taliban, and their associated forces.[24]

And finally, to sound the third chord of legitimacy, the president invoked the Western world's timeworn and long established moral framework for judging armed conflict,

> So this is a just war—a war waged proportionally, in last resort, and in self-defense. ... America's legitimate claim of self-defense cannot be the end of the discussion. To say a military tactic is legal, or even effective, is not to say it is wise or *moral* in every instance.[25]

"Law and Policy of Targeted Killing" (2010) 1 Harvard Law School National Security Journal at 149: "[w]e do so by assessing the American and Israeli experience in employing targeted killings and its legal, moral, and strategic implications."

22   President Barack Obama, "The Future of our Fight against Terrorism," National Defense University (May 23, 2013) available at <http://www.whitehouse.gov/the-press-office/2013/05/23/remarks-president-barack-obama> accessed Sept 2014.

23   Ibid (my emphasis). President Obama also suggested, "[d]on't take my word for it. In the intelligence gathered at bin Laden's compound, we found that he wrote, 'we could lose the reserves to the enemy's air strikes. We cannot fight air strikes with explosives.' Other communications from al-Qaeda operatives confirm this as well."

24   Ibid (my emphasis).

25   Ibid (my emphasis). Due to the fact that moral judgment today can take a broad host of forms, the president also intended to shore up validity by asserting "over the

To this editor and author, these formulations and arguments do not come as a surprise. My previous work at the University of Geneva was an investigation into how legitimacy can be conceptualized as a target in asymmetrical armed conflict, and posited a theory on the components of such legitimacy. Thus I have my own motivations for organizing the volume in this manner, but the chapter authors have not been asked to explicitly endorse this theory. Nonetheless, even if it is beyond the scope to offer here a full elaboration of this theory, which can be found elsewhere,[26] this Introduction is extended to provide an overview to our readers so that they can judge for themselves the validity of this theory as an organizational device, and the tools of analysis it offers.

Legitimacy as a Target

For a society to function collectively, the *power to command* must be established by the population's *will to obey*—this would include a meaningful portion of high-ranking officials to mid-level bureaucrats to low-level administrators to even the common citizenry. Most plainly, legitimacy must be bestowed from below to those on high. Guglielmo Ferrero, an Italian historian who investigated the concept of legitimacy within his work written in Geneva during the Second World War, cogently explains that:

> government is ... legitimate if power is conferred and exercised according to principles and rules accepted without discussion by those who must obey. There are still peoples who, without knowing the abstract theory of legitimacy, recognize [it] in the respect for these rules and principles [as] the source of the right to govern.[27]

The contention here is that this conferred legitimacy in the context of asymmetrical conflict—state vs non-state actors—is of particular significance.

Non-state actors clearly wish to avoid and upend the conventional battlefield where they are at an enormous disadvantage, yet these individuals and groups also

---

last four years, my Administration has worked vigorously to establish a framework that governs our use of force against terrorists—insisting upon clear guidelines, oversight and accountability that is now codified in Presidential Policy Guidance that I signed yesterday." Although President Obama forwarded this as a form of moral argument, there are two better explanations: 1) these parameters are a legal, or institutional, argumentation; 2) it is rather at a point of overlap between legality and morality. Nonetheless, it is not within the scope here to further pursue such an analysis.

26   Steven J Barela, *International Law, New Diplomacy and Counterterrorism: An Interdisciplinary Study of Legitimacy* (Routledge 2014); see also excerpts on my website <LegitimacyasaTarget.com>.

27   Guglielmo Ferrero, *The Principles of Power* (GP Putnam's Sons 1942) at 135.

wish to undermine or even topple the ruling authority. The theory of *legitimacy as a target* posits that terrorist groups attempt to achieve their strategic aims by avoiding the traditional battleground and instead try to drive a wedge between the power to command and the will to obey within the enemy state. By provoking an overreaching reaction from a government to deal with the terrorist threat, or triggering a response that is considered to be outside of the confines of the government's authority, the goal of the terrorists is for the citizenry to deem the actions an illegitimate exercise of physical force. If the determination of illegitimacy becomes widespread, meaning that officials, bureaucrats and citizens no longer orient their actions in accordance with the authority, then a society is destabilized because it cannot function or move as a unit. This conceptual wedge certainly explains why terrorism has been described as "destabilizing of legitimately constituted governments" for almost two decades by United Nations bodies.[28]

Although it is easiest to speak about legitimacy when it no longer exists, a government that is experiencing a diminishing pull toward compliance with its authority—unauthorized leaks of damaging information, high-profile resignations, mass strikes, demonstrations, and acts of civil disobedience—encounters difficulties even before a drastic break in the social order occurs.[29] Hence, legitimacy is not an all-or-nothing proposition because it can be "eroded, contested or incomplete" and, therefore, "judgements about it are usually judgements of degree."[30] The statement that today's purveyors of terror wish to disorient and immobilize their enemy to help reach their strategic goals is hardly controversial. Nonetheless, the specific intention here is to illuminate the fact that non-state actors employing terrorism attempt to reach their overarching goals through a "strategy of provocation" meant to target legitimacy.[31]

Ferrero articulates this fundamental organization of society in a way that helps provide useful imagery for our purposes. By charting the place from which command and obedience emanate we can better comprehend where the gaze of the counterterrorism analyst should be fixed. Ferrero posited,

> [i]f, in democracies as in monarchies, the authority comes from above, in monarchies as in democracies legitimacy comes from below, since only the consent of those who must obey can create it. In every regime, therefore, the

---

28   General Assembly resolutions: UN Doc A/RES/48/122 (1993); UN Doc A/RES/49/185 (1994); UN Doc A/RES/50/186 (1995); UN Doc A/RES/52/133 (1997); UN Doc A/RES/56/160 (2001); UN Doc A/RES/58/174 (2003); World Conference on Human Rights, Vienna A/CONF.157/24 (1993) Part I, chap III; Commission on Human Rights resolutions 2001/37 as well as 2004/44; and UN High Commission of Human Rights resolutions 2001/37 and 2003/37.

29   David Beetham, *The Legitimation of Power* (Palgrave Macmillian 1991) at 205–221.

30   Ibid at 20.

31   This phrase comes from the classic work on terrorism by Laqueur (n 19) at 81.

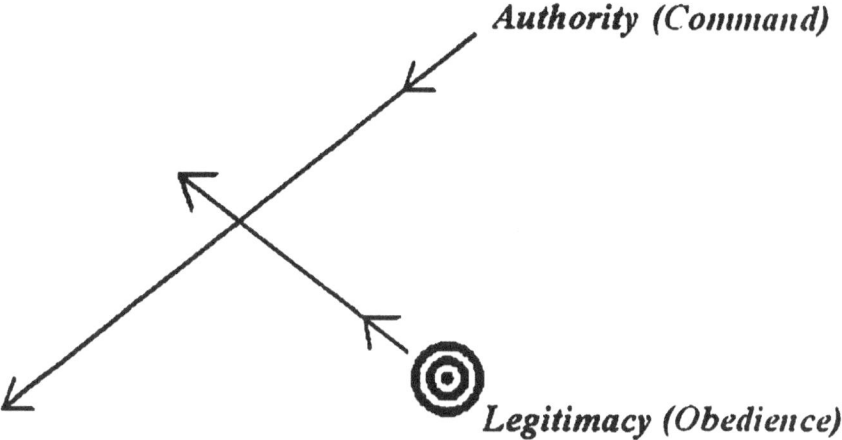

Authority (Command)

Legitimacy (Obedience)

Figure I.1   Legitimacy as a target

> plenitude of the state is realized at the intersection of two lines—one descending, which is authority, and the other ascending which is legitimacy.[32]

The intention is to highlight the idea of tacit and active endorsement, or legitimacy, as the bull's-eye on the target of attacks conducted by actors employing terrorism. It is possible to sketch how this understanding of a terrorist attack would look when applied to the imagery of power provided by Ferrero. We place a target on the bottom-up element of legitimacy since this is where terrorist acts are proposed to be directed. We can also illustrate how a diminishing legitimacy would appear—threatening a break at the essential intersection—by showing when terrorist violence is successful in goading a reaction that oversteps the bounds of authority considered valid by the population. The result would appear as shown in Figure I.1.

Of course, accepting the thesis of *legitimacy as a target* has analytical consequences. By identifying a target at which terrorists aim, this most importantly provides policy-makers and analysts with the opportunity to think defensively. Since the deliberate targeting of civilians is generally accepted as the defining characteristic of terrorism,[33] defending every civilian or civilian object would be extremely difficult, if not impossible. Nonetheless, if we understand that a response needs to be deemed legitimate to best defend this point of attack, it is possible to integrate strategic defensive thinking into the equation of counterterrorism.

One keen terrorism scholar has pointed out that while "[s]peed and force are both critical elements in a successful military campaign; it is far less clear that

---

32   Ferrero (n 27) at 171.

33   See eg Richardson (n 19) at 6. Although who exactly should be included in the term "civilian" is contentious as the United States, for example, views the killing of police and soldiers of a government it deems democratic, or at least friendly, as a terrorist act: see Farer (n 19) at 15–29.

they are necessary ingredients of a successful counterterrorism policy."[34] The reason why this is so critical is because many who confront terrorism begin with the belief that the primary objective is to kill, interrogate, and detain (often without trial) potential enemies rapidly and with overwhelming force.[35] To wit, this overly offensive approach aims to eliminate or remove all potential enemy combatants without delay or substantial caution. Such an approach confuses tactics with strategy. Overemphasizing this aggressive and tactical aspect of the conflict can lead to an undermining of the defensive strategy of protecting the target of legitimacy—safeguarding that these methods are only employed legally, morally and effectively.

There is little doubt that there will be times when the aggressive methods are necessary, just as in any ordered society. However, losing sight of defending the legitimacy of the government in such a struggle can come at great cost to the society as a whole, and plays into the hands of the purveyors of terror. Most notably, this offensive/defensive framing underscores the fact that defending legitimacy represents an essential strategic interest.

If legitimacy can indeed be examined,[36] this again raises the question of determining the components of legitimacy so that we can analyze a counterterrorism response—in this case, the use of UCAVs across borders. To do so, the model of *legitimacy as a target* has been constructed using an interdisciplinary approach

---

34   Richardson (n 19) at 101.

35   It worthwhile to note that these aggressive tactics deal directly with the most basic "personal security rights" that are found as fundamental in both the laws of war and human rights law; they have been enumerated as a "protection from summary execution, torture and other cruel and inhuman and degrading treatment, and conviction and/or punishment without due process of law" by Farer (n 19) at 14.

36   There is an important clarification that should be mentioned concerning the analysis of legitimacy during times of political struggle. Although Max Weber's work on legitimacy has achieved a classical status in the literature, it has also been described as incomplete. Most importantly, Weber offers a view of legitimacy that is not amenable to proof and cannot be tested because he sees it as based on unspecific subjective *beliefs* without identifiable content. In contrast to Weber, other scholars adopt the position that the legitimacy of a regime can be accepted or rejected on rational grounds, which actually rests on verifiable content (see eg Beetham (n 29) at 13: "the evidence is available in the public sphere, not in the private recesses of people's minds"). Most significantly, Jürgen Habermas keenly zeroed in on this shortcoming and suggested, "legitimacy is assumed to have an immanent relation to truth, the grounds on which it is explicitly based contain a rational validity claim that can be tested and criticized independently of the psychological effect of these grounds" (*Legitimation Crisis*, T McCarthy (trans) (Beacon Press 1975 [German text 1973]) at 97). Considering this significant distinction, it is important to spell out that the theory of *legitimacy as a target* necessarily rejects the Weberian view of a belief-based legitimacy, and rather embraces the Habermasian notion that it is truth-dependent and testable.

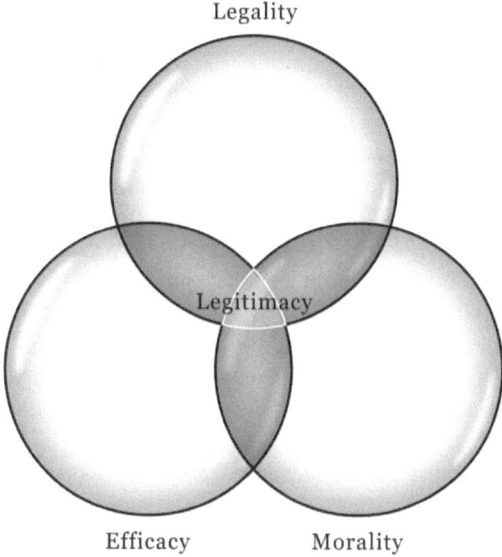

Figure I.2   The lenses of legitimacy

meant to integrate the most valuable features of differing viewpoints and methods.[37] What this means specifically is that the three proposed realms of inquiry have a shared space of overlap to capture the integrative nature of the pull toward compliance; no form of validity is presented to dominate the model at the expense of the others. Additionally, the model allows us to conceptualize and visualize how these individually valuable components relate to each other. There is no doubt that determining legality, morality, and efficacy are valid analytical pursuits in and of themselves. Yet this model explains them as interrelated, rather than independent of one another.

Consequently, in employing the Habermasian view that the content of legitimacy can be rationally tested,[38] I propose that there is a useful and proper way to conceptualize legitimacy in a manner that affords the application of a series of overlapping lenses (i.e. tools of analysis) for viewing and discussing these central issues. The model appears as set out in Figure I.2.

Accordingly, this book has three sections: formal validity (legality), axiological validity (morality), and empirical validity (efficacy). The logic and reasoning for using *legality*, *morality*, and *efficacy* as the tools of analysis for assessing legitimacy will be expanded as the chapters are presented below.

---

37   The starting point for this model is the sound interdisciplinary work on legal validity by François Ost and Michel van de Kerchove, *De la Pyramide au Réseau? Pour une Théorie Dialectique du Droit.* (Publications des Facultés Universitaires Saint-Louis 2002) at 307–383.

38   See (n 36).

Legality—Formal Validity

If there is a point of agreement among the scholars of legitimacy, it is on legality; that is, legality is considered to be one of legitimacy's central components.[39] Thus, the predominant scholars of legitimacy point to legality as a starting point. To do so in the context of counterterrorism it is important to take note of one distinctive feature of 9/11 that must shape the way we think about and analyze the response to them: these were cross-border attacks. Non-state actors stepped across international boundaries to conduct public and politically motivated violence against civilians. As a result, when citizens of the victim country, and even more critically the official members of the government itself, attempt to understand the rights and duties related to its response inevitably reaching outside of its domestic borders, *international law* is the natural, not to mention legally fitting, framework.

Of note, there has been a dramatic increase in global cooperation throughout the twentieth century—between international organizations and their state missions of diplomats, foreign officers, international civil servants, intelligence officers, military personnel, police investigators, judges, legislators, and financial regulators—all governed by international law. These rules have mushroomed following the conclusion of each world war and this dramatic growth is arguably the most significant change the international structure has experienced since the inception of the state-based system ushered in with the Peace of Westphalia in 1648.

The contention here is that this surge in international regulation has had a bearing on the shape and content of the domestic political order. To suggest that this immense growth has had absolutely no effect on the domestic legal order or the mental framework for how citizens judge the obligations of their own government in an information age is to deny the shifting terrain. It is not a question of whether any adjustment has occurred, but rather the pertinent question is: How much change has actually come about?[40]

---

39   Max Weber explained in *The Theory of Social and Economic Organization*, Henderson AM and Parsons T (trans) (The Free Press 1947): "[t]oday the most usual basis of legitimacy is the belief in *legality*, the readiness to conform with rules which are formally correct and have been imposed by accepted procedure" at 131; David Beetham (n 29) expounded: "[p]ower can be said to be legitimate in the first instance if it is acquired and exercised in accordance with *established rules*" at 16; And in the context of political violence, Cook (n 19) suggested: "legitimacy is a controversial and complex notion, involving not only *legal* and political matters," while Crenshaw (n 19) explained: "[a]ctions against the terrorists must be scrupulously legal" at 33.

40   For international lawyers who comprehend this legal framework as being a valuable source of understanding for common citizens see Antonio Cassese, *Terrorism, Politics and Law* (Polity Press 1989) at viii: "[a]n international lawyer should no longer write for rulers alone (who may or may not heed his words); he ought to write mainly for ordinary citizens: he should offer them parameters by which to judge international affairs, and analytical mechanisms for examining the intricacies of the world community";

As it is posited that growing portions of civil society turn to international law to understand and judge the legitimacy of their government's counterterrorism policies reaching across international borders, this discipline undergirds our book with nearly all of its contributors credentialed or deeply grounded within it. To lay this foundation of formal validity, one must begin with the strength behind the logic of legal positivism and lay out a jurisprudential view of the applicable law. In consequence this section analyzes the applicable legal treaties, customary law as evidenced by state practice, general principles of law, and *opinio iuris* in order to peer through the lens of legality.

There are four primary questions that are raised in international law when it comes to analyzing the lethal use of UCAVs for counterterrorism operations across borders, and each is explored in its own chapter. Those questions are: Is this use of force in the territory of another state justified? Who is a legal target in an armed conflict between a state and groups of individuals? Are there geographical limits to the application of humanitarian law—that is the laws of war? What is the proper application of human rights law when such deadly force is exercised?

Christian J. Tams and James G. Devaney launch the investigations of this book with their chapter of preliminary relevance, "*Jus ad Bellum*: Crossing Borders to Wage War against Individuals." Although the ban on the use of force across borders has been described as the "cornerstone" of the contemporary international system, the precise scope of the legal rules governing this prohibition is by no means beyond debate. The UN Charter lays down the legal framework, but the contours of the current interpretation of this shifting regime must be explored to understand where the law stands today when it comes to violent individuals and/or groups. As the use of armed drones has become a publicly recognized reality, this chapter argues that over the last 25 years the legal rules have been readjusted through interpretation to permit forcible counterterrorism under more lenient conditions. Nevertheless, the risks posed by current practice might indeed compel further readjustment to a legal regime undergoing stress as this act of force becomes more common.

The next chapter is "Who Can Be Killed?: Legal Targets in Non-International Armed Conflicts" by Patrycja Grzebyk. While combat drones are weapons and not unique from a legal point of view, there continue to be particular difficulties in identifying legal targets in these asymmetrical armed conflicts and this chapter addresses those challenges. It delves into the problems related to the principle of distinction between those engaging in hostilities (members of armed groups with a continuous combat function) and civilians who do not directly participate. Additionally, she discusses another related point of international humanitarian law

---

Philippe Sands, *Lawless World* (Viking 2005) at 7: "[i]nternational rules are now frequently seen as providing an independent benchmark against which to assess the justification of behavior—and in particular the behavior of states—which is politically or morally contentious"; see also generally Brunnée and Toope, *Legitimacy and Legality in International Law* (n 19).

concerning the elimination vs the neutralization of legitimate targets—namely the obligation to capture if possible.

Chapter 3 by Katja Schöberl is titled "Boundaries of the Battlefield: The Geographical Scope of the Laws of War." In it she discusses the fact that the use of armed drones in operations labeled as counterterrorism has raised important questions about where one can properly apply the rules of international humanitarian law. This has become of particular consequence as UCAVs have made it easier for armed conflict to be conducted well beyond the traditional combat zone, and perhaps without any territorial limit. This contribution gives an overview of the debate and subsequently analyzes the geographical scope of international and non-international armed conflicts both within and outside the territory of belligerent states. Ultimately, she looks into the possible extraterritorial application of the law of non-international armed conflicts.

To conclude this section, Gloria Gaggioli provides the contribution "Lethal Force with Drones: The Human Rights Question." She addresses the fact that armed drones are being used—and are likely to be increasingly used—in situations in which the rules on the conduct of hostilities provided for in international humanitarian law do not apply. This can occur when force is used outside of armed conflict situations, but also, in certain cases, in the context of an ongoing armed conflict. This chapter discusses what the precise situation would be when human rights law is applicable, and then explores whether lethal force with drones can actually be used within its requirements of necessity, proportionality, and precautionary measures. Overall, real doubts are raised as to whether UCAVs can indeed operate within this framework, especially considering the fact that the premeditated aim is to kill suspected offenders with no gradual use of force envisaged.

Morality—Axiological Validity

Moral debate over war has been with us for centuries and its historical filiations in the Western world are best traced through the familiar just war doctrine. Its protracted existence and constantly developing recurrence in Western thought speaks to the depth to which the moral dimensions of war run within the populations from which it sprung. Even though there has been a history of formulating a reasoning for casting aside any ethical and legal considerations in war-making,[41] classical, modern, and contemporary forms of moral (at times mixing clearly with legal) analysis are a testament to the durability and relevance

---

41   See eg the well-known *Pro Milone* speech by Cicero found in *The speeches / Cicero. Pro Milone; In Pisonem; Pro Scauro; Pro Fonteio; Pro Rabirio postumo; Pro Marcello; Pro Ligario; Pro rege Deiotaro*, Watts NH (trans) (Harvard University Press 1931); and for a contemporary example see John Yoo, "Using Force" (Summer 2004) 71, University of Chicago Law Review 729: "[r]ather than pursuing these doctrinal or moral

of such an approach.[42] However, it should be remembered that limits on warfare have come together over scores of years from an amalgamation of sources. These include Christian theologians as a moral doctrine came into being inside the church, international lawyers with principles that guided and set in motion the first texts of this discipline, and military professionals using considerations of fair play rooted in chivalry.[43] Therefore as the just war theory will serve as the overarching vehicle that will drive the axiological analysis in this section, it should be understood as a relatively imprecise term that rightfully reflects the intersection of the vital concepts of law, morals, and prudence. As such, it is a fitting tradition to be used in this work treating the equally interdisciplinary subject of legitimacy.[44]

Of course, there is a great amount of debate over whether the type of operations carried out by UCAVs for counterterrorism should always and only be categorized as war. Once armed conflict is used as the framework for analyzing the use of force, the constraints become more lenient, and there is a wider margin for the use of violence. Accordingly, the debate over this difficulty of determining the proper applicable norms will be discussed in this section.

In addition, morality is undoubtedly an extremely complex concept that creates many difficulties for analysts who wish to explore its relevance. However, it is this editor's opinion that it is in fact possible to clarify what is intended when the word morality is invoked in this book by drawing a distinction between a "morality of duty" and a "morality of aspiration."[45] These two categories represent opposite ends of the same scale, and where one leaves off for the other to begin has long dominated moral argument. A *morality of duty* starts at the bottom and "lays down the basic rules without which an ordered society is impossible," whereas a *morality of aspiration* is at the top end of this scale and is about human excellence or "the fullest realization of human powers."[46]

It is essential to understand that the difficulty in precisely specifying all the forms of a *morality of aspiration* as we mount this scale does not mean that the more basic requirements of a baseline *morality of duty* are unknowable or incomprehensible. This wide scope that morality can cover, from the most obvious

---

approaches, this Article addresses the rules governing the use of force from an instrumental perspective" at 731.

42   See eg Richard Tuck, *The Rights of War and Peace* (OUP 1999); Michael Walzer, *Just and Unjust Wars* (3rd edn, Basic Books 1977); James Turner Johnson, *Ideology, Reason and the Limitation of War: Religious and Secular Concepts 1200–1740* (Princeton University Press 1975) and *Just War Tradition and the Restraint of War* (Princeton University Press 1981); Peter Haggenmacher, *Grotius et la doctrine de la guerre juste* (Presses Universitaire de France 1983).

43   Ibid, Johnson, *Just War Tradition ...* at xxi.

44   This paragraph describing the legal, moral and prudential origins of just war comes from Barela (n 26) at 149.

45   Lon Fuller, *The Morality of Law* (Yale University Press 1964) at 3–32.

46   Ibid at 5.

demands of social living to the highest reaches of human aspiration, often leads to confusion when we speak about morality. Yet, since this book deals with the use of deadly force, the severest act that can be carried out against an individual, we will be solely concerned with the more discernible bottom end of this scale.

In addition, to speak about legitimacy in a thorough manner one needs to not only address the questions of formal validity, but also where there is conceivably overlap with the more axiological questions of morality. In other words, attention in this section will focus on what the consummate positivist H.L.A. Hart has described as the "minimum content of natural law": the unstated assumption that all human activity is properly understood as perpetuating survival.[47] Indeed, this theory posits that there is an overlap of legality and morality in the prohibition of free, unconstrained violence, and the destruction of human life (by terrorism and drones alike) poses problems that cross disciplinary boundaries. It is with this approach that this section will gaze through the lens of morality.[48]

To initiate this section is Alexis Keller's chapter "Old Ideas in New Skins: The Sixteenth Century Debate over Artillery." Here he addresses the fact that armed drones are certainly not the first time that humanity has been confronted with a technology that upsets the conventional norms of war. At the beginning of the sixteenth century, Renaissance Europe, particularly France and Italy, encountered a similar upheaval—the introduction of artillery weapons. While previous armed conflicts had been largely conducted by knights and nobles who were highly educated and followed a code of honor in their hand-to-hand combat, the fact that untrained laborers and uneducated individuals of a lower social background could suddenly launch deadly material into the air to rain down on opposing fighters sparked an important debate over the expanded warfare. Ominously, combatants were now unexpectedly struck dead from objects dropping from the sky—a zone previously unknown to be perilous—by an unknown and unseen enemy. In fact, among other changes, this revolution and the debate that surrounded it paved the way to one of the most significant developments in the theory of just war: *jus in bello.*

Next is my contribution, Chapter 6, "The Question of 'Imminence': A Historical View on Anticipatory Attacks." Although there is general agreement that an anticipatory attack can be used to thwart an imminent strike, the relevance of this norm pivots directly on how one defines the term "imminence." This chapter will discuss a revealing bifurcation in the just war doctrine that occurred in the seventeenth century on this very question. The historical moment, and the ideas advanced in that context which helped to give birth to international law, illuminate the distorting definition given to this critical standard by the Obama administration (following directly in the steps of its predecessor). Because there is believed to

---

47   HLA Hart, *The Concept of Law* (Claredon Press 1961, 1994) at 193–200. For a fuller explanation of how Hart's views can be applicable in the international realm, see my Chapter 6.

48   As there are a host of moral inquiries that can be carried out for such UCAV operations, the chapters found here should certainly not be seen as exhaustive.

be an overlap between legal positivism and natural law due to the universally shared goal of survival, it is possible to discern the untenable interpretation of "imminence" being used in the US targeted killing program today.

Chapter 7 by Avery Plaw and Carlos R. Colon, "Correcting the Record: Civilians, Proportionality and the *Jus ad Vim*," explores the question of civilian casualties with armed drones. To do so they dig deeply into the available data on injured and killed civilians in Pakistan to refute the work of two scholars who advocate using standards of proportionality that come from a set of developing rules to regulate the just use of force (short of war) away from conventional battlefields: *jus ad vim*. In essence, the formulation of *jus ad vim* is meant to offer a middle ground between those who insist on applying the framework of war to US counterterrorism operations and those who judge those operations according to the standards of law enforcement and human rights. After working through the moral claims and reported data (making this an investigation of the overlapping space between axiological validity and empirical validity), Plaw and Colon propose an alternative "elevated just war theory" framework that is considered to be more plausible, attractive, and applicable to the drone strikes in Pakistan.

Delphine Hayim's contribution, "From Just War to Clean War: The Impact of Modern Technology on Military Ethics," is Chapter 8. She posits that the clean war doctrine is progressively supplanting the just war theory as a moral and legal justification for armed conflicts. Specifically, the cleanliness of effects is replacing the legality/morality of the cause. Technologies for conducting warfare are altering the nature of conflict in that public acceptance of war is now garnered through faith in scientific advances, trust in experts, and the use of medico-surgical terminology. This chapter uses the three moral frameworks of *jus ad bellum, jus in bello,* and *jus post bellum* in an attempt to grasp to what extent this clean war doctrine leads to a redefinition, or at least to a new interpretation of, international law rules and principles. And, considering the prospects for developing entirely automated and robotized wars, the moral dimension of the clean war doctrine and the ethical problems that are closely intertwined with the emergence of (semi-) autonomous weapons are highlighted.

Efficacy—Empirical Validity

While at first glance the realm of efficacy can be mistakenly considered straightforward, this sphere is better understood as "tremendously complex and deceptively simple."[49] One of the reasons behind this is the fact that there are various terms and meanings that can be applied.[50] For example, the word "efficiency" is related, but addresses cost/benefit analysis to measure the outcome

---

49   Ost and Kerchove (n 37): "l'effectivité est une notion extrêmement complexe, faussement simple" (my translation) at 329.

50   Barela (n 26) at 44–49.

of a law on society,[51] whereas the term "effectiveness" can be seen as the capacity of a rule to bring about the behavior desired within a society.[52] However, this model is not meant to analyze the effect of a rule on society, but rather the effectiveness of government in achieving its goals with counterterrorism policy. Thus the lens is inverted to look at the government rather than society.

The first step in carrying out this inverted objective of analyzing policy is to explain that the term we apply in this model of legitimacy is *efficacy*. The working definition comes from its direct derivative: efficacious. A policy should be judged by the Oxford English Dictionary definition of the term: "that [which] produces, or is certain to produce, the intended or appropriate effect." Considering all that is at stake in the use of deadly force, this is an appropriately high standard for judging efficacy.

As a second step, it is necessary to note the fact that this definition of efficacy opens the door to different tools of measurement for arriving at an empirical validity. The method can shift depending upon the specific policy under review and it is not appropriate to pretend that there is but one applicable instrument of measurement that can be put forward. For instance, we find that some of the empirical research questioning the effectiveness of armed drones has arrived at directly opposite conclusions. In this case, their divergence can be attributed to various reasons: differing definitions of the desired outcome, a difficulty with proving lines of causality, the introduction of counterfactual claims, and a dearth of empirical evidence.[53] To avoid sweeping assertions of little utility, the authors in this section remain measured and modest in their assessments of empirical validity. Additionally, in an attempt to delineate some of the essential differences regarding efficacy, this section is divided into chapters treating the data on leadership targeting, tactical efficacy, strategic efficacy, and systematic efficacy as we look through this lens of the model.[54]

Jenna Jordan begins this section with her chapter "Data on Leadership Targeting and Potential Impacts for Communal Support." While leadership targeting has become a key feature of current counterterrorism policies, both academics and policy-makers have argued without much investigation into the historical data that the removal of leaders is an effective policy. Accordingly, this examination of the empirical data on the history of leadership removal is of real additional

---

51   Wilfrid Rumble, "Legal Realism, Sociological Jurisprudence and Mr. Justice Holmes" (1965) 26, 4 J of the History of Ideas 547, at 549.

52   Ost and Kerchove (n 37) at 329.

53   See eg Stephanie Carvin, "The Trouble with Targeted Killing" (2012) 21 Security Studies 529.

54   As this theory suggests that a growing number of citizens use international treaties as standards of assessment for whether a government is exercising force legitimately, we can understand that there will be times when there is an overlap of legality and efficacy in this section. This overlay does not dominate the section, but there will be times when this crossover analysis is key to the conclusions put forward.

value. It reveals that groups whose leaders have not been forcibly removed, in fact, have had a faster rate of decline. Further, she identifies that the data indicate that large, older, religious, and separatist organizations—like Al-Qaeda—are the most likely to survive attacks on their leadership. Looking at the overall efficacy of leadership targeting, she then turns to discuss the role that communal support plays in understanding organizational resilience since it is essential to the ability of a terrorist group to withstand attacks on its leadership. This analysis suggests that, in fact, such targeting may have the counterproductive consequence of bolstering local support for the militants cause.

Next is Chapter 10 by Marek Madej—"Tactical Efficacy: 'Notorious' UCAVs and Lawfare." This chapter starts with an overview of the current state of technological development of unmanned aerial vehicles (UAVs) in use worldwide. The actual and potential advantages and limitations of that category of military equipment are then discussed, with particular attention devoted to decisive factors for their effectiveness as a tool of disruption and elimination of targets assumed hostile (combatants, terrorists). With a crossover into the realm of legality, the question of so-called lawfare against entities using drones is considered, with an analysis of the practical implications for tactical efficacy of UCAV operations. That is to say, the creative interpretations of law employed by the dominant operator of this technology appear to undermine their tactical efficacy because the creation of more legal targets does not necessarily increase the elimination of more true combatants.

I am the author of Chapter 11—"Strategic Efficacy: The Opinion of Security and a Dearth of Data." As the primary objective of counterterrorism is to increase security for the targeted population, it must be noted that Montesquieu spotlighted this concept as "an opinion" in the eighteenth century. Noting that there is a pivotal divergence between objective and subjective security—the absence of threats vs the absence of fear—begins to elucidate the immense difficulty in putting forward a scientific measurement of strategic efficacy. Additionally, a US program cloaked in secrecy, largely taking place in some of the world's most remote regions with little or no independent journalism, and with a fluid definition of the enemy to be defeated, has led to a constellation of data that is thus far impossible to collate systematically and completely. In other words, there is a currently insurmountable impediment to definitive conclusions on this pressing question.

Robert Kolb concludes this section with his chapter "Systemic Efficacy: 'Potentially Shattering Consequences for International Law.'" Within weeks of passage of the United Nations Security Council Resolutions passed in the aftermath of the attacks of 9/11, international jurist Antonio Cassese identified a disturbing incoherence in those Resolutions. His concern was that they introduced a disruption of crucial legal categories, and his disquiet becomes amplified with the use of combat drones employing lethal force across borders. This chapter addresses the tectonic impacts between the various bodies of law that are colliding and ponders what this strain means to the global structure as we know it. Hence, even if the use of drones were shown to be tactically and strategically efficacious, the resulting damage to the international system could negate such theoretical gains. Under discussion in this work—using the realm of legality to pose central questions about efficacy—is the construction of the contemporary global system, and the legal disturbance caused by using armed drones for cross-border killing.

Creating a Drone Court: Integration via a Policy Proposal

If the legitimacy of government is a battleground in conflicts waged by non-state actors, exercising lethal force legally, morally, and effectively against those specific actors is about proceeding defensively. There is certainly a legitimacy in pursuing, and attacking if necessary, those who are directly involved in hostilities against the government. Yet doing so in a manner that is considered justifiable requires great attention and institutional checks upon power to avoid jeopardizing legitimacy.

As a way to synthesize the ideas explored in this book in the previous sections, the final section will delve into a practical policy proposal that aims to reconcile, or at least ease, this tension between counterterrorism offense and defense. That is, since deadly force has been described as an offensive tactic that will at times be necessary, we must find a way to use it in a manner that does not sacrifice the strategic battleground of legitimacy. The proposal that is put forward and explored is, accordingly, meant to integrate the three key elements of legitimacy in view of the domestic constitutional order of the predominant user of UCAVs for counterterrorism across borders—the United States.[55]

The idea under consideration is put forward by Amos N. Guiora and Jeffrey S. Brand in Chapter 13, "Establishment of a Drone Court: A Necessary Restraint on Executive Power." As Guiora served for 20 years in the Judge Advocate General's Corp of the Israeli Defense Forces and took part in targeted killing decisions during that time, and Brand is Dean Emeritus of the law faculty at the University of San Francisco School of Law, this policy proposal represents a unique composite of theory and practice. The authors assert that to minimize collateral damage and enhance narrow application of a target-specific, criteria-based drone policy, the US Congress should create a "Drone Court." Specifically, there is a concern that the combination of broad definitions of "imminence" and "legal target" has undermined legal and moral principles by at times targeting the wrong people—a result that must be described as inefficacious. They believe that a court consisting of sitting judges should be created by Congress to establish an

---

55   In 2013 there was some discussion of the idea to be proposed here with a modeling on the Foreign Intelligence Surveillance Court, or keeping the court within the executive branch (see eg Neal K Katyal, "Who Will Mind the Drones?" *New York Times* [Feb 20, 2013]   <http://www.nytimes.com/2013/02/21/opinion/an-executive-branch-drone-court.html> accessed Dec 2014). The problems with these initial proposals are extensively discussed here.

institutional check against a power that is now exclusively exercised within the executive branch. Otherwise, the concern is that international targeted killing will become completely unhinged.

Focusing on the fundamental separation of powers principle found in the US Constitution, Brand and Guiora explore the Foreign Intelligence Surveillance Court to demonstrate the pitfalls that must be avoided in order to conduct operational counterterrorism in accordance with the rule of law. They provide substantial detail for their proposal: a rebuff of full secrecy and *ex parte* hearings; who will sit as judges; structure and appeal processes; guarantees of the right to confront and cross examine; burden of proof; credible, reliable facts; the definition of "imminence"; and a possible post hoc review. In the end, fleshing out these institutional and domestic legal questions serves to demonstrate the feasibility and utility of a Drone Court. Not only would such a court create a forum for rigorous discussion of legality and morality, it would also serve to create a collection of data for measurement of empirical validity—efficacy.

Finally, Tom Farer and Frédéric Bernard provide the closing chapter to this volume, "Can UCAVs be Reconciled with Liberal Governance?: The Substantive Law of a Drone Court." Farer is a former dean at the Korbel School of the University of Denver, and Bernard is a lecturer at the University of Geneva, and thus this co-authorship represents a distinctive transatlantic alliance to defend the general principles undergirding liberal governance as they can become compromised in the reaction to a terrorist threat.

These authors endorse the creation of a Drone Court, believing it would be an important step forward, and build on the proposal in their own chapter. To move the conversation forward, this chapter digs into the substantive international law that such a court would need to apply to hard cases to suggest what defending the international rule of law might look like when contemplating the use of lethal force against individuals and across borders. The authors explore a holistic approach that the proposed court could use to address the competing bodies of law that we see colliding in this work. In doing so, they introduce valid and important questions that a Drone Court would face in order to protect liberal values while allowing the executive branch to strike legitimate targets in foreign territories.

A Final Note on Multiple Disciplines

This book is a collection of investigations from different disciplinary perspectives, and at times those disciplines are interwoven and overlapping. Therefore, it is useful to make a more declarative statement as to how one might categorize this work within disciplinary lines. It is first necessary to elucidate the terms used to describe a research project that is meant to bring together and integrate various disciplines to resolve a question that is too complex to solve through just one field of study. For our purposes, it is the distinction between *multi*disciplinarity and *inter*disciplinarity that matters most. While a *multi*disciplinary work looks at a topic from the perspective of several disciplines at one time and makes a modest attempt to join together their insights, an *inter*disciplinary project brings together a collection of viewpoints from various disciplines and draws on the diverse insights by finding ways to integrate them.[56]

Using these definitions, this book can best be described as a multidisciplinary work that at times crosses over into interdisciplinarity. This structure and design has been employed to help achieve two goals. The first is to consciously separate the work into clear disciplinary categories to shed light on the various questions at stake from important and valid perspectives. The second, made possible with the sharp distinctions in place, is to afford dialogue between the disciplines, as the chapter authors chose, in order to explore the space in which these spheres interact. To the extent that this has been accomplished to successfully illuminate worthwhile points of interplay, the contributing authors deserve credit. However, the complexity of the task undertaken also requires that responsibility falls to the editor for any failings resulting from this design.

***

It is hoped that through exploring the three varied pillars that would erect the legitimacy of drones the reader departs with an expanded purview on this developing military technology, the current usage of which is agitating our existing norms. This is not to say that a final judgment is to be rendered here on the legitimacy of armed drones for cross-border counterterrorism; rather the intent has been to set out some of the central parameters for such an assessment precisely because their legitimate use is so critical.

Beyond that, there has also been an effort to deal with some of these disturbances with a meaningful and practical proposal aimed at calming turbulent waters— that is, the establishment of a Drone Court that might reconcile the substantive international law at stake. A veritable discussion has begun about this idea,[57] and this publication aims to push it further.

As the intention behind this project was to stimulate conversation about how a government under the threat of terrorism can best defend itself strategically—not just tactically—it is believed that these valuable and diverse contributions can serve to do just that. After all, discussing the legitimacy of armed drones used for cross-border counterterrorism is significant because it is central to the outcome of the conflict itself.

---

56   Allen Repko, *Interdisciplinary Research: Process and Theory* (Sage Publications 2012) at 20–21.

57   A Guiora, J Brand, M Hakimi, M O'Connell, S Vladeck, N Rao, "Room for Debate: Should a Court Approve all Drone Strikes?" *New York Times* (April 24, 2015).

SECTION I
Through the Lens of Legality— Formal Validity

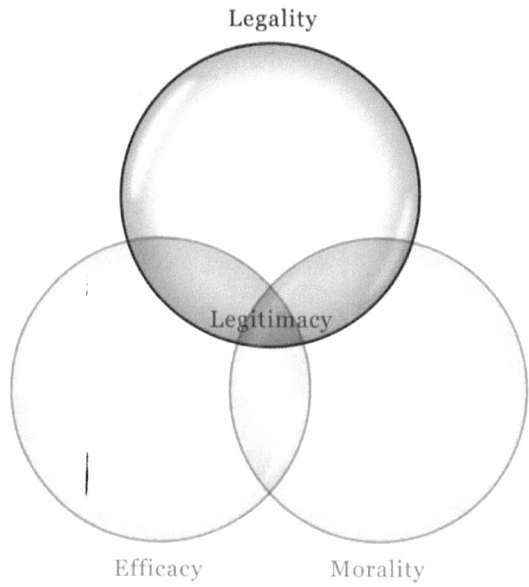

Chapter 1

Jus ad Bellum: Crossing Borders to Wage War against Individuals

Christian J. Tams and James G. Devaney

I. Introduction

Drone strikes have become an everyday reality in the war on terror. Hailed by supporters as an effective and targeted tool, drones require us to rethink questions of legitimate warfare. Legal analysis is part of that inquiry; and one would hope a relevant part. As the debate of the last decade suggests, drone strikes do not neatly fit established patterns of legal argument. In a report issued in 2010, the UN's Special Rapporteur on Extrajudicial, Summary and Arbitrary Executions at the time, Professor Philip Alston, indicated that targeted killing by drones had led to a "highly problematic blurring ... of the boundaries of the applicable legal frameworks" the result of which "has been the displacement of clear legal standards with a vaguely defined licence to kill."[1]

Whether the existing standards were as clear as that statement suggests may be open to discussion, but that the present state of the law is at best "vaguely defined" seems difficult to dispute. From human rights law to the laws of war and principles of domestic constitutional law, drone strikes implicate a great number of legal provisions and standards and quite often are not easily categorized. The present contribution does not engage with human rights law, international humanitarian law, or indeed domestic legal orders. It addresses a separate aspect of the legal analysis: the question of whether force can be used across borders at all—whether states pursuing anti-terrorist policies can use drones against targets located abroad. In the recent discourse on drones, this matter has not received the same amount of attention as questions relating to human rights or to the rules governing the conduct of hostilities. Yet they remain fundamental, and perhaps indeed of preliminary relevance: since international law excludes, subject to certain exceptions, the use of force in the international relations between states, unless a state can establish a right to use force across borders (whether by means of drone strikes or otherwise) its conduct will fall foul of an essential standard of international legality.

---

1 Philip Alston, Special Rapporteur on Extrajudicial, Summary or Arbitrary Executions, "Study on Targeted Killings," UN Doc A/HRC/14/24/Add.6 (May 28, 2010) at para 6.

While essential—perhaps even the "cornerstone" of the contemporary legal order—the rules governing recourse to force, the *jus ad bellum* in international legal terminology, are not static. Drone strikes no doubt present a fresh challenge, but they are not the first challenge to the *jus ad bellum*, which in previous decades has been tested by arguments about wars for national liberation, claims for humanitarian intervention, and more recently by debate about the war on terror. In fact, in many respects, anti-terrorist drone warfare, which is at the heart of the current debate, is but one aspect of a broader challenge to adjust the existing law to the new, or newly perceived, threat of global terrorism.

Against this background, the subsequent inquiry assesses to what extent the current *jus ad bellum* permits the use of drones against terrorists in foreign countries. More particularly, it situates debates about drone warfare within the broader debate about the *jus ad bellum*. As so much of the current debate centers on questions of human rights law and international humanitarian law, the present contribution deliberately steps back and takes a general look at the Charter regime regulating inter-state force, which it is argued continues to set the parameters of the debate. As will be shown, these parameters can evolve, and arguably have already evolved. But in order to assess the legality of drone strikes, these parameters (evolving or not) need to be engaged with—and need to be engaged with seriously. This is the aim of the present contribution. The argument is presented in two steps: Part II below introduces the general regime regulating inter-state force as set out in the UN Charter. The bulk of the analysis, in Part III, then revisits contemporary challenges to that regime and assesses whether and under what conditions, the current regime permits the use of drone strikes against terrorists. In that part, the focus is on the main exception to the use of force, namely the right to act in self-defense.

II. The General Regime Regulating Recourse to Inter-State Force

A. Basic Charter Rules

The legal rules which make up the *jus ad bellum* form the core of modern international law: the general ban on military force has been of crucial relevance in the development of the modern system of inter-state relations. It is considered the "cornerstone" of the UN Charter,[2] and "the most important principle in

---

2   *Armed Activities on the Territory of the Congo (Democratic Republic of the Congo v Uganda)* Judgment, ICJ Reports 2005: "The prohibition against the use of force is a cornerstone of the United Nations Charter" at 223, para 148; *Military and Paramilitary Activities in and against Nicaragua (Nicaragua v United States of America)* Merits, Judgment, ICJ Reports 1986 Rep 14, separate opinion of President Singh: "the very cornerstone of the human effort to promote peace in a world torn by strife" at 153; C Joyner, *International Law in the 21st Century* (Rowman & Littlefield 2005) at 165.

contemporary international law to govern inter-State conduct."[3] However, reference to this principle being the cornerstone of the UN Charter does not imply that the prohibition on the use of force is absolute or immutable. In fact, there is a wide range of issues related to the use of force that are not settled. For example, there has been significant debate regarding whether states are justified in using force to protect human rights, their nationals abroad, and, most significantly for our purposes, the use of drones against terrorist organizations operating in other states.

1. The scope of the ban on force

Mention of the UN Charter prompts closer consideration of its provisions relating to the use of force. The general prohibition on the use of force set out in Article 2(4) of the Charter obligates UN members to "refrain in their international relations from the threat or use of force against the territorial integrity or political independence of any state, or in any other manner inconsistent with the Purposes of the United Nations." Articles 42, 43, and 51 of the Charter, taken together, recognize two exceptions to the general prohibition on the use of force: enforcement measures sanctioned as part of the UN system of collective security and measures taken in self-defense.

While many specific aspects of the law remain disputed, and of course the law is by no means perfectly adhered to, the Charter rules are regularly affirmed by states and other actors as the regime governing inter-state force. When applying this regime to the problems of drone strikes, two questions need to be answered:

 i. Does a drone strike violate the prohibition against the use of force?
 ii. Can a drone strike be justified under the Charter regime?

The first question in most settings can be answered with relative ease. A drone strike amounts to the use of military force and as such implicates Article 2(4) of the UN Charter.[4] That provision purposefully—and in deliberate extension of earlier rules renouncing "war"[5]—bans not only large-scale military operations but all uses of "force." If a drone strike is directed against a foreign state, it will *prima facie* violate the prohibition on the use of force. This is true irrespective of the

---

 3 *Oil Platforms (Islamic Republic of Iran v United States of America)* Judgment, ICJ Reports 2003, at 161, dissenting opinion of Judge Elarby, at 291.

 4 See N Bhuta and others, *Targeted Killing, Unmanned Aerial Vehicles and EU Policy*, vol RSCAS PP 2013/17 (RSCAS Policy Papers 2013).

 5 Notably the Pact of Paris (or Kellogg–Briand Pact) of 1928. As Dinstein notes, "[w]hen the Charter of the United Nations was drafted in San Francisco, in 1945, one of its aims was redressing the shortcomings of the Kellogg–Briand Pact. ... Article 2(4) avoids the term 'war.' The use of force in international relations, proscribed in the Article, includes war. But the language transcends war and covers also forcible measures 'short of war.'" Y Dinstein, *War, Aggression and Self-Defence* (Cambridge University Press 2011) at 85.

motives prompting the strike, and irrespective of whether the target hit is a state agent, a member of the foreign military, or similar. Article 2(4), adopted against the backdrop of the devastating experience of World War II, deliberately opts for a comprehensive ban. While some types of breaches are mentioned expressly (namely, uses of force against the territorial integrity or political independence of other states) "any use of inter-State force by Member States for whatever reason is banned, unless explicitly allowed by the Charter."[6]

2. Exceptions and carve-outs

As noted above, this ban is not absolute. It admits of two exceptions expressly set out in the UN Charter, one of which is of crucial relevance. Moreover, the protection against military force can be waived if the targeted state consents to the use of force. To begin with the latter scenario, the ban against force is premised on the notion that coercive might is used against another state's will (and hence found to be illegal). If the "victim" consents the use of force—at least in the relationship between the states involved—is no longer coercive in character, but consensual. In line with this, international law in principle admits the consensual use of force on foreign soil. The question is less one of principle, but rather one related to the modalities for the expression of consent. In essence, consent is validly expressed if it is given by the competent and recognized organs of government *ex ante*, and in a transparent manner. This general position has been tested, without clear results, by the proverbial hard cases involving dubious forms of consent—given by puppet governments, factions involved in civil strife, and consent given ex post facto.[7] In relation to drone strikes against terrorists operating abroad, separate questions arise which eschew a clear-cut answer, and include those concerning the validity of secret (or tacit) consent, consent expressed by organs other than the government, or the impact of discrepancies between public silence and secret connivance. At this stage, all that is required is to restate that in principle, states targeted by drone strikes can consent to the use of force and thereby remove its illegality.

As stated above, with regard to the expressly recognized exceptions to the ban on force, the UN Charter envisages two settings: military force can be employed lawfully (i) if the UN Security Council has authorized it as a collective security measure under Chapter VII of the Charter; or (ii) if force is used in self-defense under Article 51 of the Charter. The first of these exceptions has so far been irrelevant to the debate about drone strikes and is unlikely to gain relevance in the near future. While there is no doubt that the Security Council would be competent to sanction the use of drones under Chapter VII, to date it has not done so— nor indeed has it contemplated doing so. Therefore the majority of the following analysis will focus on the issue of self-defense as the most likely potential justification upon which states could rely.

---

6   Ibid at 87–88.

7   *Nicaragua Case* (n 2) para 194; Alston (n 1) at 12.

As stated above, the issue of drones undoubtedly presents a fresh challenge to the *jus ad bellum*, but it is not the first such challenge. In fact the issue of drones raises many of the same factual and legal issues as the discourse on anti-terrorist self-defense. Consequently, some consideration of the early debates about anti-terrorist self-defense are instructive in this regard.

B. Early Debates about Anti-Terrorist Self-Defense

Even though the early debates on anti-terrorist self-defense did not center around drone strikes, they did involve the cross-border use of force against individuals. Like today, terrorism was a very real threat to states in the 1980s. However the legal regime governing anti-terrorist measures was of a very different character. While terrorism was denounced, and a number of sector-specific conventions were adopted addressing certain specific types of terrorist activity,[8] the international community stopped short of condemning terrorism outright.[9] Terrorism was treated as a problem of criminal law that would be addressed through state cooperation in that field. What is more, with respect to the *jus ad bellum*, when a small number of states asserted a right to respond militarily[10]—by invoking self-defense, or at times relying on the defense of necessity—their legal claims were rejected.[11] For instance, Israel's raid on the Palestine Liberation Organization (PLO) Headquarters in 1985 was condemned vigorously by the Security Council as being an "act of armed aggression ... in flagrant violation of the Charter of the United Nations."[12] Similarly, South Africa's attempt to "invent" a right of hot pursuit against offenders across borders met with almost unanimous resistance.[13]

In essence, in rejecting Israel and South Africa's claims, the international community was espousing a restrictive construction of the *jus ad bellum*, which—shaped by number of judicial decisions[14]—sought to limit the availability of

---

8   See the list of conventions available at <http://www.un.org/en/terrorism/instruments.shtml> accessed Dec 2014.

9   See P Klein, "Le droit international à l'èpreuve du terrosime" (2006) 321 Recueil des Cours 203, at 209.

10   Christine D Gray, *International Law and the Use of Force* (OUP 2008) at 136–140.

11   Klein (n 9) at 375.

12   See SC Res 573 (1985); in this respect see also the US' raid on targets in Libya in 1986 which was described by the General Assembly as "a violation of the Charter of the United Nations and of international law," GA Res. 41/38.

13   See the rejection of this doctrine in SC Res. 568 (1985), as well as SC Res. 527 (1982), SC Res. 586 (1985) and the summary provided by Gray (n 10) at 31–33.

14   *Corfu Channel Case (UK v Albania)* (Merits) [1949] ICJ Rep 4; *Nicaragua Case* (n 2). The restrictive construction was reflected in the leading works published at the same: see the commentaries on Articles 2(4), 39–43 and 51 of the UN Charter in JP Cot and A Pellet (eds) *La Charte des Nations Unies* (Economica 1985) (contributions by Virally, Cohen, Jonathan, Fischer, and Cassese) and in B Simma (ed) *Die Charta der Vereinten Nationen* (CH Beck 1991) (contributions by Randelzhofer and Frowein).

military force in international relations and which dominated debates until the 1990s. As far as self-defense is concerned, this restrictive view rested on three assumptions, which remain important to the current debate on drone warfare, and which thus merit setting out.

First, under the text of the Charter, self-defense is available against "armed attacks" only. While the ban on force was construed to cover all uses of force (including relatively minor incidents), the notion of "armed attack," and this notion was understood to cover "the most grave forms of the use of force" only, which were distinguished "from other less grave forms."[15] This threshold was crucial with regards to extraterritorial anti-terrorist violence in that such attacks were likely to be of lesser intensity than attacks carried out by state forces, and as such more likely to fail to pass the required threshold.[16] What is more, while some states seemed prepared to recognize that, under the so-called accumulation doctrine, a series of "continuous pin-prick assaults" could be "appraise[d] ... in their totality as an armed attack,"[17] most states considered that each armed attack in and of itself would have to reach the threshold limit.[18] This construction meant that terrorist attacks would typically not trigger a right of self-defense.

A second normative choice reinforced the first, and further limited the availability of self-defense in response to terrorist violence. Under the traditional approach, self-defense was rationalized as an inter-state defense that would only be available against armed attacks by another state. This did not mean that force had to be used by foreign armies. However, where irregular armed bands—or indeed terrorists—operated from foreign territory, their conduct was attributed to the territorial state only under relatively narrow conditions, namely if the armed group was acting under the "effective control" of the foreign state.[19] The "effective control" test—devised in the International Court of Justice's (ICJ) judgment in the *Nicaragua* case, widely accepted at the time,[20] and subsequently affirmed (and generalized) in Articles 5

---

15   *Nicaragua Case* (n 2) at para 191.

16   G Guillaume, "Terrorisme et droit international" (1989-III) 215 Recueil des Cours 406.

17   See, for instance, Dinstein (n 5) at 230–231.

18   See C Wandscher, *Internationaler Terrorismus und Selbstverteidigungsrecht* (Duncker & Humblot GmbH 2006) 170; Barry Levenfeld, "Israel's Counter-Fedayeen Tactics in Lebanon: Self-Defense and Reprisal Under Modern International Law" 21 Columbia Journal of Transnat'l Law 1.

19   *Nicaragua Case* (n 2) paras 109, 115, 195; cf. the dissenting opinion of Judge Jennings at 533.

20   See the separate opinion of Judge Kooijmans in *Legal Consequences of the Construction of a Wall (Advisory Opinion)* 2004 available at <http://wwwicj-cijorg/icjwww/idocket/imwp/imwpframehtm> accessed April 2012 at para 35 stating that this inter-state interpretation of self-defense "has been the generally accepted interpretation for more than 50 years"; see also CJ Tams, "Note Analytique-Swimming with the Tide or Seeking to Stem It-Recent ICJ Rulings on the Law of Self-Defence" 18 Revue quebecoise de droit int'l 275, at 278–280.

and 8 of the International Law Commission's (ILC) text on State Responsibility[21]—required effective control to be exercised over the specific attacks in question. Article 51 of the UN Charter, textually requiring only "an armed attack," was thus read to require an "armed attack by a State"—a plausible, but by no means uncontestable interpretation that would subsequently come under pressure.

Finally, under the restrictive interpretation of the *jus ad bellum*, self-defense was limited by a "functional argument." In other words, it was accepted that "self-defence [was] not an open-ended instrument but only has the aim of repelling armed attacked attacks and provisionally guaranteeing the security of States."[22] This functional approach to the right of self-defense meant that not only did the conditions of necessity and proportionality apply, there was necessarily a temporal link between the measures taken in self-defense and the armed attack, commonly referred to at this time as the requirement of "immediacy."[23] As such, force had to be taken to stop ongoing attacks, or perhaps (according to one of the big debates) to avert imminent future attacks (so-called anticipatory self-defense[24]). As regards the end point, measures of self-defense—aimed at repelling attacks—were necessarily finite, and in fact meant to be limited in time. Notably, self-defense under this traditional understanding did not justify punitive or other "belated" measures that were taken "after the event and when the harm has already been inflicted" and as such could not be classified as means of protection.[25]

All three factors restricted the scope of self-defense as a justification permitting recourse to force; and restricted it in ways that remain highly relevant to the current debate on drones. They were based on a plausible interpretation of international law that sought to restrain options for the lawful use of force. This restrictive approach could draw on the language of the Charter and highlighted the development of the *liberum jus ad bellum* of nineteenth-century international law into a *jus contra bellum*. At the same time, it failed to accommodate demands to permit unilateral uses of force for ostensibly benign purposes (such as the defense of human rights) or as part of a robust policy of repression against criminal conduct

---

21   James Crawford, *The International Law Commission's Articles on State Responsibility: Introduction, Text and Commentaries* (Cambridge University Press 2002) commentary to Art 8(3)–(5).

22   Enzo Cannizzaro, "Contextualizing Proportionality: Jus ad Bellum and Jus in Bello in the Lebanese War" 88 International Review of the Red Cross 779, at 782.

23   See Tarcisio Gazzini, *The Changing Rules on the Use of Force in International Law* (Manchester University Press 2005) at 143.

24   See for example Abdul Ghafur Hamid, "The Legality of Anticipatory Self-Defence in the 21st Century World Order: A Re-Appraisal" (2007) 54 Netherlands International Law Review 441; Terry D Gill, "The Temporal Dimension of Self-Defence: Anticipation, Pre-emption, Prevention and Immediacy" (2006) 11 Journal of Conflict and Security Law 361; Ian Brownlie, *International Law and the Use of Force by States* (Clarendon Press 1963) at 257.

25   Derek Bowett, "Reprisals Involving Recourse to Armed Force" (1972) 66 American J of Int'l Law 1, at 3.

(such as terrorism).[26] Not surprisingly, then, the restrictive approach was by no means unanimously accepted, and it would come under increased pressure toward the end of the twentieth century, when the tougher policy against international terrorism would begin to influence the interpretation of the *jus ad bellum*. These developments form the subject of the next section.

III. The Anti-Terrorist Challenge to the *Jus ad bellum*

A. The Restrictive Regime Under Pressure

Revisiting, if only briefly, the debates of the 1980s is useful for the fact that it becomes immediately clear how much things have changed. For the past two decades, states have taken a more robust line against terrorism. This has resulted in the emergence of a body of "anti-terrorism law,"[27] which spans wide areas of international relations, from increased duties of cooperation[28] to strict Security Council sanctions.[29] Unsurprisingly, the more robust anti-terrorist policy is also evident in debates about the *jus ad bellum*. In fact, these debates could perhaps be seen as the vanguard of an "anti-terrorist challenge." Drone strikes may perhaps be seen as the vanguard of that vanguard, but they also need to be seen as part of a

---

26   As a potential exception, there was debate, especially during the late 1970s and 1980s, whether international law should permit military conduct in support of anti-decolonization movements. That particular form of "just war" enjoyed considerable support, especially among the G-77, even though it remained firmly opposed by Western States; for a brief summary see CJ Tams, "Prospects for Humanitarian Uses of Force" in A. Cassese (ed), *Realizing Utopia: The Future of International Law* (OUP 2012).

27   Ben Saul, "Attempts to Define 'Terrorism' in International Law" (2005) 52 Netherlands International Law Review 57; Larissa Van den Herik, "The Security Council's Targeted Sanctions Regimes: in Need of Better Protection of the Individual" (2007) 20 Leiden Journal of International Law 797; Daniel Moeckli, "Emergence of Terrorism as a Distinct Category of International Law" (2009) 44 Tex Int'l Law J 157.

28   See for example Giuseppe Nesi, *International Cooperation in Counter-terrorism: The United Nations and Regional Organizations in the Fight Against Terrorism* (Ashgate 2007); chapters 7, 13, and 14 of Michael J Glennon and Serge Sur, *Terrorisme et droit international: Terrorism and international law* (Martinus Nijhoff Publishers 2008); and R Higgins, "The General International Law of Terrorism" in R Higgins and M Flory Higgins (eds), *Terrorism and International Law* (Routledge 1997) in particular Part II: "Cooperation Against Terrorism."

29   See for example SC Res 1368 and 1373 (concerning 9/11), SC Res 1438 (Bali), SC Res 1530 (Madrid); SC Res 1377 ("Declaration on the Global Effort to Combat Terrorism"); SC Res 1456 ("Declaration on the Issue of Combating Terrorism"). In the latter document, the Council, acting under Chapter VII, eg "[c]ondems in the strongest terms all acts of terrorism irrespective of their motivation, whenever and by whomsoever committed, as one of the most serious threats to peace and security"; SC Res 1456, second preambular paragraph. For a full list of SC Res see <www.un.org/terrorism/sc-res.shtml>.

broader debate that has seen the restrictive regime, seeking to limit options for the lawful use of unilateral force, come under pressure.[30] This pressure has made itself felt in renewed assertions of a right of humanitarian intervention, the increased willingness of states to accept targeted uses of force as a means to rescue nationals abroad, and in these debates about means to suppress terrorism.

As a preliminary point, it is worth noting that while faced with a new challenge, the UN Charter continues to set the parameters of the *jus ad bellum*. Articles 2(4), 42, 43 and 51 are still central to the debate owing to states' continuing reluctance to even consider formal change to the existing legal rules. In particular, the prohibition on the use of force as set out in Article 2(4) of the UN Charter, often said to be dead or outdated,[31] has not been seriously questioned by relevant numbers of states.[32] Rather than softening Article 2(4), the debate has centered on exceptions to the ban on force, and in particular on whether the claim that anti-terrorist force, whether exercised through drone strikes or otherwise, should be qualified as self-defense. It is to this debate that we now turn.

B. State Practice as an Agent of Change?

If the debate is a real one, then it is mainly because state practice has evolved. To put things provocatively, positions advocated by Israel and South Africa in the 1980s (rejected by most other states at that time) have now become much more acceptable. That the *jus ad bellum* should recognize a right to use self-defense against terrorists operating abroad, even where they were not acting under the direction and control of a foreign state under the narrow test devised in the *Nicaragua* judgment, is not a universally accepted position today. However it does enjoy much more support than it did in the 1980s. The debate about drones, in this respect, is part of a wider challenge to the *jus ad bellum* which began well before the advent of drone policies. This practice needs to be briefly reviewed as it informs the interpretation of the rules governing anti-terrorist drone warfare; and it is well illustrated by the following five examples of forcible anti-terrorist action:

> *First*, the international community's response to the 9/11 attacks provides the most prominent example.[33] While not controlled or directed by any foreign

---

30 Tams (n 26).

31 See for instance Michael J Glennon, "Why the Security Council Failed" (2003) 82 Foreign Aff 16; Michael J. Glennon, "How International Rules Die" (2005) 93 Geo Law Journal 939; Gregory M Travalio, "Terrorism, International Law, and the Use of Military Force" (2000) 18 Wis Int'l LJ 145 and the summary of debates in Gray (n 10).

32 See Michael Wood, "Law on the Use of Force: Current Challenges" (2007) 11 The Singapore Yearbook of International Law 1; CJ Tams, "The Use of Force Against Terrorists" (2009) 20 Eur J of Int'l Law 359, at 375.

33 See Thomas M Franck, "Terrorism and the Right of Self-Defense" American Journal of International Law at 839; C Stahn, "'Nicaragua is Dead, Long Live Nicaragua'—

state, the attacks were generally considered to be "armed attacks" in the sense of Article 51 of the Charter. In Resolution 1368 of 12 September, the Security Council expressly affirmed the United States' right to use self-defense.[34] Similar language appeared in the preamble of Security Council Resolution 1373 (2001) and in Resolutions adopted by the North Atlantic Treaty Organization (NATO) and the Organization of American States (OAS) member states.[35]

Second, Turkey has consistently invoked the right to self-defense in relation to its forcible measures against Kurdistan Workers' Party (PKK) bases in Northern Iraq, including during the ground operation known as Operation Sun in 2008.[36] The reaction of other states was largely sympathetic with some expressly voicing their support for Turkey's position.[37]

Third, the forcible actions of the US in response to the 1998 terrorist attacks on its embassy in Dar es Salaam[38] did not provoke notable criticism from other states, with a small number speaking out in support of the US's right to self-defense in this particular situation.[39]

Fourth, non-Western states like Russia and Iran have equally espoused a broad right to use force extraterritorially against terror networks in response to prior attacks. In line with this view, Russia used force against bases on Georgian

---

The Right to Self-Defence under Art 51 UN Charter and International Terrorism" Terrorism as a Challenge for national and international Law: Security versus Liberty at 827; since the US began this action there has been significant academic debate regarding whether the US had overstretched the limits of self-defense: on that point see Gray (n 10) Chapter 4; CJ Tams, "Light Treatment of a Complex Problem: The Law of Self-Defence in the Wall Case" (2006) 16 Eur J of Int'l Law 963.

34   SC Res 1368 (2001) (Preamble); also reproduced in (2001) 40 ILM 1277.

35   For the text of the relevant documents see (2001) 40 ILM 1267 and 1273 respectively.

36   See T Ruys, "Quo Vadit Jus ad Bellum? A Legal Analysis of Turkey's Military Operations against the PKK in Northern Iraq" (2008) 12 Melbourne Journal of International Law 334.

37   See the comments of the Belgian and Dutch foreign ministers, Questions et réponses écrites, Chambre des représentants de Belgique (Feb 25, 2008) QRVA 52 010, 1357 available at <http://www.dekamer. be/QRVA/pdf/52/52K0010.pdf> accessed Oct 2014; Maxime Verhagen, "Beantwoording vragen van het lid Van Bommel over een Turkse invasie in Noord-Irak" Ministerial statement (March 3, 2008) as cited in ibid.

38   J Lobel, "The Use of Force to Respond to Terrorist Attacks: The Bombing of Sudan and Afghanistan" (1999) 24 Yale Journal of International Law at 537.

39   See G Wettberg, *The International Legality of Self-Defence Against Non-State Actors* (Peter Lang International Academic Publishers 2007) at 157–59; for a summary of responses: UN Doc S/1998/780 (August 20, 1998); see also K Drozdiak, "European Allies Back US Strikes; Japan Says It 'Understands'" *Washington Post* (August 21, 1998) A20.

territory in 2004,[40] and Iran repeatedly mounted raids against the Mojahedine Khalq Organization (MKO) terror group operating within Iraq in 1999.[41]

Fifth, and relatedly, Israel has continued to support a broad right of self-defense, in 2003 and 2006 responding to terrorist attacks with military action north of Damascus and Lebanon respectively.[42] The reaction of the international community to Israel's actions in both 2003 and 2006 was mixed, with many states condemning Israel's actions as being disproportionate[43] while not necessarily contesting Israel's right to take such action in self-defense per se.[44]

These examples are illustrative only. They provide evidence of a new willingness of states—mostly Western, but also including Iran and Russia—to take forcible action against terrorists operating abroad. And of course even states that have not acted themselves have hardly ever protested, thus perhaps manifesting a willingness to accept the conduct in question. While the picture is no doubt mixed, all things considered, it can be said that state practice reveals an "opening up" of the right of self-defense to apply to forcible measures taken against terrorist entities in a manner that would not have been accepted under earlier, more restrictive, interpretations of the *jus ad bellum*.

C. US Practice with Drones in Particular

The advent of concerted drone policies needs to be seen against this background. Detailed information on drone strikes in recent times can be found in Chapter 9 by Jenna Jordan and Chapter 10 by Marek Madej in this edition, and therefore it is not necessary to set out precise figures here. However, a brief review is perhaps convenient, if only to draw out parallels between drone strikes and other forms of anti-terrorist action. As is well known, the United States is the most regular user

---

40   For further information see Wandscher (n 18) at 194.

41   See UN Doc S/1999/781 for Iran's legal explanation of the measure.

42   See Y Ronen, "The 2006 Conflict in Lebanon" (2008) 9 Yearbook of International Humanitarian Law at 362–93; see also Secretary-General's Report on UNIFIL, UN Doc S/2006/560 (July 22, 2006); A Zimmermann, "The Second Lebanon War: Jus ad Bellum, Jus in Bello and the Issue of Proportionality" (2007) 11 Max Planck Yearbook of United Nations Law at 99.

43   See the statements of Russia, Argentina, Japan and France: UN Doc S/PV.5488 (July 13, 2006), UN Doc S/PV.5489 (July 14, 2006), UN Doc S/PV.5493 (July 21, 2006), UN Doc S/PV.5493 (Resumption 1) (July 21, 2006).

44   During the Security Council meeting of July 14, 2006, only China and Qatar expressly rejected Israel's reliance on self-defense: see UN Doc S/PV 5489. In contrast, Israel's right was recognized in ibid.: Argentina (9), Japan (12), the UK (12), Peru (14), Denmark (15), Slovakia (16), Greece (17), France (17) and the USA (17). The UN Secretary General also acknowledged Israel's right to defend itself against Hezbollah attacks under UN Charter, Art 51: see UN Doc S/PV.5492 (July 20, 2006) 3; UN Doc S/PV.5498 (July 30, 2006) at 3.

of drones, repeatedly using them to target members of Al-Qaeda and "associated forces"[45] in Afghanistan, Iraq,[46] Pakistan, Yemen, and Somalia (regardless of whether such attacks take place on an "active battlefield"[47]), and even in Northern African countries such as Libya[48] and Mali.[49] At some point, and especially in relation to Pakistan, the consent of the host state may have justified this drone policy in terms of the *jus ad bellum*.[50] However, whether that consent continues to be given is difficult to discern.[51] Regardless, the United States claims that, even without the consent of those states involved, it is acting in accordance with international law relying on its inherent right to self-defense.[52]

---

45   See for instance John Brennan, Speech at Woodrow Wilson Center, "The Efficacy and Ethics of U.S. Counterterrorism Strategy" (April 30, 2012) available at <http://www.wilsoncenter.org/event/the-efficacy-and-ethics-us-counterterrorism-strategy> accessed Dec 2014.

46   See for instance <http://www.theguardian.com/world/2014/jun/28/us-flying-armed-drones-in-iraq> accessed Dec 2014.

47   Brennan (n 45).

48   Arguably sanctioned by SC Res 1973, UN Doc S/RES/1973 (March 17, 2011).

49   Additional detailed information on drone strikes by region is available at <http://www.thebureauinvestigates.com/category/projects/drones/> accessed Dec 2014; see also Michael J Boyle, "The Costs and Consequences of Drone Warfare" (2013) 89 Int'l Affairs 1, at 2.

50   Leaked cables confirm that former Pakistani presidents, prime ministers and generals had expressly given their consent to the United States (see "strong evidence" cited by Special Rapporteur on the Promotion and Protection of Human Rights and Fundamental Freedoms While Countering Terrorism, "Interim Report to the General Assembly on the Use of Remotely Piloted Aircraft in Counter-Terrorism Operations" UN Doc A/68/389 (Sept 18, 2013) at 53; Tim Lester, "WikiLeaks: Pakistan Quietly Approved Drone Attacks, US Special Units" CNN.com (Dec 1, 2010) available at <http://articles.cnn.com/2010–12–01/us/wikileaks.pakistan.drones_1_drone-attacks-predator-strikes-interior-minister-rehman-malik?_s=PM:US> accessed Dec 2014; "Kayani Wanted More Drone Strikes in Pakistan," *The Express Tribune with the Int'l New York Times* (May 20, 2011) available at <http://tribune.com.pk/story/172531/wikileaks-kayani-wanted-more-drone-strikes> accessed Dec 2014; Amanda Hodge, "Pakistan Allowing CIA to Use Airbase for Drone Strikes" *The Australian* (Feb 19, 2009) available at <http://www.theaustralian.com.au/news/pakistan-permits-cia-base-for-strikes/story-e6frg6t6- 1111118893683> accessed Dec 2014, despite publicly speaking out against such forcible measures (Zeeshan Haider, "Pakistan Denies Accord with U.S. on Drone Attacks" *Reuters* (Nov 17, 2008)). However, more recently Pakistan's consent to such action has been called into question by its statements denouncing the use of drones on its territory and calling for an immediate end to US measures in this regard (see for instance "Pakistan Summons US Ambassador to Protest Against Latest Drone Killings," *The Guardian* (June 8, 2013) available at <http://www.theguardian.com/world/2013/jun/08/pakistan-us-drone-killings> accessed Dec 2014.

51   See <http://www.thebureauinvestigates.com/category/projects/drones/>; Special Rapporteur on Extrajudicial, Summary or Arbitrary Executions, "Armed Drones and the Right to Life," 91, UN Doc A/68/382 (Sept 13, 2013).

52   See remarks of Secretary of State John Kerry (Aug 1, 2013) available at <http://www.state.gov/secretary/remarks/2013/08/212626.htm> accessed Dec 2014.

This claim is based on a particular understanding of the UN Charter rules on self-defense. The US position on the matter has evolved over time, but its current position is encapsulated in a series of speeches given by senior members of the Obama administration. In these, the United States recognizes that "[i]nternational legal principles constrain our ability to act unilaterally," but argues that "the use of force in foreign territory would be consistent with these ... principles ... after a determination that the [foreign state] is unable or unwilling to deal effectively with a threat to the United States."[53] Against that background, drone strikes are claimed to be justified if directed against "leader[s] of Al-Qaeda or an associated force who presen[t] an imminent threat of violent attack" against the United States.[54] With regard to the notion of "imminence," the US drone policy is premised on a more flexible understanding of "imminence": one that has to be "broadened in light of the modern-day capabilities, techniques, and technological innovations of terrorist organizations."[55] Consequently:

> the condition that an operational leader present an "imminent" threat of violent attack against the United States does not require the United States to have clear evidence that a specific attack on U.S. persons and interests will take place in the immediate future. Given the nature of, for example, the terrorist attacks on September 11, in which civilian airliners were hijacked to strike the World Trade Center and the Pentagon, this definition of imminence, which would require the United States to refrain from action until preparations for an attack are concluded, would not allow the United States sufficient time to defend itself.[56]

Arguably going even further, the US government has claimed that it faces a "continuing, imminent threat," triggering a continuing right to use self-defense to avert attacks.[57] The key point that emerges from this brief summary of practice is that the US has sought to defend the legality of drone strikes through a particular

---

53   Attorney General Eric Holder, speech at Northwestern University School of Law, Chicago (March 5, 2012) available at <http://www.justice.gov/opa/speech/attorney-general-eric-holder-speaks-northwestern-university-school-law> accessed Oct 2014.

54   Ibid.

55   John Brennan, "Remarks at Harvard Law School: Strengthening Our Security by Adhering to Our Values and Laws" (Sept 16, 2011) available at <http://www.whitehouse.gov/the-press-office/2011/09/16/remarks-john-o-brennan-strengthening-our-security-adhering-our-values-an> accessed Oct 2014.

56   US Department of Justice, "Lawfulness of a Lethal Operation Directed against a U.S. Citizen Who Is a Senior Operational Leader of Al-Qa'ida or an Associated Force" (Nov 8, 2011) available at <http://www.fas.org/irp/eprint/doj-lethal.pdf> accessed Oct 2014.

57   President Obama, "Remarks by the President at the United States Military Academy Commencement Ceremony" (May 28, 2014) available at <http://www.whitehouse.gov/the-press-office/2014/05/28/remarks-president-united-states-military-academy-commencement-ceremony> accessed Dec 2014.

interpretation of the right to self-defense. Against this background, the following sections examine the legality of such claims under *the jus ad bellum*.

D. Assessing Recent Practice

Given the diverse nature of state practice that we have just examined, it is remarkable that, by and large, states seeking to justify forcible action taken against terrorist entities have relied on their right to self-defense. States have not asserted a free-standing right to enforce international law against terrorists, or reverted to old arguments about the legality of armed reprisals, but rather considered it as part of the law of self-defense.[58] This at some level may be read as a confirmation of the *jus ad bellum*—the basic structure of which remains intact. However, reliance on self-defense, in Christine Gray's words, increasingly amounts to a "ritual incantation of a magic formula."[59] As Daniel Bethlehem perceptively notes, "[t]he reliance by States on self-defence in virtually every conceivable circumstance" leads to "normative drift, as attempts have been made to stretch the concept."[60] Given such widespread invocation of the right to self-defense we are forced to consider what remains of the traditional interpretation explored earlier, starting with the previously strict rules on attribution.

1. What remains of strict rules on attribution?

As a result of recent practice many commentators have questioned whether the strict test of attribution and the restrictive approach to the *jus ad bellum* prevalent in the 1980s is still valid,[61] with some proposing a new reading of self-defense that would abolish the need for attribution as the Court had envisaged in its *Nicaragua* judgment, allowing states to invoke self-defense and use force against terrorists irrespective of the involvement of any state.[62] Interestingly, the ICJ—which typically lacks jurisdiction to address questions of self-defense and has never had

---

58   Dinstein (n 5) at 247.

59   Gray (n 10) at 119.

60   D Bethlehem, "International Law and the Use of Force: the Law as It Is and as It Should Be," written evidence to the Select Committee on Foreign Affairs—Minutes of Evidence (June 8, 2004), available at <www.publications.parliament.uk/pa/cm200304/cmselect/cmfaff/441/4060808.htm> accessed Oct 2014, at para. 21.

61   A Cassese, "Terrorism is also Disrupting some Crucial Legal Categories of International Law" (2001) 12 Eur J of Int'l Law 993; Stahn (n 33) at 827; SD Murphy, "Self-Defense and the Israeli Wall Advisory Opinion: An Ipse Dixit from the ICJ?" (2005) 99 American J of Intl Law 62; R Van Steenberghe, "Self-Defence in Response to Attacks by Non-State Actors in the Light of Recent State Practice: A Step Forward?" (2010) 23 Leiden J of Intl Law 183.

62   See SD Murphy, "Terrorism and the Concept of Armed Attack in Article 51 of the UN Charter" (2002) 43 Harv Int'l Law J 41; Christopher Greenwood, "International Law and the Pre-emptive Use of Force: Afghanistan, Al-Qaida, and Iraq" (2003) 4 San Diego Int'l Law Journal 7, at 17.

an opportunity to pronounce on drones—has been called upon to address that question in two recent decisions, namely the *Wall* and *Armed Activities* cases.

Initially the Court appeared to stand by its earlier position. In the *Wall* case the Court stated simply "Article 51 of the Charter ... recognizes the existence of an inherent right of self-defence in the case of armed attack *by one State against another State*" and as such Israel's measures designed to prevent attacked by terrorist entities operating from the occupied territories were not justified.[63] However only one year later in the *Armed Activities* case the Court rejected Uganda's reliance on self-defense as justification for measures it had taken against irregular forces operating from the Democratic Republic of Congo (DRC) because such acts could not be attributed to the DRC.[64] Curiously, however, the Court then stated that it was unclear "whether and under what conditions contemporary international law provides for a right of self-defence against large-scale attacks by irregular forces."[65] The Court's pronouncements in the *Wall* and *Armed Activities* cases were the subject of significant criticism from individual judges for their lack of reasoning and for not taking into account recent state practice. In this regard, Judges Burgenthal, Kooijmans, and Simma argued that self-defense was available to states "even if [attacks against that state] cannot be attributed to the territorial State."[66] The uncertainty created by the Court's pronouncements, together with the considerable practice of states, led to serious questions being raised as to whether a more lenient approach to the *jus ad bellum* had developed.

The practice summarized above indeed seems to suggest that this is the case: too numerous are assertions of a right of self-defense in response to attacks that, while armed, are not "effectively controlled" by another state. The contours of this new interpretation of the law are not yet fully clear, and in fact there are two competing perspectives of change.[67] The first argues that any armed attack, whether by a state or another group (such as a terrorist organization) triggers a right of self-defense.[68] The second retains the traditional understanding of self-defense as a justification for the use of force between states, but recognizes that special rules exist with regards to the attribution of uses of force by terrorist entities[69]—to the point where

---

63 *The Wall Advisory Opinion*, para 139.

64 *Armed Activities Case* (n 2) at para 147.

65 Ibid at para 147.

66 Ibid, separate opinion of Judge Simma, at para 12; and separate opinion of Judge Koiijmans at para 30 and the declaration of Judge Burgenthal in *The Wall Advisory Opinion*, para 6.

67 Tams (n 32); Kimberley N Trapp, "The Use of Force against Terrorists: A Reply to Christian J. Tams" (2009) 20 Eur J of Int'l Law at 1049; T Ruys and S Verhoeven, "Attacks by Private Actors and the Right of Self-Defence" (2005) 10 Journal of Conflict and Security Law at 289; Ashley S Deeks, "Unwilling or Unable: Toward a Normative Framework for Extraterritorial Self-Defense" (2012) 52 Va J Int'l Law 483.

68 Kimberley N Trapp, "Back to Basics: Necessity, Proportionality, and the Right of Self-defence Against Non-State Terrorist Actors" (2007) 56 Intl and Comparative Law Q 141.

69 See Tams (n 32) at 385; Ruys and Verhoeven (n 67).

self-defense can be used against terrorists operating on foreign state territory when the foreign state is unable or unwilling to take action against the terrorists.[70]

As has been noted elsewhere,[71] of these two forms of "opening up" the law of self-defense, the second is perhaps more easily brought in line with the traditional regime: it accepts that, under the *jus ad bellum,* forcible action abroad is illegal because it is used against a foreign state—and thus violates Article 2(4) of the UN Charter, international law's cornerstone rule.[72] In line with this, the justification for using force abroad, even where the targets are terrorists, should also be construed as an inter-state defense.

That said, differences ought not to be exaggerated: the key distinction is not between the two ways of "opening up" self-defense, but between them (seen as part of a general trend) and the traditional understanding requiring a strict rule of attribution. And with respect to this real divergence, state practice indeed seems to have acted as a catalyst of change. For the debate about drone warfare, this change is relevant. It means that terrorist attacks—unlike under the traditional understanding of the *jus ad bellum*—can today amount to armed attacks in the sense of Article 51 of the UN Charter, even where they have not been effectively controlled by a foreign state. This means that, as a matter of principle, self-defense becomes available against terrorist attacks; and that drone strikes, subject to the further conditions addressed in the following, can *in principle* be viewed as justified measures of self-defense.

2. What remains of the threshold requirement?

For self-defense to be available, under the Charter regime, it is not sufficient for terrorist attacks to involve the use of force; they need to reach the threshold of an armed attack. As noted above, under the traditional understanding this presupposed a qualified breach, or indeed, in the words of the ICJ, a "most grave for[m] of the use of force."[73] Recent practice suggests that this element of the traditional approach has also come under pressure, raising a number of questions.

For instance, it is important to note that while wavering over the question of state versus non-state attacks,[74] notwithstanding recent state practice, international courts have affirmed the need for some form of threshold criterion. On a number of occasions, the ICJ has reaffirmed the distinction between "armed attacks" on

---

70   CJ Tams and JG Devaney, "Applying Necessity and Proportionality to Anti-Terrorist Self-Defence" (2012) 45 Israel Law Review at 91; CJ Tams, "Armed Conflict and Terrorist Organisations" in LJ van den Herik and Nico Schrijver (eds) *Counter-Terrorism Strategies in a Fragmented International Legal Order: Meeting the Challenges* (Cambridge University Press 2013); see also the terminology used in speeches formulating the US policy on drones, for example, the speech of AG Holder (n 53).

71   For a focused debate see Tams (n 32) and Trapp (n 67).

72   See section II above.

73   *Nicaragua Case* (n 2) at para 191.

74   Section II(B) above.

the one hand, and other ("lesser") uses of force on the other.[75] For instance in the *Armed Activities* case (where the Court had left open whether states could respond to "attacks by irregular forces") the Court stated that the use of force could only be undertaken against "large-scale attacks."[76] The Eritrea-Ethiopia Boundary Commission has also endorsed the distinction.[77] As such, jurisprudence suggests that the distinction between armed attacks and lesser uses of force remains important.[78]

Practice is difficult to bring into line with this jurisprudence, which by and large affirms the line of the traditional understanding. The large-scale military anti-terrorist operations by states like Israel and Turkey seemed to have been triggered by incidents that were not themselves of a large-scale nature; however, these incidents were part of long-term campaigns.[79] The US drone policy goes beyond that: it seeks to apply self-defense to an almost "granular" level, where targeted measures against one person are designed to eliminate a threat posed by that one person.

The international community's relatively positive response to forcible actions by Turkey or Israel (as summarized above) could perhaps be rationalized as an implicit endorsement of the accumulation doctrine mentioned above. It may be arguable that, in an age of low-level but long-term conflict, international law should accept the possibility of "aggregating" (or "accumulating" for that matter) a series of small-scale attacks and, taken together, treat it as a large-scale armed attack sufficient to trigger the right to use self-defense.[80] A passage from the ICJ's judgment in the *Oil Platforms case* seems to admit as much. In this case the Court noted that "even taken cumulatively" a series of armed incidents in that case did not amount to an armed attack on the United States (thus perhaps implying that one should "take cumulative" a series of armed incidents).[81] Yet the trend toward recognition of the accumulation doctrine is far from consolidated—in fact there is little express mention of it in official documents. And perhaps this caution is justified, for the implications of admitting the doctrine would be considerable. Who is to decide whether consecutive attacks are sufficiently linked so as to be accumulated? What time frame is required for a series of attacks to be treated as consecutive? Would attacks have to come from the same state? How are attacks by different, but allied, actors to be qualified?

---

75 See for instance *Oil Platforms Case* (n 3) at paras 51 and 62.

76 *Armed Activities Case* (n 2) at para 147.

77 Award on Ethiopia's Jus ad bellum Claims 1–8, (2006) ILM 45 at 430.

78 Gray (n 10) at 144; L Shore et al, "The Legality Under International Law of Targeted Killings by Drones Launched by the United States" in *New York City Bar Association Report, June 16, 2014* (2014) at 78; cf. Alston (n 1) at 40.

79 Cannizzaro (n 22).

80 On the face of it, the argument supporting the doctrine no doubt sounds plausible: to take a hypothetical example, should a state really be precluded from defending itself forcibly against 20 small-scale attacks killing three nationals each, where self-defense would be permissible against one massive attack killing 60 citizens?

81 *Oil Platforms Case* (n 3) para 67; for comments on the doctrine see also *Land and Maritime Boundary between Cameroon and Nigeria (Cameroon v Nigeria: Equatorial Guinea intervening)* Judgment, ICJ Reports 2002, para 303; and *Armed Activities Case* (n 2).

And, most importantly, does the right of self-defense apply as long as a series of pinprick assaults continues? These questions point to potential implications of the accumulation of events doctrine, which as yet have hardly been broached, let alone addressed sufficiently.[82] They suggest that—while the accumulation doctrine may be on the rise, and while it is certainly not implausible—the *jus ad bellum* remains premised on the idea of a threshold criterion, setting armed attacks apart from other violent incidents that cannot simply be aggregated.

While the accumulation doctrine may stretch the right of self-defense, the US policy on drones presents a challenge of a different kind. It is based on the most general of links between attacks extending over a considerable period of time. What is more, in its focus on individual members of terrorist organizations, it moves away from the concept of a large-scale attack conduct by an organized entity—whether a state or a non-state actor—and views self-defense as a means of responding against threats posed by individuals. This might perhaps be said to illustrate the risks of an accumulation doctrine writ large. But more importantly, by singling out the risks posed by members of a terrorist organization (which then can bet targeted), the US policy seems to turn the right of self-defense into a law enforcement measure, or a police power, available against individuals. This construction is perhaps the end point of the "opening up" of self-defense. It is clear that the US drone policy accepts—indeed argues for—such an opening up, and a number of states at least tacitly seem to accept the underlying claim. However, even our brief summary of arguments suggests that it marks a conceptual break with the traditional understanding of self-defense, and the need for a large-scale armed attack in particular.

While the US policy has no doubt started a process of debate, in our assessment, at present it reflects the position of one state that has not been tested seriously, and that marks a break with the traditional construction of self-defense which, even throughout the "anti-terrorist" challenge to the *jus ad bellum,* remained intact. On the basis of this traditional understanding, drone strikes can be used against terrorist organizations, but only in response to armed attacks that reach the threshold, that is those that can be qualified as "large-scale attacks."[83] In contrast, they are not a

---

82   For a different view, but with important limitations, see Claus Kress "Self-Defence, Conflict Qualification, and the Geographical Scope of Targeting Powers in (Transnational) Non-International Armed Conflicts" in Bhuta et al: "Though controversial, the better view is to construe the concept of a non-State armed attack so as to require large-scale transnational violence by a non-State organization. It is suggested that the term 'large-scale' is, in this specific context, to be construed in accordance with the intensity element of the concept of non-international armed conflict. Under the lex lata, there is room to argue that the intensity threshold may be reached through an accumulation of lower scale acts of force. Importantly, however, the different components of a continuous non-State armed attack must emanate from the same non-State organizational source and the intervals between the several lower-scale attacks must not be too significant. If those limits are respected, it would seem fair to take the accumulation of events into account in determining whether a non-State armed attack has occurred" at para 4.

83   *Armed Activities Case* (n 2) at para 147.

lawful means of responding against small-scale terrorist attacks or waves of small-scale attacks either. While the law may be in the process of development, in our view there is insufficient evidence suggesting that the threshold criterion had been given up, or that self-defense had been turned into a police power capable of being used against individuals.

3. The scope of the right: boundless self-defense?

While much of international legal scholarship has focused on issues such as the armed attack requirement, the scope of the right of self-defense is equally affected by "normative drift."[84] To elaborate, if in recent practice references to self-defense have at times become mere "ritual incantations"[85] then it is because a number of states have asserted a broad license to use force against terrorists that bears little resemblance to the traditional understanding of self-defense as a defensive concept designed to justify temporary responses against armed attacks.

4. Anticipation, imminence, preemption?

Expansive tendencies are manifold, but two in particular stand out. The first concerns the temporal scope of the right to self-defense. As is clear from the summary of practice, states have increasingly claimed a right to use self-defense in order to thwart future attacks. In so doing, they have gone beyond traditional debates about anticipatory self-defense.[86] US defense strategies published soon after the 9/11 attacks adopted a particularly broad approach, asserting a right to pre-empt future threats from emerging.[87] The underlying normative claim has however met with rather concerted resistance, being rejected by, among others, the ICJ (in *Armed Activities*[88]), important UN Reform Reports,[89] and by other

---

84   Daniel Bethlehem, "Self-Defense Against an Imminent or Actual Armed Attack by Nonstate Actors" (2012) 106 American J of Intl Law at 769.

85   Gray (n 10).

86   Section II(b) above.

87   White House, The National Security Strategy of the United States of America (Oct 2002) available at <http://www.state.gov/documents/organization/63562.pdf> accessed Oct 2014; and US Department of Defense, "The National Defense Strategy of the United States of America" (March 2005), available at <http://www.defense.gov/news/mar2005/d20050318nds1. pdf> accessed Oct 2014; for extensive debates see Reisman and Armstrong, "The Past and Future of the Claim of Preemptive Self-Defense" (2006) 100 American J Intl Law 525; Hoffmeister, "Preemptive Strikes: a New Normative Framework" (2006) 44 *Archiv des Völkerrechts* 187; Murphy, "The Doctrine of Preemptive Self-defense," (2005) 50 Villanova Law Rev 699.

88   *Armed Activities Case* (n 2) in its judgment, the Court held that Uganda's assertion of a right to defend security interests "does not allow the use of force by a State to protect perceived security interests ... Other means are available to a concerned State, including, in particular, recourse to the Security Council"; at para 148.

89   Kofi Annan, "In Larger Freedom: Towards Security, Development and Human Rights for All" Report of the Security General of the United Nations, UN Doc A/59/2005 (paras 124–125); as well the High Level Panel Report, UN Doc A/59/565 at para 188.

states.[90] The US no longer uses this terminology. At the same time, however, the international community now seems quite willing to accept that force can be used to preclude future attacks if the responding state possesses reliable information of their imminence. The traditional view (never easy to apply in practice) that the right to self-defense would only emerge once the armed attack had begun has lost support.[91] The debate has become one about the proper interpretation of "imminence."

In a way, there has been some convergence: there is at least agreement on the proper categorization. However, this is a relatively thin level of agreement. For example, President Obama's reference to a "continuing, imminent threat," while accepting the notion of "imminence," marks a clear shift away from the traditional interpretation of the law, as well as from the (pre-drone) anti-terrorist practice of states described in section III(b) above. The point may be illustrated by contrasting US statements on drones with the careful attempt by a group of British lawyers to distill principles governing the use of force in self-defense in a post-9/11 setting. With respect to imminence these so-called Chatham House Principles note that a "forcible response [against an imminent, future attack] lies at the limits of an already exceptional legal category, and therefore requires a correspondingly high level of justification."[92] Against that background, there would have to be "circumstances of irreversible emergency" for self-defense against an imminent attack to be lawful: "any further delay in countering the intended attack will result in the inability of the defending state effectively to defend itself against the attack."[93]

These guidelines can be taken to reflect a balanced assessment of the law, one that recognizes the need for self-defense to permit responses against imminent attacks, but that understands "imminence" to require a specific threat on the verge of materializing. In that respect, it reflects the rationale of the Webster Formula—there must have been an "instant and overwhelming" need for action that left "no moment for deliberation."[94] It may well be that some of the US drone strikes could be justified under those terms. However, the assertion of a "continuing, imminent threat" would seem very difficult to bring in line with any of the accepted understandings

---

90   Pars pro toto, see the UK Attorney General's statement that "international law permits the use of force in self-defence against an imminent attack but does not authorise the use of force to mount a pre-emptive attack against a threat that is more remote"; Statement in HL Debs, April 21, 2004, vol. 660, cols. 370–371.

91   For example, some form of anticipatory self-defense against imminent attacks is accepted both in "In Larger Freedom" and in the "UN High Level Panel Report" (n 89).

92   Chatham House Principles, principle 4, commentary.

93   Chatham House Principles, principle 4.

94   On this issue see Timothy Kearley, "Raising the Caroline" (1999) 17 Wis Intl Law J 325; Maria Benvenuta Occelli, "Sinking the Caroline: Why the Caroline Doctrine's Restrictions on Self-Defense Should Not Be Regarded as Customary International Law" (2003) 4 San Diego Int'l Law J 467; James A Green, "Docking the Caroline: Understanding the Relevance of the Formula in Contemporary Customary International Law Concerning Self-Defense" (2006) 14 Cardozo J Int'l & Comp L 429.

of imminence and cannot be taken to reflect the current law. The better view is that, contrary to assertions in the Department of Justice White Paper quoted above, "clear evidence that a specific attack ... will take place in the immediate future"[95] remains necessary for self-defense to be lawful. (For a historical investigation into the question of "imminence" see Chapter 6 by Steven J. Barela.)

5. Necessity and proportionality; the defensive character of self-defense?

The normative drift toward some form of anticipatory self-defense also raises issues with regard to the traditional constraints on the use of force in self-defense, namely the principles of necessity and proportionality. The application of necessity and proportionality is no less essential when considering self-defense against terrorist entities than it is in the traditional inter-state setting.[96] In relation to anti-terrorist practice generally (whether through drones or otherwise), necessity in particular has played a relevant role, as typically one would expect the territorial state to take action against terrorists operating on its territory—in fact, international law expects it to do so. Only if no such territorial action is forthcoming (e.g. because the territorial state is unable or unwilling to take action) could self-defense be considered necessary.[97] With respect to drones, Special Rapporteur on Extrajudicial, Summary or Arbitrary Executions Christof Heyns has argued that "whether a State is unable or unwilling to take action, the State acting in self-defence might be required to request such action before the commencement of acts taken in self-defence, to establish that it is necessary."[98]

Perhaps more fundamentally, normative drift is evident with regard to the defensive character of the right to self-defense in recent drone practice. In short, recent wide-ranging and protracted drone programs bear little resemblance to a right of supposedly defensive character, which has traditionally held that force can only be used in self-defense within a relatively short period of time after the initial attack lest it lose its defensive character and instead constitute an unlawful reprisal.[99] But perhaps in this respect, the US policy on drones is in fact in line with other, recent invocations of self-defense: one need only look at examples such as Iran's pursuit of Kurdish fighters into Iraq for purposes other than repelling an attack: to arrest the individuals in question for evidence.[100] As such, a considerable part of the recent anti-terrorist practice seems to have stretched the idea that self-defense should be defensive in character. The defensive character of measures is regularly

---

95   US Department of Justice (n 56).

96   Tams and Devaney (n 70).

97   Special Rapporteur on Extrajudicial, Summary or Arbitrary Executions, "Armed Drones and the Right to Life," 91, UN Doc A/68/382 (Sept 13, 2013) at 94.

98   Ibid at 91.

99   Mary Ellen O'Connell, "Unlawful Killing with Combat Drones: A Case Study of Pakistan, 2004–2009" in S Bronitt, M Gani and S Hufnagel (eds), *Shooting to Kill: Socio-Legal Perspectives on the Use of Lethal Force* (Hart 2012) at 278.

100   Wettberg (n 39) at 151.

proclaimed (hence the attempt to portray measures as necessary to thwart imminent attacks) but there is a risk of merely paying lip-service to the need to act in self-defense while in fact pursuing much broader aims.[101] This risk can never be fully avoided, but the recent practice summarized in the preceding sections evidences a willingness to use a defensive title to pursue attacks designed to root out threats. In response, it is worth citing another statement by the Special Rapporteur on Extrajudicial, Summary or Arbitrary Executions, who recently noted that:

> The right to self-defence persists only for so long as it is necessary to halt or repel an armed attack and must be proportionate to that aim. In determining what is necessary to bring an attack to an end and what is a legitimate objective for self-defence, however, States are not entitled to continue to act in self-defence until the absolute destruction of the enemy is achieved, such that the enemy poses no long- term threats.[102]

While it is in some ways understandable that states wish to rely upon, and further adapt, the right to self-defense in relation to the use of drones, there must come a point at which the right can be stretched no further without losing its basic character. In fact, were one to apply a strict interpretation of the right to self-defense to contemporary drone practice, it is clear that much of this practice could not be justified as self-defense in international law. While states have undoubtedly attempted to reshape the right to self-defense, it has become increasingly clear that there is little room for maneuver left.

IV. Conclusion

Much of the legal debate about drones focuses on questions of human rights law and the law of armed conflict. The preceding considerations suggest that this focus is too restrictive. The use of drones raises equally complex questions under the *jus ad bellum*; and perhaps these ought in fact to take priority: before discussing whether a state's drone program could be brought in line with, say, its obligations under applicable human rights treaties, it is crucial to assess whether the state was at all entitled to use force abroad. Seen from that perspective, there is no principled difference between the use of force by means of drones and other the modalities of using force. As such, the preceding discussion has drawn in rather large measure from legal arguments developed prior to the use of drones on the current scale.

As we have tried to make clear, the use of drones abroad *prima facie* violates the UN Charter's prohibition on military force as set out in Article 2(4). If validly given, consent may remove the wrongfulness of that conduct under the *jus ad bellum* (though perhaps not under human rights law). In the absence of consent,

---

101   P Klein, "Vers la reconnaissance progressive d'un droit à des représailles armées?" in O Corten et al, *Le droit international face au terrorisme* (Pedone 2002) at 249.

102   Report of the Special Rapporteur on Extrajudicial ..., (n 98) at 90.

states pursuing drone warfare have to rely on the right of self-defense recognized in the UN Charter. While Article 51 provides a clear starting point for the debate, State practice in the last two decades and the military fight against terrorism, in particular, have resulted in considerable uncertainty about the conditions and limits of self-defense. As the fight against terrorism has been militarized, including through the increasing use of drones, the law has responded. Even in an area as crucial as military force it has clearly not been static but has been adapted and adjusted in a process of contestation and claims.

As we have shown, in important respects state practice has opened up the right to self-defense, which now no longer requires an armed attack by a state, but can be taken against attacks by terrorists operating abroad, and which now permits responses against imminent attacks. This may have been an overdue adjustment of the law that takes account of the need to fight terrorists effectively. However, as our discussion has shown, the application of self-defense to terrorism remains highly contentious. More specifically, our analysis suggests that—notwithstanding the more "liberal" invocation of self-defense in recent practice—international law continues to impose three important limitations on the use of self-defense:

i. armed attacks must be of a significantly large scale to trigger a right to self-defense;
ii. responses can only be directed against future attacks if these are imminent and present a "circumstance of irreversible emergency"; and
iii. the action taken must be necessary and proportionate as well as being defensive in nature, having the sole objective of bringing to an end the immediate threat.

None of these criteria are self-explanatory. No doubt Daniel Bethlehem is right to emphasize that the "principle is sensitive to the practical realities of the circumstances that it addresses, even as it endeavors to prohibit excess and the egregious pursuit of national interest."[103] And indeed our analysis highlights the remarkable ability of the *jus ad bellum* to develop and change. However, there are some respects in which the normative drift in the right to self-defense, required in order to justify the use of drones as part of the right to self-defense, ought to be of concern. For instance, it seems clear that in opening up the right to self-defense, in particular in relation to the temporal limitation and the defensive character of the right, it becomes much more open to abuse. This willingness to broaden the right to self-defense in the face of recent practice in relation to the use of drones risks accepting reprisals thinly-veiled as measures taken in self-defense. In fact, it seems clear that if the legal justifications for the use of drones (most often the "ritual incantation" of an ever-broader right to self-defense) as as put forward by some states were accepted wholesale the right to self-defense as it was once understood would cease to have any meaning and there would be little to stop states running roughshod over the Charter's system of collective security.

---

103   Bethlehem (n 84) at 773.

Chapter 2

Who Can Be Killed?: Legal Targets in Non-International Armed Conflicts

Patrycja Grzebyk*

I. Introduction

Remotely piloted aircraft are becoming a common fixture in contemporary armed conflicts, especially in case of conflicts of asymmetrical nature.[1] The benefits of using this type of weapons are clear: limited risk exposure is accompanied by the ability to track the enemy-controlled territory for hours, select a target, and chose the most suitable time of attack. It is therefore hardly a surprise that many states have implemented unmanned flight systems, and are now working on their advancement.[2] Even the United Nations decided to use reconnaissance and surveillance drones within its peacekeeping operations, despite preliminary doubts raised by some states.[3] However, what is most controversial is the use

* This project was financed by the Polish National Center of Science based on the decision number DEC-2013/11/D/HS5/01413.

1 On the history of use of drones in combat, see Victor Hansen, "Predator drone attacks" (2011–2012) 46 New Eng. Law Rev On Remand, 27, at 28 ff; Mary Ellen O'Connell, "Seductive Drones: Learning from a Decade of Lethal Operations" (2011) 21 JLIS 116, at 118–119; Brendan Gogarty and Isabel Robinson, "Unmanned Vehicles: A (Rebooted) History, Background and Current State of the Art" (2011) 21 JLIS 1, at 9 ff; William C Marra and Sonia K McNeil, "Understanding 'the Loop': Regulating the Next Generation of War Machines" (2013) 36 Harv JL&Pub Pol 1139, at 1161 ff.

2 See Sebastian Wuschka, "The Use of Combat Drones in Current Conflicts—A Legal Issue or a Political Problem?" (2011) 3 Goettingen J Int'l Law 891, at 892; Stuart Casey-Maslen, "Pandora's Box? Drone strikes under *ius ad bellum, ius in bello* and international human rights law" (2012) 94 IRRC 597, at 598–9, 608; Chris Cole, "Is Drone Proliferation about to Explode?" Drone Wars UK (May 25, 2012) <http://dronewars.net/2012/05/25/is-drone-proliferation-about-to-explode/> accessed Sept 2014.

3 The first drones (Falco drones, produced by the Italian enterprise Selex ES, a unit of the Italian defense group Finmeccanica) were used by the UN peace-keeping forces in the MONUSCO operation in the eastern part of Democratic Republic of Congo (DRC) to monitor the border between DRC and Rwanda. See eg Kenny Katombe, "U.N. forces use drones for first time, in eastern Congo" Reuters (Dec 3, 2013) <http://www.reuters.com/article/2013/12/03/us-rop-congo-democratic-drones-idUSBRE9B20NP20131203≥ accessed Sept 2014. See also <http://www.unmultimedia.org/tv/unifeed/2013/12/drc-drones-launch/> accessed Sept 2014.

of combat drones armed with missiles and bombs.[4] Germany for instance has at first put a hold on acquiring armed drones, considering the serious ambiguities pertaining to the legality of their use.[5] The hesitation is by no means universal. The United States is the largest operator of armed drones, but weapons of this type are also present in the military equipment of China, Israel, and Iran.[6]

Introduction of a new type of weapon or means of warfare usually generates discussion. Often, it also provokes outcries to prohibit it. This is particularly true if the new weapon is unprecedentedly effective in killing. In the Middle Ages, the crossbow was perceived as a weapon of mass destruction, a threat to the existence of the human species. From today's perspective, it may seem ridiculous and exaggerated. The question is: Are the drones the crossbow of our times?

The legal problems related to the usage of drones are twofold. Firstly, they pertain to the legal regulations as to the use of force (particularly if there is no consent of the state in the territory of which the attack is taking place).[7] Secondly, they involve the regulations with regard to human rights law (when the attack is

---

4   Combat drones can be equipped with eg Hellfire missiles; guided bomb units (eg Paveway 12) and lighter missiles. They are also able to launch weapons of mass destruction. See Meredith Hagger and Tim McCormack, "Regulating the Use of Unmanned Combat Vehicles: Are General Principles of International Humanitarian Law Sufficient?" (2011) 21 JLIS 74, at 85 ff. Even more controversies arise with regard to fully autonomous drones which would decide on their own for the launching an attack. However at this moment their use remains in the realm of fiction.

5   See eg Sven Heymann, "German military deploys lethal drones in Afghanistan" World Socialist Web Site (March 26, 2013) <http://www.wsws.org/en/articles/2013/03/26/dron-m26.html> accessed Sept 2014; "'We reject illegal killings': Germany suspends drone purchase" RT (Nov 15, 2013) <http://rt.com/news/germany-halts-drone-purchase-763/> accessed Sept 2014; Karolina Libront, "Niemiecka awantura o drony" Portal Geopolityka. org, (June 29, 2013) <http://www.geopolityka.org/analizy/2324-niemiecka-awantura-o-drony> accessed Sept 2014. One of the most controversial statement in German debate was of Thomas de Maizière, former Minister of Defense, who claimed that drones are ethically neutral, see Robert Birnbaum, "Die Drohne als Henker und Richter" *Der Tagespiegel* (April 21, 2013) <http://www.tagesspiegel.de/meinung/essay-die-drohne-als-henker-und-richter/8097512.html> accessed Sept 2014. See also the assessment of drone warfare from an ethical point of view in: Joseph Pugliese, "Prosthetics of Law and the Anomic Violence of Drones" (2011) 20 Griffith Law Rev 931.

6   Hin-Yan Liu, "Categorization and legality of autonomous and remote weapons systems" (2012) 94 IRRC 628, at 634.

7   See eg European Parliament Resolution on the Use of Armed Drones (2014/2567(RSP), para E, available at <http://www.europarl.europa.eu/sides/getDoc. do?pubRef=-//EP//TEXT+MOTION+P7-RC-2014–0201+0+DOC+XML+V0//EN&language=en> accessed Sept 2014. Especially Pakistani politicians indicated that use of drones in their territory violates Pakistan's sovereignty; see the interview with Masood Khan, Pakistan's permanent representative to the United Nations for Radio Free Europe of April 24, 2014, <http://www.rferl.org/content/drones-pakistan-un/25206076.html> accessed Sept 2014.

outside the scope of armed conflict)[8] and international humanitarian law (when the attack is within the scope of armed conflict). In this paper, I focus on the legality of employment of unmanned combat aerial vehicles (UCAV) within non-international armed conflicts (NIACs), and in particular on the choice of legitimate human targets and rules on their elimination.

A NIAC, according to the International Criminal Tribunal for the former Yugoslavia (ICTY) jurisprudence, is a conflict in which there is "protracted armed violence between governmental authorities and organized armed groups or between such groups within a State."[9] When a third-party state intervenes with its armed forces on the side of governmental authorities, it does not change the qualification of the conflict, that is, the conflict does not become international. Consequently, drones operations in, for example, Afghanistan (since June 2002) are considered a part of warfare in the non-international armed conflicts that is taking place there, because they were/are performed with consent of local authorities.[10] Doubts arise as to qualification of situations in Somalia, Yemen, or Pakistan as a NIAC.[11] Yet these doubts have been disregarded by the United States, which claims to be engaged in a global armed conflict with Al-Qaeda and its associates. (For a full discussion of this legal question see Chapter 3 by Katja Schöberl.) As a result, the United States applies the law of non-international armed conflicts also to attacks on the territory of these countries, no matter if there is a NIAC on their territory

---

8   According to Sarah Kreps, 98 per cent of the estimated 465 non-battlefield targeted killings undertaken by the United States since November 2002 were carried out by drones, Sarah Kreps, "The Foreign Policy Essay: Preventing the Proliferation of Armed Drones" Lawfare (April 13, 2014) <http://www.lawfareblog.com/2014/04/the-foreign-policy-essay-preventing-the-proliferation-of-armed-drones/> accessed Sept 2014.

9   ICTY, *Prosecutor v Dusko Tadic*, Decision on the defense motion for interlocutory appeal on jurisdiction, Oct 2, 1995, IT-94-1-AR72, at para 70. See also Article 8(2)(f) of the Rome Statute, 2187 UNTS 90.

10   See eg Robin Geiß and Michael Siegrist, "Has the armed conflict in Afghanistan affected the rules on the conduct of hostilities?" (2011) 93 IRRC 11, at 13 ff.

11   On possible qualification of the situation in Pakistan, Somalia, and Yemen as non-international armed conflict (during different periods), see Benjamin R Farley, "Targeting Anwar Al-Aulaqi: A Case Study in U.S. Drone Strikes and Targeted Killing" (2011–2013) 2 Nat'l Sec Law Brief 57, at 62–63; Kevin Jon Heller, "'One Hell of a Killing Machine' Signature Strikes and International Law" (2013) 11 J Int'l Criminal Justice 89, at 92; Louise Arimatsu and Mohbuba Choudhury, "The Legal Classification of the Armed Conflicts in Syria, Yemen and Libya" International Law PP 2014/01, Chatham House (March 2014) 1, at 29 ff, available at <http://www.chathamhouse.org/sites/default/files/home/chatham/public_html/sites/default/files/20140300ClassificationConflictsArimatsuChoudhury1.pdf> accessed Sept 2014; Human Rights Watch, "'Between a drone and Al-Qaeda.' The Civilian Cost of US Targeted Killings in Yemen" October 2013, available at <http://www.hrw.org/reports/2013/10/22/between-drone-and-al-qaeda> accessed Sept 2014 84; also in part Wuschka (n 2) at 904; cf. Donna R Cline, "An Analysis of the Legal Status of CIA Officers Involved in Drone Strikes" (2013) 15 San Diego Int'l Law J 51, at 53; Michael W Lewis, "Drones and the Boundaries of the Battlefield" (2011–2012) 47 Tex Int'l Law J 293, at 295.

between governmental forces and any armed group.[12] Without further discussion of the scope of the battlefield and validity of American argumentation, this chapter investigates whether the rules of international humanitarian law (IHL) governing the choice of human targets are applied in contemporary drone attacks.

In the first part of the chapter, I analyze if drones should be considered a special kind of means of warfare from the point of view of humanitarian law. In the second part, I present the problems related to choosing legal human targets within NIACs. In the third part, I analyze the issues arising with regard to the elimination of such targets (proportionality, precautions, "kill or capture"). The last part is devoted to the problem of accountability for drone strikes performed in violation of IHL.

II. Means of Warfare of Special Concern or Just the Modern Crossbow?

There is no convention that prohibits the use of combat drones,[13] and the fact that combat drones are used extensively and many states are willing to acquire them may indicate that there is also no customary rule forbidding their use.[14] However, the Lotus principle—according to which restriction on state's independence cannot be presumed[15]—does not fully apply to states' right to use drones, as law of armed conflicts clearly provides that parties to the conflict are not unlimited in their choice of means and methods of warfare.[16]

---

12   Harold Hongju Koh, Speech: "The Obama Administration and International Law" Annual Meeting of the American Society of International Law, (March 25, 2010) available at <http://www.state.gov/s/l/releases/remarks/139119.htm> accessed Sept 2014; Milena Sterio, "The United States' Use of Drones in the War on Terror: The (Il)Legality of Targeted Killings under International Law" (2012–2013) 45 Case W Res J Int'l Law 197, at 201 ff; Susan Breau and Marie Aronsson, "Drone attacks, International Law, and The Recording of Civilian Casualties of Armed Conflict" (2012) 35 Suffolk Transnat'l L Rev 255, at 266–270; Andrew C Orr, "Unmanned, Unprecedented, and Unresolved: The Status of American Drone Strikes in Pakistan under International Law" (2011) 44 Cornell Int'l Law J 729, at 742 ff. See also *Hamdan v Rumsfeld*, 548 US 557 (2006), at 566–567.

13   The only non-proliferation regimes applied to drones today (and still in a limited manner) are the Missile Control Technology Regime (1987), the Wassenaar Arrangement (1995), the Intermediate-Range Nuclear Forces Treaty (1987), and the Treaty on Conventional Armed Forces in Europe (1992). See more in Kreps (n 8); Hagger and McCormack (n 4) at 84–85.

14   Ibid Hagger and McCormack (n 4) at 85.

15   Permanent Court of International Justice, "The case of the SS 'Lotus'" (Sept 7, 1927) Collection of Judgments, Series A—no. 10, 18.

16   Article 35(1) of the First Additional Protocol to Geneva Conventions of 1949 (AP I), 1125 UNTS 3.

From the point of view of IHL, there is no difference between qualifying UCAVs as weapons or just as the platform from which weapons are deployed,[17] as UCAVs may nonetheless constitute "means and methods of warfare."[18] Certainly, drones are not a standard platform, because they are unmanned and thus they do not endanger the life or health of their operator. However, the fact that they are controlled remotely has no impact on the applicability of basic principles of humanitarian law, such as the principles of distinction,[19] proportionality,[20] and obligation to take precautions.[21] Additionally, Article 35 of the first Additional Protocol to Geneva Conventions of 1949 (AP I) stipulates that all sorts of weapons, projectiles, material or methods of warfare which cause superfluous injury or unnecessary suffering are prohibited, as are methods and means of warfare which may cause widespread, long-term, and severe damage to natural environment. The subsequent Article 36 also provides that development, acquisition, or adoption of a new weapon, but also of new means or methods of warfare, requires determination if its employment does not violate any applicable rule of international law. The International Court of Justice in its advisory opinion on *Legality of the Threat or Use of Nuclear Weapons* of May 8, 1996 confirmed that the entire law of armed conflict "applies to all forms of warfare and to all kinds of weapons, those of the past, those of the present and those of the future."[22] Articles 35 and 36 of AP I apply in international armed conflict. However, according to the International Committee of the Red Cross (ICRC), also in NIACs, the customary rule on prohibition of use

---

17   See more on the qualification of drones as weapons or weapons systems in Hin-Yan Liu (n 6) at 629.

18   ICRC commentary to the Article 35 AP I emphasizes that "the words 'methods and means' include weapons in the widest sense, as well as the way in which they are used," Yves Sandoz, Christophe Swinarski, Bruno Zimmermann, *Commentary on the Additional Protocols of 8 June 1977 to the Geneva Conventions of 12 August 1949* (Martinus Nijhoff Publishers 1987) at 398.

19   See Articles 48, 51(1) of AP I, Article 13 of the Second Additional Protocol Geneva Conventions of 1949 (AP II); rules 1, 11, 12 and 71 as recognized in the study of the ICRC on customary international humanitarian law (CIHL), Jean-Marie Henckaerts and Louise Doswald-Beck, *Customary International Humanitarian Law, Volume I: Rules* (Cambridge 2005) at 3 ff. See also International Court of Justice (ICJ), Advisory Opinion, Legality of the Threat or Use of Nuclear Weapons (May 8, 1996) at para 78. More on principle of distinction, Michael W Lewis and Emily Crawford, "Drones and Distinction: How IHL Encouraged the Rise of Drones" (2012–2013) 44 Geo J Int'l Law 1127. Nils Melzer, *Targeted Killing in International Law* (OUP 2008) at 300 ff.

20   Articles 51(5)(b) and 57(2)(a)(iii) of AP I and Article 8(2)(b)(iv) of the Rome Statute refer only to international armed conflicts. However, see also rule 14 of CIHL, similar to the mentioned provisions, which must be applied to both international and non-international armed conflicts.

21   Rules 15–24 of CIHL applied in international and non-international armed conflicts (compare with Articles 57–58 of the AP I).

22   ICJ, Nuclear Weapons Advisory Opinion (n 19) at para. 86.

of means and methods of warfare which are of a nature to cause superfluous injury or unnecessary suffering and weapons which are by nature indiscriminate must be applied.[23] Additionally, taking into account that drones are used in all types of conflicts (international and non-international), their position as new means of warfare definitely warrants an assessment of legality.

In order to determine if the use of combat drones is compliant with IHL, it is necessary to consider not only drones as such, but also a combination of drones with specific bombs or missiles. In the case of combat drones, the projectiles they employ differ in no significant manner from other types of projectiles typically used in armed conflicts.[24] Shrapnel dispersion is smaller than from a bomb dropped by a plane or under mortar fire,[25] thus theoretically drone attacks should not cause greater collateral damage than other means of warfare currently deemed acceptable; in fact, drone use should cause fewer incidental civilian casualties compared to the use of conventional weapons. Moreover, with the option of observing the potential target for multiple hours, the drone-employing party may take greater precautions and choose the target, place, and time of attack with greater precision, which further improves the protection of civilians.[26] At first glance, drones therefore appear to be more discriminate than other means of warfare, especially if we take into account lower flight speeds and greater opportunities for visual inspection of the area.[27] Drones are considered a precise weapon;[28] more importantly, they are considered extremely effective in destruction of terrorist networks.[29] Overall, it could be argued that drones are the perfect weapon in terms of humanitarian law.

---

23   See Rules 70–71 of CIHL.

24   See eg Philip Alston, *Study on Targeted Killings, Addendum to Report of the UN Special Rapporteur on Extrajudicial, Summary or Arbitrary Executions*, UN Doc A/HRC/14/24/Add.6 (May 28, 2010) at para 79: "a missile fired from a drone is no different from any other commonly used weapon, including a gun fired by a soldier or a helicopter or gunship that fires missiles." However, in my opinion, for assessment of discriminate or indiscriminate character of means of warfare it is essential to verify if the use of different platforms impacts the precision of the attack.

25   Daniel Byman, "Why Drones Work: The Case for Washington's Weapon of Choice" (2013) 92 Foreign Affairs 32.

26   Jakob Kellenberg, Philip Spoerri, "International Humanitarian Law and New Weapon Technologies," 34th Round Table on current issues of international humanitarian law, San Remo, September 8–10, 2011, (2012) 94 IRRC 809, at 812–813. See also Laurie R Blank, "'After Top Gun': How Drone Strikes Impact the Law of War" (2011–2012) 33 U Pa J Int'l Law 675, at 687 and Marco Sassòli, "Autonomous Weapons and International Humanitarian Law: Advantages, Open Technical Questions and Legal Issues to be Clarified" (2014) 90 Int'l Law Study 308, at 335–336.

27   Wuschka (n 2) at 896.

28   Hagger and McCormack (n 4) at 86.

29   Byman (n 25).

Yet the potential advantages of drones are difficult to verify in practice. The states that employ combat drones publish no official data on the number of drone attacks and the damage they inflicted. It is therefore difficult to find reliable data and thus to address the suspicion that drone use is inherently indiscriminate, taking into account large civilian losses. On the one hand, in an armed conflict it is difficult to expect precise statistics on the victims of each type of weapon used (drones, rifles of various types, mortars, helicopters, airplanes, etc.). However, in light of Article 36 AP I, states are obliged to reveal reliable data in order to assess legality of a weapon or means of warfare under IHL rules. Since drones have been introduced into the field as new means or a method of warfare, states are obliged to reveal statistical data at least at the beginning of their use. This is one reason why it is particularly disturbing that the US Senate stripped the provision from an intelligence bill in April of 2014 to require the president to make public the number of people killed or injured by the use of targeted lethal force when inflicted by remotely piloted aircraft.[30] For now, scholars are forced to use statistics prepared by private entities such as the Bureau of Investigative Journalism (BIJ), which examines drone strikes performed by the United States in Pakistan, Yemen, and Somalia. Data provided by the BIJ show that civilian deaths amount to approximately 25 percent of all killings in the case of Pakistan, and nearly 0 percent in the case of Somalia. Having in mind that civilians constitute usually 30 percent to 90 percent of all war victims depending on the conflict,[31] statistics concerning drone strikes are not particularly shocking. Information provided by media or NGOs (non-governmental organizations) focusing on killings of innocent victims (wedding guests or children)[32] distorts this general picture to a certain degree.[33]

Other scholars provide more alarming data. Some claim that killing one militant results in the death of even 50 civilians.[34] The research carried out by Leila Nadya Sadat demonstrates that civilian losses that resulted from the North Atlantic Treaty Organization (NATO) intervention in former Yugoslavia in 1999

---

30 See Mark Mazetti, "Senate Drops Bid to Report on Drone Use" *New York Times* (April 28, 2014) available at <http://www.nytimes.com/2014/04/29/world/senate-drops-plan-to-require-disclosure-on-drone-killings.html> accessed Sept 2014.

31 See Adam Roberts, "Lives and Statistics: Are 90% of war victims are civilians?" (2010) 52 Survival 115.

32 One controversial drone strikes took place in Dec 2013 and was executed in Yemen. In this strike, 14 wedding guests were killed and 22 others injured: see Hakim Almasmari, "Yemen says U.S. drone struck a wedding convoy, killing 14" CNN (Dec 13, 2013) available at <http://edition.cnn.com/2013/12/12/world/meast/yemen-u-s-drone-wedding/> accessed Sept 2014.

33 Available figures are presented in Chapter 7 by Avery Plaw, Chapter 9 by Jenna Jordan and Chapter 10 by Marek Madej in this volume.

34 See eg Mary Ellen O'Connell, "Unlawful Killing with Combat Drones: A Case Study of Pakistan, 2004–2009," (2010) ND Law School Legal Studies Res Paper No 09–43, available at <http://ssrn.com/abstract=1501144> accessed Sept 2014.

(495 civilians killed and 820 wounded as a result of 10,000 strike sorties and the release of more than 23,000 air munitions) are far fewer than in the case of drone strikes.[35] Similarly, research completed by Larry Lewis (who had access to classified information) indicates that drone strikes were ten times more deadly to Afghan civilians than those performed by fighter jets.[36]

Certainly, reports indicating high rates of civilian losses can be criticized as exaggerated, claiming that their authors took an uncritical approach to data provided by Taliban or, for example, Pakistani media.[37] However, practice definitely shows that the precision of drones is a myth. High civilian losses can be a result of the great distance from a potential target, which makes target selection much easier, and abuse more likely.[38] The ceiling of combat drones is also higher compared to surveillance drones, so the screen picture is less clear.[39] These facts must be taken into account when legality of drones is discussed. Yet it must be clearly stated that the answer to the question of whether the use of drones is or is not prohibited by IHL depends on the analysis not only of the design and intended purpose of drones, but also of the manner in which they are used.[40] As Michael Schmitt rightly notes, using discriminate weapons in indiscriminate ways is illegal.[41] Even if drones are not inherently unlawful, constant engagement in illegal actions can impact their legality,[42] that is, the legality of their use, especially if their main effect is to terrorize the civilian population.[43] It is therefore essential to verify how and against whom drones are usually used. If their main victims are civilians, as some NGO's claim,[44] now is high time to demand that drones be

---

35   Leila Sadat, "Second Annual Katherine B. Fite Lecture: Drone Wars and the Nuremberg Legacy" (2012) 45 Stud. Transnat'l Legal Policy 9, at 39.

36   Larry Lewis, "Drone Strikes in Pakistan: Reasons to Assess Civilian Casualties," CNA Analysis & Solutions, April 2014. See also Larry Lewis and Sarah Holewinski, "Civilian Protection for an Evolving Military," (2013) 4 PRISM 57. See also Spencer Ackerman, "US drone strikes more deadly to Afghan civilians than manned aircraft– adviser" *The Guardian* (July 2, 2013) <http://www.theguardian.com/world/2013/jul/02/us-drone-strikes-afghan-civilians> accessed Sept 2014.

37   See eg Farhat Taj, "Drone Attacks: Challenging Some Fabrications" *Daily Times* (Jan 2, 2010) available at <http://archives.dailytimes.com.pk/editorial/02-Jan-2010/analysis-drone-attacks-challenging-some-fabrications-farhat-taj> accessed Sept 2014.

38   Kellenberg and Spoerri (n 26) at 812; cf. Blank (n 26) at 702.

39   Heller (n 11) at 106.

40   Kathleen Lawand, *A Guide to the Legal Review of New Weapons, Means and Methods of Warfare. Measures to Implement Article 36 of Additional Protocol I of 1977* (ICRC 1996) at 10.

41   Michael N Schmitt, "Drone Attacks Under the Jus ad Bellum and Jus in Bello: Clearing the Fog of Law" (2010) 13 Yearbook of Int'l Humanitarian Law 311, at 321.

42   Hin-Yan Liu (n 6) at 643.

43   Sadat (n 35) at 12.

44   See eg Amnesty International, "'Will I Be Next?' US Drone Strikes in Pakistan" (2013) available at <http://www.amnestyusa.org/research/reports/will-i-be-next-us-drone-

improved before they are used any further. A ban on drones seems a mere fantasy, considering how attractive they have proved to politicians (eg Barack Obama)[45] and the military sector.

III. Who Can be Targeted in Non-International Armed Conflicts?

As noted above, when using drones, the principle of distinction must be applied. Thus, the distinction must be made between legal targets and protected persons like civilians. In international armed conflict, there is a clear division into civilians and combatants; a civilian is every person who cannot be qualified as a combatant.[46] Combatants can be attacked and they cannot be prosecuted for taking part in hostilities. Civilians are protected, which means that they cannot be targeted, but they cannot take part in hostilities either. If they do, they can be punished and they lose their protection for the duration of their engagement in hostilities.

However, in NIAC, the distinction between those engaged in hostilities and civilians is one of the most controversial issues in IHL. The notion of combatants does not apply, so it is impossible to use a negative definition of a civilian (*per analogiam* to international armed conflicts). States were unwilling to establish the status of combatants for members of non-state armed groups, as this would effectively give them the right to legally take part in hostilities. However, at the same time the states complained that limitation of legal targets only to persons taking direct part in hostilities and only for the time of engagement is untenable. States noted that in asymmetrical conflicts, the opponent pretends to be a civilian while in fact engaging in combat ("farmers by day and fighters at night"). Thus it is not reasonable to uphold the restriction that only allows them to target the persons who at a given time are actively engaged in a hostile act.

Yet the same states are obsessed with the zero casualty paradigm.[47] Apparently, the life of a foreign civilian is less valuable than the life of that state's soldier.[48] Shaped by these tendencies, law of armed conflict nowadays is moving toward exclusion of more and more groups from the scope of a term "civilian," or at least from the protection given to civilians. In consequence, the pool of potential legitimate targets is expanding. The guidance issued by the ICRC under the title "Interpretative Guidance on the Notion of Direct Participation in Hostilities under

---

strikes-in-pakistan> accessed Sept 2014.

45   See eg Ryan J Vogel, "Drone Warfare and the Law of Armed Conflict," (2010–2011) 39 Denver J Int'l Law & Policy 101, at 105.

46   See Article 4A of the Third Geneva Convention of 1949, 75 UNTS 135, and Article 50 of AP I.

47   Markus Wagner, "Beyond the Drone Debate: Autonomy in Tomorrow's Battlespace" (2012) 106 Proceedings of Am Society Int'l Law 80, at 83.

48   Kellenberg and Spoerri (n 26) at 815.

International Humanitarian Law" (Interpretative Guidance) proved this by creating a status of a member of an organized armed group within a NIAC.[49]

The first group of legitimate targets in a NIAC are the members of a state's armed forces. Humanitarian law definitely does not prohibit targeting the armed forces of a state that is party to a conflict. That is the theory. In practice, in light of the fact that no other state's forces have the right to take part in hostilities, anyone who engages in an attack against the state's armed forces can be prosecuted and even sentenced to death based on national law.[50] In consequence, IHL does not prohibit killing members of a state's armed forces, but it allows them protection by accepting penalization of attacks against them. In terms of drone use, it must be noted that a large proportion of the operations involving drones is performed by civilians (Central Intelligence Agency (CIA) officers or private military contractors) who are not members of armed forces and thus do not have the privilege of combatancy.[51]

When drone operators perform attacks in armed conflicts, their work can be qualified as direct engagement in hostilities. In effect, they could be legally attacked, but only during the time of involvement in hostilities. In my opinion, nothing in conventional law prohibits attacks on drone bases located in the United States, Germany, or Saudi Arabia on behalf of an armed group with which the United States is engaged in fighting, even if no hostilities take place in the territory of these mentioned states. The decisive factor is that operators of drones who work in these bases are engaging in hostilities within an armed conflict. However, the distance between them and the areas where the hostilities are taking place can make them feel immune to attacks.

In theory, civilian operators of drones can be punished for their role in strikes based on domestic law of the states where attacks are launched or executed.[52] If they are engaged in war crimes, they can also be prosecuted by third-party states if their jurisdiction covers war crimes committed in NIACs.[53] These options remain purely

---

49   "Interpretative Guidance" were adopted by the Assembly of the ICRC on February 26, 2009, available with commentary in (2008) 90 IRRC 991. Reference can be found not only in academic articles, but also in UN official documents such as the Report of the Special Rapporteur (Ben Emmerson) on the promotion and protection of human rights and fundamental freedoms while countering terrorism, UN Doc A/68/389 (Sept 18, 2013) passim.

50   AP II only encourages state's authorities to grant at the end of hostilities the broadest possible amnesty to persons who have participated in the armed conflict, or those deprived of their liberty for reasons related to the armed conflict, whether they are interned or detained (Article 6(5)).

51   Cline (n 11) at 110.

52   See Blank (n 26) at 705; Cline (n 11) at 112; Alston (n 24) 71. Cf. Jordan J Paust, "Self-Defense Targetings of Non-State Actors and Permissibility of U.S. Use of Drones in Pakistan" (2010) 19 J of Transnat'l Law & Policy 237, at 277–278.

53   See ICRC study on Rule 157 (available at <http://www.icrc.org/customary-ihl/eng/docs/v1_cha_chapter44_rule157#refFn_22_7> accessed Sept 2014) for references to national case law concerning prosecution of war crimes committed in non-international armed conflicts.

theoretical due to the locations of the drone bases and the certainty of no cooperation with regard to extradition—although attempts have been made to this effect.[54]

Involvement of CIA operatives in attacks within armed conflicts provokes questions about their lack of adequate training in the use of force, including insufficient knowledge about IHL rules concerning targeting.[55] However, some analysts note that there is no evidence to prove that the strikes launched by the CIA are more detrimental to civilians than those of Pentagon. Shifting drone warfare exclusively to militaries (which was announced by the American authorities) does not therefore guarantee better compliance with the laws of war.[56] Nonetheless, the division of roles between the CIA and special military forces is blurring, which can impact the concept of combatancy in the future.[57]

The second group of legitimate targets are the civilians taking direct part in hostilities. Additional Protocols to Geneva Conventions for the Protection of War Victims of 1949 stipulate that civilians lose their protection when they directly participate in hostilities.[58] According to *Interpretative Guidance on the Notion of Direct Participation in Hostilities under International Humanitarian Law*, in order to qualify certain actions as "direct participation in hostilities," and thus in order to deprive a civilian of protection, three requirements must be satisfied.[59]

- Firstly, action must be likely to adversely affect the military operations or military capacity of a party to an armed conflict or, alternatively, to inflict death, injury, or destruction on persons or objects protected against direct attack. It means that the act has to achieve a threshold of harm.
- Secondly, there must be a direct causal link between the act and the harm likely to result either from that act, or from a coordinated military operation of which that act constitutes an integral part. Commentaries to *Interpretative Guidelines* note that the harm in question must be brought about in one causal step.
- Thirdly, the act must be specifically designed to directly cause the required threshold of harm in support of a party to the conflict and to the detriment of another.

---

54   Hagger and McCormack (n 4) at 94.

55   Sterio (n 12) at 212.

56   Michael Hirsh, "Is the CIA Better Than the Military at Drone Killings?" National Journal (Feb 25, 2014) available at <http://www.nationaljournal.com/magazine/is-the-cia-better-than-the-military-at-drone-killings-20140225> accessed Sept 2014.

57   See Columbia Law School–Human Rights Clinic, Center for Civilians in Conflict, "The Civilian Impact of Drones: Unexamined costs, unanswered questions" (2012) 13–14, available at <http://web.law.columbia.edu/human-rights-institute/counterterrorism/drone-strikes/civilian-impact-drone-strikes-unexamined-costs-unanswered-questions> accessed Sept 2014.

58   Article 51(3) of AP I, Article 13(3) of AP II and Rule 5 of CIHL.

59   Interpretative Guidance (n 49) at 995–996.

Even with these requirements, it is quite difficult to differentiate between support for a party to the conflict that does not deprive a civilian of protection and engagement in hostilities that puts a person in a position of legal target. Little controversy arises with regard to qualification of taking part in military action, armed fighting, provision of ammunition and explosives to fighters, transmission of sensible information to them, or being their guide as a direct participation in hostilities.[60] However, according to the ICRC commentary to *Interpretative Guidelines*, production and transport of weapons (unless it is a part of specific military operation), financing of armed groups, recruitment and training of fighters, helping fighters in escape, or supplying them with all other services can be qualified only as indirect participation in hostilities, so there is a lack of a direct causal link.[61] All these examples meet with opposition.[62]

The cold reaction to *Interpretative Guidance* was a result of the opinion that the guidance does not reflect common *opinio iuris*, and, for example, the American practice shows that at least the American understanding of the direct participation in hostilities is much broader. Americans perceive drug lords as legal targets because they provide financial support to armed organizations and enable them to function.[63] Americans eliminate those who provide tactical training, but also other facilitators, such as passport forgers, bomb makers, recruiters, fundraisers who are linked with some terrorist group, and even authors of enemy propaganda magazines. A direct causal link between those facilitators and any harm is very difficult to prove, but nowhere in conventional law is there a reference to the one step test. There are also different theories of causation that can be applied in armed conflicts. The United States prefers to talk about elimination of terrorists (as it is engaged in a global war on terror). However, planning or preparation of terrorist acts designed be executed outside the battlefield in Afghanistan or Pakistan does not necessarily qualify as taking direct part in hostilities in the framework of NIAC, because these terrorist acts (and the preparation for them) are unrelated to such hostilities.[64] Even if the activities of terrorists are perceived as taking part in hostilities, the terrorists are to be eliminated only at the time when they are

---

60　International Criminal Tribunal for the Former Yugoslavia (ICTY), *Prosecutor v Strugar* (July 17, 2008) IT-01–42-A, Judgment, at para 177.

61　Interpretative Guidance (n 49) at 1021–1022.

62　See Michael Schmitt, "The Interpretative Guidance on the Notion of Direct Participation in Hostilities: A Critical Analysis" (2010) 1 Harvard Nat'l Security J 5. It must be noted that states did not define the term "direct participation in hostilities" on purpose for broader acceptance of AP; see Emily Camins, "The past as prologue: the development of the 'direct participation' exception to civilian immunity" (2008) 90 IRRC 853, at 877.

63　See "A Report to the Committee on Foreign Relations," United States Senate, One Hundred Eleventh Congress, First Session (August 10, 2009) available at <http://www.gpo.gov/fdsys/pkg/CPRT-111SPRT51521/html/CPRT-111SPRT51521.htm 1 May 2014> accessed Sept 2014.

64　See Alston (n 24) at 64. Cf. Cheri Krarner, "The Legality of Targeted Drone Attacks as U.S. Policy" (2011) 9 Santa Clara J Int'l Law 375, at 378.

engaging in hostilities. Killing them as a punishment for their previous attacks would be against the rules of IHL.[65]

The third group which was identified in the *Interpretative Guidelines* as a separate category of those who can be legally attacked are members of organized armed groups of a party to the conflict who have a "continuous combat function."[66] In consequence, those who assume only political, administrative, or other non-combat functions may not be lawfully targeted unless and until they directly participate in hostilities.[67] The American practice shows however that elimination of the main financial officers of an armed group is perceived by the United States as having as much value as the elimination of commanders.

The creation of a continuous combat function (which is not explicitly indicated in treaty law)[68] is in practice a formation of a separate status within NIAC.[69] A person who would be encompassed by this term can be legally eliminated and at the same time is deprived both of a civilian's privilege of protection and of a combatant's right to take part in hostilities. In theory, it should be more difficult to qualify someone as a person with an uninterrupted combat role than as a person engaged directly in hostilities.[70] In practice, evidence showing a relationship with an organized armed group that actively takes part in hostilities could be perceived as sufficient proof

---

65   Cf. Amos Guiora, *Legitimate Target: A Criteria-Based Approach to Targeted Killing* (OUP 2013) at 2.

66   Interpretative Guidance (n 49) at 1002 ff. Cf. Yoram Dinstein, *The Conduct of Hostilities under the Law of International Armed Conflict* (Cambridge University Press 2004) at 29; and UN Doc A/68/389 (n 48) at para 69.

67   Interpretative Guidance (n 49) at 1007.

68   The words of Chief Justice Barak are worth invoking here. In his analysis of temporal scope of deprivation of protection of civilian who is taking part in hostilities, including members of terrorist groups, he noted that in case of "gray cases," "customary international law has not yet crystallized": The Supreme Court of the State of Israel Sitting as the High Court of Justice, *The Public Committee against Torture in Israel et al v The Government of Israel et al* (Dec 11, 2005) HCJ 769/02, at para. 40.

69   See more Nobuo Hayashi, "Continuous Attack Liability Withour Right of Fact of Direct Participation in Hostilities–The ICRC Interpretative Guidance and Perils of a Pseudo-Status" in Joanna Nowakowska-Małusecka (ed) *Międzynarodowe Prawo Humanitarne. Antecedencje i wyzwania współczesności. (International Humanitarian Law. Antecedences and Challenges of the Present Time)* (Oficyna Wydawnictwa Branta 2010); Jens David Ohlin, "Is Jus in Bello in Crisis?" (2013) 11 J Int'l Criminal Justice 27, at 28; Alston (n 24) at 65. Cf. the judgment of the Supreme Court of the State of Israel (n 68) where the Court stated that we are moving towards the function paradigm instead of the status paradigm (see eg para 31 or 35); see also Helen Keller and Magdalena Forowicz, "A Tightrope Walk Between Legality and Legitimacy: An Analysis of the Israeli Supreme Court's Judgment on Targeted Killing" (2008) 21 Leiden J Int'l Law 185, at 207. However, it must be noted that in comparison to law applied in international armed conflict which establishes a complete division of population on combatants and civilians, rules applied in non-international armed conflict do not directly prohibit distinction of other than civilian status.

70   Cline (n 11) at 99.

justifying elimination. This solution is extremely dangerous for the population living in the territory where hostilities are taking place. The creation of this status has completely changed the approach to targeting: Instead of executing attacks only "for such time" during which a person is engaged in hostilities, attacks can be performed at "all times" against members of armed groups.[71]

The ICRC stresses that those who disengage permanently from combat function regain their civilian protection.[72] However, the burden in reality lies on a person who was previously a member of this armed group. The state can always claim that a person was only pretending, and intelligence data confirm that they were still in contact with former colleagues. Even the ICRC is of the opinion that an individual recruited, trained, and equipped by an organized armed group to continuously and directly participate in hostilities can be perceived as having uninterrupted combat function, even if they have not performed any hostile act.[73]

This would mean that states have a tremendous excuse at their disposal. It would be sufficient to prove that a person was trained to take part in hostilities in order to demonstrate that they can now be eliminated. This continual quasi-combatant status implies that not only high-profile commanders can be targets, but also ordinary executors of orders.

Also of importance, drone strikes are not always "personality" strikes, aimed at a specific person whose identity was verified and where is clear evidence of their continuous combat function within an organized armed group or direct engagement in hostilities. Americans forces execute also "signature" strikes, which target suspicious persons in the territory controlled by the enemy, without knowing these persons' identity, or in which suspicious locations are attacked without confidence of who is there.[74] Humanitarian law certainly does not require the decision-maker to know the identity of the target, but it does demand knowledge about their participation in hostilities (or having a continuous combat function).[75]

What is most alarming is that for a drone strike to be ordered, it suffices if a pattern of behavior of specific people is noticed.[76] The United States treats all military-age males in a strike zone as combatants unless it is proved after their death that they were innocent.[77] Unsurprisingly, American authorities can claim

---

71   Alston (n 24) at 65.

72   Interpretative Guidance (n 49) at 1036 ff.

73   Ibid at 1007.

74   Sadat (n 35) at 35.

75   See also Human Rights Watch (n 11) at 87; Christof Heyns, Report of the Special Rapporteur on extrajudicial, summary or arbitrary executions UN Doc A/68/382 (Sept 13, 2013) at para 72.

76   Arianna Huffington, "'Signature Strikes' and the President's Empty Rhetoric on Drones" *Huff Post Politics* (Oct 7, 2013) <http://www.huffingtonpost.com/arianna-huffington/signature-strikes-and-the_b_3575351.html> accessed Sept 2014.

77   Jo Becker and Scott Shane, "'Secret "Kill List' Proves a Test of Obama's Principles and Will" *The New York Times* (May 29, 2012) available at <http://www.nytimes.

that civilian casualties are minimal or none at all.[78] This practice not only broadens the limits of the definition of legitimate targets, but also clearly violates IHL and is highly discriminative toward men. It is not acceptable (yet it does happen nowadays) to treat the fact that men are carrying weapons[79] or are moving in the direction of battlefield[80] as a proof of direct participation in hostilities or continuous combat function, unless there is other intelligence refuting their engagement in hostilities. The proximity of civilians to the hostilities does not deprive them of protection. The tendency to reverse the burden of proof, forcing a person to prove that they are a civilian, completely undermines IHL and cannot be accepted.[81]

All attempts to broaden humanitarian rules on the choice of legitimate targets are still considered too restrictive by the United States. As a result, when it is impossible to submit evidence that a person was to continually perform a combat function or was taking direct part in hostilities, the argument of legitimate self-defense (including an anticipatory one) is used. Self-defense is invoked to justify the right to eliminate every person that threatens the security of a state's citizens.[82] This approach mixes two separate branches of international law, that is, IHL (*jus in bello*) and use of force law (*jus ad bellum*).[83] While use of force law explains when

---

com/2012/05/29/world/obamas-leadership-in-war-on-al-qaeda.html?pagewanted=all&_r=0> accessed Sept 2014.

78   Scott Shane, "C.I.A. Is Disputed on Civilian Toll in Drone Strikes" *The New York Times* (Aug 11, 2011) <http://www.nytimes.com/2011/08/12/world/asia/12drones.html?pagewanted=all> accessed Sept 2014.

79   See also ICTY, Trial Chamber, *Prosecutor v Blagoje Simić et al*, IT-95-9-T, (Oct 17, 2003) at para 659.

80   Heller (n 11) at 98–99. See also Daniel Klaidman, "Drones: The Silent Killers" *Newsweek* (May 28, 2012) <http://www.newsweek.com/drones-silent-killers-64909 accessed September 4, 2014> accessed Sept 2014.

81   Sadat (n 35) at 27.

82   Koh (n 12); Charlie Savage, "US Law May Allow Killings, Holder Says" *The New York Times* (March 5, 2012) available at <http://www.nytimes.com/2012/03/06/us/politics/holder-explains-threat-that-would-call-for-killing-without-trial.html?_r=0> accessed Sept 2014. See also Paust (n 52) at 261 ff; Elinor June Rushforth, "There's an App for That: Implications of Armed Drone Attacks and Personality Strikes by the United States Against Non-Citizens, 2004–2012" (2012) 29 Arizona J Int'l & Comp Law 623, at 629 ff. For critiques of this approach see eg Marry Ellen O'Connell, "Remarks: The Resort to Drones under International Law" (2010–2011) 39 Denver J Int'l Law & Policy 585, at 599; Guiora (n 64) at xi. Compare with US Department of Justice, Office of Legal Counsel, *Memorandum for the Attorney General. Re: Applicability of Federal Criminal Laws and the Constitution to Contemplated Lethal Operations Against Shaykh Anwar al*-Aulaqi, (July 16, 2010) at 21–22. In this OLC Memorandum, from one side there is an indication that al-Aulaqi's activities in Yemen posed "a 'continued and imminent threat' of violence to United States persons and interests" and at the same time there is a passage focusing on his active participation in hostilities which deprived him of protection of Common Article 3 of the Geneva Conventions.

83   Jasmine Moussa, "Can *jus ad bellum* override *jus in bello*? Reaffirming the separation of the two bodies of law" (2008) 90 ICRC 963; Robert D Sloane, "The Cost of

force can be used in the inter-state relations (outside the territory of the state using force), IHL explains who can be targeted when the use of force is justified. The fact that the state is authorized to use force in self-defense does not entitle it to violate rules of law of war. The danger of mixing these branches lies, for example, in the problems related to proportionality, which is understood differently in these two branches of law.[84] In use of force law, proportionality of a state's actions is measured in reference to the incident (*casus belli*) that would justify a resort to force (if there was an armed attack resulting in a right to self-defense of a victim state) and also in reference to the end that state wants to achieve (establishment of security and peace). In consequence, minor border clashes not resulting in a considerable number of victims cannot be responded to by an invasion on the territory of an enemy state (even if all targets would be legitimate in the light of humanitarian law, like barracks, arms stocks, combatants, etc.). In contrast, proportionality in *jus in bello* is assessed by a comparison of an anticipated concrete and direct military advantage of the attack with possible civilian losses. The reason for the resort to force (eg self-defense or aggression) or the final aim that a party to the conflict wants to achieve have no meaning for judging the military action as proportional or disproportional in the light of IHL. It must be also noted that justifying attacks on civilians by invoking a right to use force stands in contradiction with a prohibition of reprisals against civilians.[85]

IV. How to Eliminate a Legitimate Target?

The fact that a person is a civilian and does not take any part in hostilities does not assure them of immunity from attacks. According to the IHL, an attack which results in proportional civilian losses in comparison to military advantage can be legally justified.[86] Only those attacks that are clearly excessive in relation to the concrete and direct overall military advantage are criminalized.[87] The perspectives

---

Conflation: Preserving the Dualism of Jus ad Bellum and Jus in Bello in the Contemporary Law of War" (2008) 34 Yale J Int'l Law 47; François Bugnion, "Just Wars, Wars of Aggression and International Humanitarian Law" (2002) 84 IRRC 3.

84   Michael Newton and Larry May, *Proportionality in International Law* (OUP 2014). See, however, attempts to mix the two branches of law, Guiora (n 64) at 33–34.

85   See Rule 147 of CIHL and Article 33 of GC IV, Articles 20, 51, 52, 53, 54, 55, 56 of AP I.

86   Articles 51(5)(b), 57 of AP I, Rule 14 of CIHL. See also Marco Sassòli, Antoine A. Bouvier and Anne Quintin, *How Does Law Protect in War? Cases, Documents and Teaching Materials on Contemporary Practice in International Humanitarian Law. Volume I Outline of International Humanitarian Law* (3rd ed ICRC) Chapter 9.

87   See article 8(2)(b)(iv) of the Rome Statute (Intentionally launching an attack in the knowledge that such attack will cause incidental loss of life or injury to civilians or damage to civilian objects or widespread, long-term and severe damage to the natural environment which would be clearly excessive in relation to the concrete and direct overall military advantage anticipated).

on what is excessive differ a lot among commanders (as a consequence of their training, field experience, national military history, etc.)[88] and even more between states. Three examples—American, German, and British—demonstrate the scale of differences among states in understanding proportionality in targeting. As a result of American strike against Baitullah Mehsud, 83 people were killed, among them 45 civilians (including children). This attack was perceived by American authorities as completely legal, taking into account the high-priority target, which Mehsud was.[89] The German case is an attack (not performed by drone) of September 2009 in Kunduz, Afghanistan, on two fuel tankers taken by Taliban insurgents. In consequence of the attack, 90 civilians were killed, including children and elderly persons.[90] A completely different standard was applied by British forces, who decided that an attack in which four civilians died in addition to two legitimate targets was disproportionate.[91] Every time a case arises where proportionality must be assessed, it provokes heated discussion with regard to the civilian losses.[92] Not surprisingly, courts admit that proportionality must be judged on a case-by-case basis. This is clearly not very helpful for commanders in their daily work.[93]

Before every attack, feasible precautions must be taken. A UCAV operator must verify that the target is a military one, and guarantee that civilian losses will be minimized. In case of doubt as to the status of the target, the attack should be abandoned.[94] The question is how big the doubt must be to trigger the obligation to abandon an attack (in other words, what kind of information or amount of it is

---

88   ICTY, Final Report to the Prosecutor by the Committee Established to Review the NATO Bombing Campaign Against the Federal Republic of Yugoslavia (June 8, 2000) at para 19, 50. See also Carolin Wuerzner, "Mission impossible? Bringing charges for the crime of attacking civilians or civilian objects before international criminal tribunals" (2008) 90 IRRC 907, at 922.

89   Casey-Maslen, (n 2) at 613.

90   European Center for Constitutional and Human Rights, "Kunduz" available at <http://www.ecchr.de/index.php/KUNDUZ_CASES.html> accessed Sept 2014. "German prosecutors drop case against Kunduz airstrike colonel" DW (April 1, 2012) available at <http://www.dw.de/german-prosecutors-drop-case-against-kunduz-airstrike-colonel/a-5483181–1> accessed Sept 2014; "Kunduz Bombing: German Court Drops Case Over Civilian Deaths" *Spiegel Online International* (Dec 11, 2013) <http://www.spiegel.de/international/germany/court-says-germany-not-responsible-for-damages-in-afghanistan-attack-a-938490.html> accessed Sept 2014.

91   Casey-Maslen (n 2) at 613.

92   As an example, an incident of 2009 in Pakistan can be mentioned: a Taliban leader (a legal target) was killed, but also his wife, his wife's parents, his uncle, a lieutenant, and seven bodyguards (who should be also perceived as legal targets if their role was protection of the Taliban leader). Mary Ellen O'Connell considered this attack as disproportionate (n 33); whereas Paust considered the same incident as perfectly proportional (n 52) at 275.

93   HCJ 769/02 (n 68), para 81.

94   Article 57(2)(b) of AP I; Rule 19 of CIHL.

sufficient to determine the status of the target).[95] The ICRC notes that in case of any doubt, the commander must always be guided by the interests of civilians.[96] The slightest doubt should force the soldier to cancel or postpone an attack. However, some states prefer to talk about significant or substantial doubt.[97] No matter which approach is accepted, it is obvious when a drone operator is not certain who or what they are seeing on their screen, they should abort the strike.[98]

This rule must be applied especially with regard to signature strikes. Absence of confidence as to the intentions of armed personnel must result in giving them protection; it is completely understandable that during an armed conflict every person who wants to protect their life and family is forced to carry a weapon. Any other interpretation means endangering the entire (or at least the entire male) population of the area of conflict. In theory, usage of drones offers more opportunities to take precautions. Yet the tools of identification are not as perfect as the public tends to believe. There were attacks in which Afghan allied commanders[99] were attacked, and even attacks in which American soldiers were victims.[100] Evidently, the image on the operator's screen is not sufficiently clear to allow full identification of targets. What the public needs to understand is that the video provided by a drone is not usually clear enough to detect someone carrying a weapon, even on a sunny day with no clouds.[101] The argument that, for example, use of cruise missiles in that situation would not result in fewer deaths is misleading: No matter what kind of weapon is used, if there is no clarity who

---

95   See ICTY, Judgment, *Prosecutor v Galic*, IT-98–29-T (Dec 5, 2003) at para 58 where the Tribunal stated: "In determining whether an attack was proportionate it is necessary to examine whether a reasonably well-informed person in the circumstances of the actual perpetrator, making reasonable use of the information available to him or her, could have expected excessive civilian casualties to result from the attack."

96   Sandoz, Swinarski and Zimmermann (n 18) at 625–626.

97   See Practice Relating to Rule 10. Civilian Objects' Loss of Protection from Attack, available at <http://www.icrc.org/customary-ihl/eng/docs/v2_rul_rule10> accessed Sept 2014.

98   A lot of controversy arose as a result of the statements of Brandon Bryant, former drone operator for the US Air Force, who described a situation in which his commander mistook a child for a dog, see Nicola Abé, "Dreams in Infrared: The Woes of an American Drone Operator" *Spiegel Online International* (Dec 14, 2012) available at <http://www.spiegel.de/international/world/pain-continues-after-war-for-american-drone-pilot-a-872726.html> accessed Sept 2014.

99   David Zucchino, "US Report faults Air Force drone crew, ground commanders in Afghan civilian deaths" *Los Angeles Times* (May 29, 2010) <http://articles.latimes.com/2010/may/29/world/la-fg-afghan-drone-20100531> accessed Sept 2014.

100   "Afghanistan: US servicemen killed in first drone 'friendly fire' incident" *The Telegraph* (April 12, 2011) available at <http://www.telegraph.co.uk/news/worldnews/asia/afghanistan/8445063/Afghanistan-US-servicemen-killed-in-first-drone-friendly-fire-incident.html> accessed Sept 2014.

101   Heather Linebaugh, "I worked on the US drone program. The public should know what really goes on" *The Guardian* (Dec 29, 2013) <http://www.theguardian.com/commentisfree/2013/dec/29/drones-us-military> accessed Sept 2014.

will be killed, the strike cannot be executed despite guarantees of its precision. It is impossible to be precise without knowing what or who is under attack.

Another controversial dilemma is whether there is an obligation to capture rather than kill a legal target. From the general principle of humanitarianism, Jean Pictet derived that, whenever possible, combatants must avoid causing injuries.[102] This approach was adopted in the *ICRC Guidelines*, according to which, even in the case of an attack on a legitimate target, parties to the conflict must consider if the kind and degree of force does not exceed what is actually necessary to accomplish a legitimate military purpose in the prevailing circumstances.[103] In consequence, the enemy should be captured if possible, rather than killed. Many lawyers disagree with this approach. They argue that nowhere in the conventions is there such an obligation.[104] In consequence, if a person or an object is a legitimate target they/it can be eliminated with any legal means or method of warfare. However, this dilemma is much more complicated if we apply human rights law, which continues to apply in armed conflicts.[105] According to human rights (HR) law, the use of lethal force must always be gradual.[106] That is why some states (mainly the United States and Israel) strongly oppose extraterritorial application

102 Jean Pictet, *Development and Principles of International Humanitarian Law* (Martinus Nijhoff Publishers 1985) at 751.

103 Interpretative Guidance (n 49) at 996, 1040.

104 Thomas Michael McDonell, "Sow What You Reap? Using Predator and Reaper Drones to Carry Out Assassinations or Targeted Killings of Suspected Islamic Terrorists" (2012) 44 Geo Wash Int'l Law Rev 243, at 273; Richard S Taylor, "The Capture versus Kill Debate: Is the Principle of Humanity Now Part of the Targeting Analysis When Attacking Civilians Who Are Directly Participating in Hostilities?" (2010) 6 Army Law 203, at 204.

105 See ICJ, Advisory Opinion, Legal Consequences of the Construction of a Wall in the Occupied Palestinian Territory (July 9, 2004) para. 106. See the summary of the debate on the relation between the two regimes (HR and IHL) in Karima Bennoune, "Toward a Human Rights Approach to Armed Conflict : Iraq 2003" (2004–2005) 11 UC Davis J Int'l Law & Policy 171, at 179 ff; Robert Kolb, "Aspects historiques de la relation entre le droit international humanitaire et les droits de l'homme" (1999) 37 Canadian Yearbook Int'l Law 57. Françoise J Hampson, "The relationship between international humanitarian law and human rights law from the perspective of a human rights treaty body," (2008) 90 IRRC 549, at 559 ff; Hans J Heintze, "On the relationship between human rights law protection and international humanitarian law" (2004) 86 IRRC 789. On the main differences between regimes see A Hansen, "Preventing the Emasculation of Warfare: Halting the Expansion of Human Rights Law into Armed Conflict" (2007) 194 Mil Law Rev 1, at 6; Barry A Feinstein, "The Applicability of the Regime of Human Rights in Times of Armed Conflict and Particularly to Occupied Territories: The Case of Israel's Security Barrier" (2005–2006) 4 NW J Int'l HR 238, at 245 ff; Louise Doswald-Beck, "International Humanitarian Law: A Means of Protecting Human Rights in Time of Armed Conflict" (1989) 1 African J Int'l & Comp Law, 595, at 615; Cordula Droege, "Elective affinities? Human rights and humanitarian law" (2008) 90 IRRC 501, at 521 ff.

106 See also Lewis (n 11) at 300 ff. See also Chapter 4 by Gloria Gaggioli in this volume.

of HR and act as persistent objectors to prevent establishment of a customary rule that would oblige all states to apply HR standards in armed conflicts. Despite the fact that humanization of law of armed conflict constantly is taking place,[107] in the framework of armed conflict IHL allows killing legitimate targets without giving them a chance to ask for pardon (although certainly those who lay down their arms are protected). If it is completely legal to use ruses to surprise the enemy,[108] it is illogical to expect that in every case forces will disclose their location in order to fulfill exigencies of gradual application of lethal force. However, the explanation by some states that they will not gradually apply force because they do not know what to do with detained persons cannot be accepted.

V. Blurred Responsibility

Performing a lethal attack with a drone requires coordination between an intelligence officer, a commander, the drone's operators (usually one person is responsible for the flight of the drone, while another one monitors the cameras and yet another is in contact with ground troops and commanders in the place where hostilities are taking place), and a legal adviser. Poor coordination among them can result in civilian losses or even killing their own troops.[109] The question arises: Who is responsible for a miscalculated attack? Even if it is known and clear who made the decisions leading up to a specific attack, the sheer number of persons involved in the attack opes the option of invoking another person's error. The commander can claim that they relied on credible intelligence; the intelligence officer may indicate that their source lied or they misinterpreted the facts; the drone's operator may defend themselves by saying they were only performing an order and did not know that the order was unlawful, especially if it was approved by a legal adviser. If this scenario plays out, all persons engaged in the decision regarding the attack would avoid responsibility.

It has been assessed that a majority of mistaken strikes were a result of wrong intelligence.[110] The Vietnam experience (where many of those who were put on the kill list were victims of personal quarrels, having nothing in common with the

---

107   Theodor Meron, *The Humanization of International Law* (Martinus Nijhoff Publishers 2006) 1 ff.

108   Article 37(2) of AP I.

109   See Rod Nordland, "U.S. Airstrike Kills 5 Afghan Soldiers" *The New York Times* (March 6, 2014) <http://www.nytimes.com/2014/03/07/world/asia/united-states-airstrike-kills-afghan-soldiers.html?ref=world> accessed Sept 2014; Dexter Filkins, "Operators of drones are faulted in Afghan deaths" *The New York Times* (May 29, 2010) <http://www.nytimes.com/2010/05/30/world/asia/30drone.html> accessed May 2014.

110   "Researcher: Most Civilian Drone Deaths 'From Faulty Information'" VOA News (March 15, 2013) available at <http://www.voanews.com/content/resaercher-most-civilian-drone-deaths-from-faulty-information/1622442.html> accessed Sept 2014.

armed conflict)[111] should make every commander extremely cautious before taking a decision on killing anyone based on intelligence that has not been thoroughly vetted.[112] It must nonetheless be noted that even the highest standards of precautions do not prevent all mistakes.[113] Nowadays the problem lies not in the lack of information but the overload of information that cannot be properly analyzed.[114]

Certainly the role of drone operator can be compared to that of a pilot of a manned aircraft, and responsibility can be attached in a similar manner.[115] However, there are some practical and legal obstacles that can undermine efforts to attach responsibility to the operator. Accusation of a war crime, for example, murdering/ aiming at civilians, requires proving a conduct element such that there was an act that caused death (i.e. the drone strike) and a consequence element (i.e. the death of a person).[116] These elements are relatively easy to prove. However, for the accusation to be valid, the circumstance element has to be proved as well: The qualification of the victim as a civilian has to be confirmed, and thus appropriate investigation must be undertaken. It has to be verified if the targeted persons were taking part in hostilities or were members of organized armed group and had continuous combat function.[117] The problem is that strikes are carried out in remote locations, and it is usually impossible to verify who was actually killed.[118] Even if the strikes in Pakistan were performed with consent of Pakistani authorities, in case of drone attack the militants cut off the areas and removed their dead, leaving it to the media to determine who was killed.[119] The media and NGOs cannot be relied on to ensure independent verification.[120] Given these circumstances, the principle of *in dubio pro reo* always forces the court to find the suspect innocent.

Moreover, to prosecute anyone for a war crime or a crime against humanity (which in case of drone strikes would be also possible, given the widespread and systematic nature of attacks), existence of *mens rea* on the part of the perpetrator must be proven, that is, it must be proven that the perpetrator knew and intended

---

111   Noel Sharkey, "Automating Warfare: Lessons Learned from the Drones" (2011) 21 JLIS. 140, at 152.

112   Guiora (n 64) at 37 ff.

113   See eg Dakota S Rudesill, "Precision War and Responsibility: Transformational Military Technology and the Duty of Care Under the Laws of War" (2007) 32 Yale J Int'l Law 517, at 536 ff.

114   Blank (n 26) at 714.

115   William Boothby, "Some legal challenges posed by remote attack" (2012) 94 IRRC 579, at 593.

116   Heller (n 11) at 107.

117   Paust (n 52) at 277.

118   Byman (n 25). The empirical facts alter the calculations of efficacy, thus the unavailability of this information is explored in Chapter 11 of this volume by Steven J. Barela.

119   It must be stressed that in non-international armed conflicts, states are also obliged to search for, collect, and evacuate the dead (Rule 112 of CIHL); see also Rules 113–117. Analysis of this obligation in Breau and Aronsson (n 12) at 286 ff.

120   Hagger and McCormack (n 4) at 95; A/68/389 (n 48) at para 41 ff.

to kill a civilian person. Given the legal controversies related to the clarification of who is a legitimate target in non-international armed conflict, the accused could claim that they did not knowingly attack a civilian, because they were taught a different definition of civilian than is applied by the court, and so a mistake of fact or law could be invoked.[121]

VI. Conclusion

The use of drones is on the rise. They are employed both for civilian purposes (e.g. for air purification, mail delivery) and for military ones. The process appears impossible to stop. It is however necessary to note that their current employment is often in contradiction of international humanitarian law, because they are deployed in heavily populated areas or against persons who should never become military targets. It may be argued that customary rules on selecting legitimate human targets are in the process of crystallizing, taking into account the debate provoked by the *Interpretative Guidance*, as well as American and Israeli practice, which goes even further than the already permissive *Interpretative Guidance* allows. Contemporary practice of targeting by drones within non-international armed conflict is definitely expanding the pool of targets. It is caused by the fact that strikes are increasingly easy to execute, and drone operators and the commanders taking decisions on the strikes are usually thousands of miles from potential victims, so they may have a feeling of immunity. Attempts have been made to demonstrate that even where there were suspicions of intentional killing of civilians, it would be very difficult to prove guilt, given the standards of fair trial and the principle of *in dubio pro libertate*, which so often gets abused in war crimes trials.

I am doubtful whether Americans would actually find it acceptable for other states to apply the same principles to drone use—yet the American practice may lead to the emergence of customary laws that would reduce the categories of protected persons to a level below any acceptable minimum. IHL is nowadays used by states to justify doubtful cases of strikes. This warrants the conclusion that this branch of law ceased to guarantee protection for those who just happened to live in area of hostilities or in proximity to "terrorists". Consequently, both the drones themselves and the rules governing their use must be improved to become just an effective crossbow of our times instead of a weapon of mass destruction for civilians not taking part in hostilities.

---

121   Heller (n 11) at 108.

Chapter 3

Boundaries of the Battlefield: The Geographical Scope of the Laws of War

Katja Schöberl

I. Introduction

Since a missile was test-fired from a drone in February 2001, drones have not only been used for intelligence, surveillance, and reconnaissance purposes, but armed with explosive ordnances and employed for targeting during combat operations.[1]

To this day, the majority of drone strikes have been conducted within conventional theaters of armed conflict,[2] frequently referred to as "hot battlefields."[3] For example, the United States of America (US) and the United Kingdom of Great Britain and Northern Ireland have increasingly used drones in Afghanistan.[4] The North Atlantic Treaty Organization militarily intervened in Libya, not only through conventional, but also remotely piloted, aircraft.[5]

The use of drones is, however, not limited to conventional armed conflicts, but extends to so-called cross-border counterterrorism operations, whose legal

---

1 Drones are also referred to as "remotely piloted aircraft" (RPA) or "unmanned combat aerial vehicle" (UCAV). The terms are used interchangeably. However, the expression "remotely piloted aircraft" is preferred by some in order to distinguish them from autonomous weapons.

2 B Emmerson, "Report of the Special Rapporteur on the Promotion and Protection of Human Rights and Fundamental Freedoms while Countering Terrorism, Interim Report on the Use of Remotely Piloted Aircraft in Counter-Terrorism Operations" UN Doc A/68/389, (Sept 18, 2013) at para 59.

3 Notions such as "theaters of war," "zones of conflict" or "hot battlefields" lack legal meaning and should be considered strictly descriptive terms. See also K Anderson, "Targeted Killing and Drone Warfare: How We Came to Debate Whether There Is a 'Legal Geography of War'" in P Berkowitz (ed), *Future Challenges in National Security and Law* (Online Essay Series 2011) 14.

4 Emmerson, "Interim Report" (n 2) at para 29. For a current overview of targeted killings by remotely piloted aircraft and other aerial platforms in Afghanistan, see for example United Nations Assistance Mission in Afghanistan (UNAMA), "Report on the Protection of Civilians in Armed Conflict" (Annual Report 2013) 46.

5 Emmerson, "Interim Report" (n 2) at para 36. See also Human Rights Council, "Report of the International Commission of Inquiry on Libya" UN Doc A/HRC/19/68 (March 8, 2012) at para 83.

qualification is disputed. For instance, the first known US drone strike killed Abu Ali Al-Harithi, the alleged plotter of the USS Cole bombing and head of Al-Qaeda in Yemen in 2002. Another 80 targeted killing operations, partly conducted through drone strikes, are estimated to have been carried out by the United States in Yemen since then.[6] In Pakistan, US drone strikes against alleged members of terrorist organizations, such as Al-Qaeda and the Taliban, began in 2004.[7] The US is furthermore suspected of being engaged in extensive covert counterterrorism operations in Somalia (a first reported drone strike is believed to have been targeted at a group of alleged Al-Shabaab members in June 2011).[8] While Israel is assumed to have used drones in its counterterrorism operations in Gaza, that is, during Operation Cast Lead (2008–2009) and Operation Pillar of Defense (2012), it has not acknowledged its use of drones or expressed its position on the legality of their use in targeted killing operations.[9]

Although the US has partly acknowledged the general existence of cross-border counterterrorism operations,[10] it routinely classifies information about specific drone strikes. This lack of transparency and accountability is frequently criticized, especially by human rights lawyers, non-governmental organizations, and scholars.[11] The US government has, however, recently outlined its general views on the legality of drone attacks in a series of public statements.[12] While the

---

6   For an analysis of US drone strikes in Yemen, see Human Rights Watch, "Between a Drone and Al-Qaeda. The Civilian Cost of US Targeted Killings in Yemen" (October 2013).

7   For an overview of the US drone campaign in Pakistan, see The New America Foundation, "Drone Wars Pakistan: Analysis" available at <http://natsec.newamerica.net/drones/pakistan/analysis> accessed Dec 2014; The Long War Journal, "US Airstrikes in Pakistan: Charts 2004–2014" available at <www.longwarjournal.org/pakistan-strikes.php> accessed Dec 2014; or International Human Rights and Conflict Resolution Clinic at Stanford Law School and Global Justice Clinic at NYU School of Law, "Living Under Drones: Death, Injury, and Trauma to Civilians From US Drone Practices in Pakistan" (Sept 2012).

8   M Mazzetti and E Schmitt, "US Expands its Drone War into Somalia" *The New York Times* (July 2, 2011).

9   Emmerson, "Interim Report" (n 2) at para 39 and 50.

10   The existence of covert counter-terrorism operations in Somalia and Yemen was declassified by President Obama in June 2012, while drone operations in Pakistan have been publicly acknowledged by the Obama administration since 2009. See Emmerson, "Interim Report" (n 2) at para 46; and L Panetta, "Remarks at the Pacific Council on International Policy" (May 18, 2009) in which he referred to drone strikes to confront and disrupt Al-Qaeda leadership in Pakistan as "the only game in town."

11   See, for example, American Civil Liberties Union (ACLU) and others, Statement of Shared Concern regarding US Drone Strikes and Targeted Killings, Letter to President Obama (April 11, 2013).

12   See especially H Koh, "The Obama Administration and International Law" Annual Meeting of the American Society of International Law (March 25, 2010); J Brennan, "Strengthening our Security by Adhering to our Values and Laws" Harvard Law School (Sept 16, 2011); J Johnson, "National Security Law, Lawyers and Lawyering in the Obama Administration" Yale Law School (Feb 22, 2012); E Holder, "Remarks" Northwestern

term "global war on terror" was abandoned in 2009, the US continues to consider itself engaged in an ongoing armed conflict against Al-Qaeda, the Taliban, and associated forces based on its inherent right to self-defense and the domestic Authorization for the Use of Military Force (AUMF).[13]

In the US government's view, this armed conflict is global in scope and provides authority for the use of lethal force even outside of "hot battlefields," including through the use of drones. "Fighters" in this armed conflict, as well as civilians directly participating in it, are believed to carry the armed conflict wherever they move. A US Department of Justice White Paper leaked to the press confirms the US government's public position, restating that any US operation against Al-Qaeda and associated forces would be part of the non-international armed conflict even if it were to take place away from a zone of active hostilities.[14]

The US' (legal) position on the use of lethal force in cross-border counterterrorism operations is heavily criticized,[15] while concerns are voiced within the United States that drone strikes risk becoming "Obama's Guantánamo."[16] This criticism by international legal practitioners and scholars primarily relates to the US' choice of an "armed conflict" rather than a "law enforcement" paradigm since 9/11.[17] The distinction between both approaches is of particular importance

---

University School of Law (March 5, 2012); and President Obama, "Remarks" National Defense University (May 23, 2013). See also "Fact Sheet: US Policy Standards and Procedures for the Use of Force in Counterterrorism Operations Outside the United States and Areas of Active Hostilities" (May 23, 2013).

13 For the debate on a possible revision, renewal or repeal of the AUMF in relation to the end of US combat operations in Afghanistan, see ICRC, "IHL and Contemporary Challenges Series" AUMF Debate (March/April 2014) <http://intercrossblog.icrc.org> accessed Dec 2014.

14 US Department of Justice, "Lawfulness of a Lethal Operation Directed Against a US Citizen who is a Senior Operational Leader of Al-Qa'ida or an Associated Force" (White Paper) at 3. See also Amnesty International, "The Devil in the (Still Undisclosed) Details, Department of Justice 'White Paper' on Use of Lethal Force against US Citizens Made Public" AMR/51/006/2013 (Feb 2013). For the recent court decision ordering the US government to release redacted versions of documents setting forth its reasoning on the lawfulness of targeted killings of US citizens through drone strikes, see US Court of Appeals for the Second Circuit, *The New York Times Company et al v United States*, Case No 13–422-cv (April 21, 2014).

15 For a summary of the principal areas of legal controversy, see Emmerson, "Interim Report" (n 2) at para 51. See also B Emmerson, "Report of the Special Rapporteur on the Promotion and Protection of Human Rights and Fundamental Freedoms while Countering Terrorism" UN Doc A/HRC/25/59 (March 11, 2014) at para 70 as well as <http://unsrct-drones.com> accessed Dec 2014.

16 JB Bellinger III, "Will Drone Strikes Become Obama's Guantanamo?" *Washington Post* (Oct 3, 2011).

17 For an overview of different approaches in legally classifying the US conflict with Al-Qaeda since 9/11, see N Lubell, "The War (?) Against Al-Qaeda" in E Wilmshurst (ed), *International Law and the Classification of Conflicts* (OUP 2012) at 421–454. See also R Ehrenreich Brooks, "War Everywhere: Rights, National Security Laws, and the Law of Armed Conflict in the Age of Terror" (2004) 153 Univ of Penn Law Review 675.

for the use of armed force. While in accordance with the laws of war, parties to an armed conflict may direct attacks against opposing enemy forces based on their status, under international human rights law states may only use lethal force for law enforcement purposes where absolutely necessary to protect human life.

More specifically, the US is criticized for having distorted the "boundaries of the battlefield" to an extent that they virtually no longer contain armed conflict in space or time. An important part of the current debate on the legality of the use of drones thus revolves around the geographical scope of the laws of war.[18]

Nevertheless, it should be noted that the debate about the geographical scope of the laws of war predates the drone discourse. The International Criminal Tribunal for the former Yugoslavia (ICTY) systematically addressed the notion for the first time when dealing with the various armed conflicts having taken place on the territory of the former Socialist Federal Republic of Yugoslavia. The ICTY's focus, however, was on the geographical scope of application of the laws of war within the territory of a state party to an armed conflict. It thus primarily analyzed to what extent the laws of war apply to acts committed on the territory of a belligerent state, but removed from hostilities.

The geographical scope of the laws of war has subsequently become the subject of attention and debate with respect to cross-border counterterrorism operations since 9/11. It initially played a significant role in arguments on the legality of detaining terrorist suspects, including in Guantánamo Bay, and is now of significant importance to the debate on the legality of targeted killings,[19] including through the use of drones.

II. Classification of Armed Conflicts

The 1949 Geneva Conventions and their 1977 Additional Protocols, which contain the most essential rules of international humanitarian law,[20] are applicable to two

---

18   See also RJ Vogel, "Droning On: Controversy Surrounding Drone Warfare is Not Really About Drones" (2013) XIX Brown Journal of World Affairs Volume 1. See more generally N Lubell and N Derejko, "A Global Battlefield? Drones and the Geographical Scope of Armed Conflict" (2013) 11 J of Int'l Criminal Justice 65; MW Lewis, "Drones and the Boundaries of the Battlefield" (2012) 47 Texas Int'l Law J 293; and R Heinsch, "Unmanned Aerial Vehicles and the Scope of the 'Combat Zone': Some Thoughts on the Geographical Scope of Application of International Humanitarian Law" (2012) 25 J of Int'l Law of Peace and Armed Conflict 184.

19   For a definition of "targeted killings," see P Alston, "Report of the Special Rapporteur on Extrajudicial, Summary or Arbitrary Executions, Study on Targeted Killings" UN Doc A/HRC/14/24/Add.6 (May 28, 2010) at para 7.

20   The terms "laws of war," "laws of armed conflict" and "international humanitarian law" are used interchangeably by most. They refer to rules which seek to limit the effects of armed conflict by protecting persons who are not or no longer participating in hostilities

different types of situations: (1) international and (2) non-international armed conflicts (NIACs).[21]

A. Traditional International and Non-International Armed Conflicts

International armed conflicts (IAC), to which all Geneva Conventions and Additional Protocol I (AP I) apply,[22] are provided for in Common Article 2. Traditionally, they are understood to cover all cases of resort to armed force between states, regardless of their duration and intensity,[23] even if the existence of a state of war is denied. They furthermore include all cases of declared war, occupation, and armed conflicts in which peoples are fighting against colonial domination, alien occupation, and racist régimes in the exercise of their right of self-determination.[24]

In the case of a NIAC, Common Article 3 applies to "case[s] of armed conflict not of an international character occurring in the territory of one of the High Contracting Parties." It requires all parties to the conflict to apply certain minimum standards regarding persons no longer taking part in hostilities. Additional rules under customary international humanitarian law, especially with respect to the conduct of hostilities, apply once a situation is governed by Common Article 3.[25] NIACs need to be distinguished from internal disturbances and tensions, such as riots and isolated and sporadic acts of violence to which international humanitarian

---

and by restricting the means and methods of warfare (*jus in bello*). They do not cover rules which govern the lawfulness of the resort to armed violence (*jus ad bellum*).

21   For a typology of armed conflicts, see S Vité, "Typology of Armed Conflicts in International Humanitarian Law: Legal Concepts and Actual Situations" (2009) 91 Int'l Review of the Red Cross 69; and J Pejic, "The Protective Scope of Common Article 3: More than Meets the Eye" (2011) 93 Int'l Review of the Red Cross 1.

22   As treaty law AP I is only applicable to international armed conflicts between parties to the protocol. For an overview of the current state of ratification of GC I-IV as well as AP I and II, see <www.icrc.org/ihl> accessed Dec 2014.

23   The ICRC Commentary to Common Article 2 notes that "[a]ny difference arising between two States and leading to the intervention of armed forces is an armed conflict within the meaning of Article 2, even if one of the Parties denies the existence of a state of war," J Pictet, *Commentary on the Geneva Conventions of 12 August 1949*, vol I (ICRC 1952) 32; Cf. International Law Association, "Final Report on the Meaning of Armed Conflict in International Law" (Aug 2010) at 29, which argues that state practice and *opinio iuris* support the position that hostilities must reach a certain level of intensity to qualify as an armed conflict, even in case of international armed conflict. For arguments in favor of maintaining the absence of a requirement of threshold of intensity for international armed conflicts, see ICRC, "International Humanitarian Law and the Challenges of Contemporary Armed Conflicts" 31IC/11/5.1.2 (Oct 2011) 7.

24   Article 1(4) of AP I, which extends the law of international armed conflicts to national liberation wars, is not considered to reflect customary international humanitarian law: see L Moir, *The Law of Internal Armed Conflict* (Cambridge University Press 2002) at 90.

25   S Radin, "Global Armed Conflict? The Threshold of Extraterritorial Non-International Armed Conflicts" (2013) 89 Int'l Law Studies 696, at 706.

law does not apply. State practice and international jurisprudence have illustrated Common Article 3's threshold of application. Most importantly, the ICTY has established that NIACs require both:

(1) a certain threshold of violence, and
(2) a sufficient degree of organization of the conflict parties.[26]

Additional Protocol II (AP II) develops and supplements Common Article 3, but has a more restricted scope of application. Additional Protocol II only applies to NIACs that take place in the territory of a High Contracting Party between its armed forces and organized armed groups which, under responsible command, exercise such control over a part of its territory as to enable them to carry out sustained and concerted military operations and to implement the Protocol. Unlike Common Article 3, Additional Protocol II thus applies only to armed conflicts fought between a state and an organized armed group (fulfilling certain criteria), but not to conflicts between organized armed groups without the state's involvement. Furthermore, given that Article 1 AP II refers to armed conflicts "which take place in the territory of a High Contracting Party between *its* armed forces and dissident armed forces or other organized armed groups" (emphasis added), a transnational application of AP II is commonly rejected.[27]

NIACs in general were long understood as armed conflicts "which are in many respects similar to an international war, but take place within the confines of a single country"[28] and thus frequently referred to as "internal armed conflicts" or "civil wars"[29]. Yet, the emergence of "extraterritorial armed conflicts" has challenged Common Article 3's application to conflicts outside a state's own territory.

B. "Extraterritorial Armed Conflicts" (?)

In recent years, states have progressively been engaged in armed hostilities with organized non-state armed groups operating outside of the state's own territory.

---

26 ICTY, *Prosecutor v Duško Tadić a/k/a 'Dule'* (IT-94-1-AR72), Appeals Chamber Decision on the Defense Motion for Interlocutory Appeal on Jurisdiction (Oct 2, 1995) at para 70. For a summary of the ICTY's interpretation of both criteria, see ICTY, *Prosecutor v Ramush Haradinaj, Idriz Balaj and Lahi Brahimaj* (IT-04-84-T), Trial Chamber Judgment (April 3, 2008) at para 39.

27 However, the possibility of AP II's application to "spill-over" and "multinational" armed conflicts is increasingly admitted. For an argument regarding the case of Afghanistan, see A Bellal, G Giacca and S Casey-Maslen, "International Law and Armed Non-State Actors in Afghanistan" (2011) 93 Int'l Review of the Red Cross 47, at 59.

28 J Pictet, *Commentary on the Geneva Conventions of 12 August 1949*, vol II (ICRC 1960) 33.

29 For a distinction between the concepts of "civil war" and "non-international armed conflict," see A Cullen, "Key Developments affecting the Scope of Internal Armed Conflict in International Humanitarian Law" (2005) 183 Military Law Review 66, at 79.

This trend is generally explained by both an increase in non-state armed groups' ability to conduct operations across borders and (technological) advances, for example, the development of drones, allowing states to attack members of such groups even in remote areas. Under an armed conflict paradigm, these hostilities do not easily fit into either category of IAC or NIAC. While this has led to some calls for the recognition of a new category of armed conflict,[30] whose substantive content as well as scope of application would need to be determined, the majority opinion seems to reaffirm the validity of the current binary system and to argue within the existing classification regime.[31]

Most commentators thereby reject an interpretation of existing law according to which conflicts between states and transnational organized armed groups can be classified as IAC. They agree that the concept of IAC is limited to hostilities between states and defined by the parties involved rather than their geographical expansion.[32] Current state practice and *opinio iuris* also do not support the application of the law of IAC to cross-border conflicts between states and organized non-state armed groups.[33]

Following 9/11, the US government held the view that its conflict with Al-Qaeda was not covered by the laws of IAC given that Al-Qaeda is not a state party to the Geneva Conventions. However, it proceeded to argue that given its transnational scope the conflict was also not governed by the laws of NIAC, more specifically Common Article 3. The US Supreme Court was hence asked to classify the US conflict with Al-Qaeda in its *Hamdan* decision of 2006.[34] It held

---

30   See RS Schöndorf, "Extra-State Armed Conflicts: Is There a Need for a New Legal Regime?" (2014) 37 NYU J of Int'l Law and Politics 1; and G Corn, "Hamdan, Lebanon and the Regulation of Hostilities: The Need to Recognise a Hybrid Category of Armed Conflict" (2007) 40 Vanderbilt J of Transnat'l Law 295.

31   See C Kreß, "Some Reflections on the International Legal Framework Governing Transnational Armed Conflicts" (2010) 15 J of Conflict and Security Law 245.

32   D Fleck, "The Law of Non-International Armed Conflict" in D Fleck (ed) *The Handbook of International Humanitarian Law* (3rd ed, OUP 2013) at 581, 585.

33   Cf. Supreme Court of Israel (High Court of Justice), *The Public Committee against Torture in Israel (PCATI) and the Palestinian Society for the Protection of Human Rights and the Environment v Israel,* 769/02, (Dec 14, 2006) at para 18. In its *Targeted Killings* judgment, the Israeli Supreme Court classified the armed conflict between Israel and various "terrorist" organizations active in the West Bank and the Gaza Strip as an international armed conflict because it crosses the borders of the state. However, the Supreme Court's approach has largely been rejected.

34   Hamdan, a Yemeni national, was captured during hostilities between the US and Taliban forces in Afghanistan and transferred to Guantánamo Bay in June 2002. After having been charged with conspiracy to commit offenses triable by a military commission, Hamdan challenged the commission's authority to try him. Given that Hamdan argued that the commission's procedures violated international law, the Supreme Court was called upon to decide on the application of international humanitarian law. See US Supreme Court, *Hamdan v Rumsfeld* 548 US 557 (2006).

that the term "'armed conflict not of an international character' ... bears its literal meaning and is used here in contradistinction to a conflict between nations."[35] The US Supreme Court thus supported the application of Common Article 3 to the US' conflict with Al-Qaeda and did so without explicitly limiting its geographical scope to Afghanistan. Since the Supreme Court's *Hamdan* decision, the US government has consistently classified its conflict with Al-Qaeda, the Taliban, and associated forces as a NIAC without territorial confines.

The US construction of a global non-international armed conflict has provoked extensive criticism. Many legal commentators caution against the classification of hostilities in areas as distinct as Afghanistan, Pakistan, or Yemen as a single armed conflict and call for a differentiated approach.

Firstly, they argue that, while certain cross-border effects may be reconcilable with the concept of a NIAC, its criteria (i.e. the threshold of violence and degree of organization of the conflict parties) have to be fulfilled within at least one territory. According to these commentators, the threshold of violence has to be assessed by analyzing the frequency and severity of armed attacks within a defined space and cannot be established by connecting the dots and adding various hostilities across the globe.[36] Doubts are also expressed as to whether Al-Qaeda possesses the degree of organization necessary to consider it a single party to a global NIAC.[37] Most commentators are furthermore reluctant to accept that violence conducted by different armed groups linked to each other (e.g. Al-Qaeda, the Taliban, and associated forces) can be added up in order to fulfill either criterion.[38]

Secondly, legal scholars note that a global application of international humanitarian law undermines existing protections under international human rights law, most importantly with respect to the use of force by implementing more permissive rules on targeting.[39]

Thirdly, they caution against the spreading of international humanitarian law to "neutral" countries.[40] If the laws of war apply on a global battlefield, "fighters" with a continuous combat function, as well as civilians directly participating in hostilities, are arguably targetable across the globe, including in non-belligerent, neutral states. The rules on the conduct of hostilities, for example, the principle of proportionality, would hence be applicable worldwide and potentially put

---

35   Ibid at 562.

36   Emmerson, "Interim Report" (n 2) at para 63.

37   M Sassòli, "Transnational Armed Groups and International Humanitarian Law" (2009) HPCR Occasional Paper Series 6, at 10.

38   On the interpretation of the criteria of "intensity" and "degree of organization" in case of extraterritorial non-international armed conflicts, see Radin "Global Armed Conflict" (n 25) at 718.

39   ME O'Connell, "The Choice of Law Against Terrorism" (2010) 4 J of Nat'l Security Law and Policy 343, at 350. See also Chapter 4 by Gloria Gaggioli in this volume.

40   The law of neutrality only applies in case of international armed conflict. The term should thus be understood to only factually refer to states who are not party to a non-international armed conflict and placed in quotations throughout this chapter.

civilians and civilian objects in non-belligerent states at risk, possibly without their knowledge.[41] For example, civilians sitting at an outdoor café in Paris are generally not aware that they may be affected by a strike—which raises the concern whether states should employ methods causing fear of the unpredictable similar to terrorists attacking unsuspecting civilians.[42]

Finally, commentators argue that if a global armed conflict existed and were to follow "fighters" and civilians directly participating in hostilities wherever they moved, it would be almost impossible for them to disengage from the armed conflict and regain protection.[43] They could hence no longer simply walk away from the battlefield, but would have to establish by other means that they no longer exercise a continuous combat function or directly participate in hostilities.

The majority opinion therefore seems to uphold the view that no global NIAC has come into existence and prefers a case-by-case approach to analyzing the various hostilities taking place in different parts of the world, either by the resort to drones or other means. Most accept that the US could indeed be (or have been) involved in different armed conflicts—if the conditions of either international or non-international armed conflicts have been met.

According to this majority view, the initial conflict between the US-led coalition and the Taliban in Afghanistan was an IAC because the Taliban then represented the de facto Afghan government and thus acted on behalf of another state.[44] However, the ensuing armed conflict between the new Afghan government and organized non-state armed groups, such as the Taliban, is classified as a NIAC.[45]

---

41 C Heyns, "Report of the Special Rapporteur on Extrajudicial, Summary or Arbitrary Executions" UN Doc A/68/382 (Sept 13, 2013) at para 64.

42 J Daskal, "The Geography of the Battlefield: A Framework for Detention and Targeting Outside the 'Hot' Conflict Zone" (2013) 161 Univ of Penn Law Review 1165, at 1196.

43 Lubell and Derejko, "Global Battlefield" (n 18) at 82.

44 More generally, the amount of control required by a state over a non-state armed group to establish that it acts on behalf of the state remains disputed. The International Court of Justice (ICJ) held in the 2007 *Bosnian Genocide*-case that while an "overall control" test was unsuited for the purposes of state responsibility, "it may well be that the test is applicable and suitable" to classify armed conflicts. See ICJ, *Application of the Convention on the Prevention and Punishment of the Crime of Genocide (Bosnia and Herzegovina v Serbia and Montenegro)* (Feb 26, 2007) ICJ Reports 43, at para 404.

45 Given the length of the transitional process leading to the establishment of a new Afghan government requesting international assistance in its fight against the Taliban insurgency, the exact moment when the conflict transformed into a non-international armed conflict is disputed. The following propositions are made: (1) the establishment of an interim authority headed by Hamid Karzai in December 2001; (2) the appointment of Hamid Karzai by the Loya Jirga as president of the transitional authority in June 2002; (3) the adoption of a new constitution in January 2004; (4) the election of Hamid Karzai as president of Afghanistan in October 2004; and (5) the parliamentary election in 2005. See Bellal, Giacca and Casey-Maslen, "Armed Non-State Actors" (n 27), at 52.

To the extent that hostilities between the US (which acts at the invitation and in support of the Afghan government) and Al-Qaeda, the Taliban, or associated forces are part of this NIAC, they are thus considered covered by the law of NIAC—even though they do not take place on US territory, but on Afghan territory.

The extent of Pakistan's consent to US drone operations against Al-Qaeda and the Taliban on Pakistani territory is disputed. If the US were to conduct hostilities against organized non-state armed groups on Pakistani territory without Pakistan's consent, an IAC between the US and Pakistan can be assumed, despite the lack of hostilities between both states.[46] Yet, if Pakistan consented to such strikes, most agree that no IAC can be said to have come into existence. Accordingly, hostilities between the US and Al-Qaeda or the Taliban on Pakistani territory are rather to be analyzed within a NIAC framework. Because of their comparatively limited impact, single drone strikes most probably do not fulfill the criterion of "protractedness" and thus do not trigger a distinct NIAC.[47] However, the law of non-international armed conflict applies to drone strikes in Pakistan if the Afghan conflict is considered to have "spilled" onto Pakistani territory. Alternatively, the law of NIAC applies to US drone strikes if conducted in support of Pakistan's fight against the Pakistani Taliban.[48]

Outside these international and non-international armed conflicts, the use of drones in hostilities against organized non-state armed groups is arguably not governed by international humanitarian law, but international human rights and domestic law.

The Yemeni government, for example, has announced that the US routinely seeks its consent, on a case-by-case basis, for lethal drone operations on its territory.[49] An IAC can thus not be assumed. If one accepts that single US drone strikes do not trigger a NIAC between the US and Al-Qaeda in Yemen and that the United States is not acting in support of the Yemeni government in its NIAC, the law of NIAC is also not applicable to such strikes. Outside a "global non-international

---

46   The role of a lack of consent in classifying armed conflicts is subject to (scholarly) debate. According to some commentators, the separation of *jus ad bellum* and *jus in bello* demands that armed conflicts are classified based on the occurrence of factual hostilities only, see Lubell, "War (?)" (n 17) at 432.

47   ICTY, *Haradinaj* (n 26) at para 49, which notes that "the criterion of protracted armed violence has ... been interpreted in practice, including by the *Tadić* Trial Chamber itself, as referring more to the intensity of the armed violence than to its duration." It continues to note that "Trial Chambers have relied on indicative factors relevant for assessing the 'intensity' criterion, none of which are, in themselves, essential to establish that the criterion is satisfied," such as the number, duration and intensity of individual confrontations, the type of weapons and other military equipment used, and the number of casualties.

48   See LR Blank and BR Farley, "Characterizing US Operations in Pakistan: Is the United States Engaged in an Armed Conflict?" (2011) 34 Fordham Int'l Law J 151.

49   See Emmerson, "Interim Report" (n 2) at para 52.

armed conflict" doctrine, the legality of these strikes is hence to be assessed based on international human rights law, and not international humanitarian law.

Nonetheless, both approaches, that is, the construction of a single global NIAC and the creation of multiple NIACs with a cross-border dimension, call for an analysis of the geographical scope of NIACs.

III. Geographical Scope of Non-International Armed Conflicts

The Geneva Conventions and their Additional Protocols do not contain specific rules on their geographical scope of application in case of either international or non-international armed conflict.[50]

A. Traditional Non-International Armed Conflicts

According to the general rules of public international law regarding treaties, which are today codified in the Vienna Convention on the Law of Treaties (VCLT), a treaty is binding upon each treaty party in respect of its entire territory, unless a different intention appears from the treaty or is otherwise established (Article 29 VCLT). If one argues that the Geneva Conventions' objective is to govern the humanitarian consequences of armed conflict, it indeed seems sensible to assume that they apply to the entire territory of a state engaged in armed conflict. After all, such consequences cannot always be confined to a certain part of a belligerent's territory.

Common Article 3's application to the entire territory of a High Contracting Party may nevertheless be challenged. In cases where hostilities are limited to some parts of a territory,[51] Common Article 3's application to areas remote from the battlefield may lead to counterintuitive results. For example, the killing of a "fighter" in a government-controlled, peaceful capital, based on international humanitarian law, seems inappropriate to some.[52]

The ICTY was asked to address the geographical scope of both international and non-international armed conflicts in its *Tadić* decision because the defendant argued that no armed conflict, either "internal" or international, existed at the time and place the alleged offences were committed.[53] The Appeals Chamber preliminarily established that "the temporal and geographical scope of both

---

50   See more detailed K Schöberl, "The Geographical Scope of Application of the Conventions" in A Clapham, P Gaeta, M Sassòli (eds), *The 1949 Geneva Conventions: A Commentary* (OUP forthcoming).

51   See, for example, the armed conflict between the Sri Lankan armed forces and Tamil Tigers having taken place mostly in the northern and eastern regions of Sri Lanka.

52   M Sassòli and LM Olson, "The Relationship between International Humanitarian and Human Rights Law where it Matters: Admissible Killing and Internment of Fighters in Non-International Armed Conflicts" (2008) 90 Int'l Review of the Red Cross 599, at 613.

53   ICTY, *Tadić* (n 26).

internal and international armed conflicts extends beyond the exact time and place of hostilities."[54] With respect to NIACs, the Tribunal held that Common Article 3 applies "outside the narrow geographical context of the actual theatre of combat operations"[55] because it also protects all those taking not (or no longer) active part in hostilities.

With reference to Article 2(1) and (2) Additional Protocol II, the Appeals Chamber furthermore noted that international humanitarian law of NIAC is applicable to all persons affected by an armed conflict and deprived of their liberty for reasons related to an armed conflict. According to the Tribunal, the relatively loose nature of this language suggests a broad geographical scope. It thus argued that "[t]he nexus required is only a relationship between the conflict and the deprivation of liberty, not that the deprivation occurred in the midst of battle."[56] The ICTY eventually concluded that "[u]ntil that moment [i.e. a general conclusion of peace/a peaceful settlement], international humanitarian law continues to apply in the whole territory of the warring States or, in the case of internal conflicts, the whole territory under the control of a party, whether or not actual combat takes place there."[57] The Tribunal's reference to "the whole territory under the control of a party" can be understood as a reminder of the possible parties to a NIAC. Contrary to IAC, territory may thus not be under state control, but under the control of an organized non-state armed group.[58]

In applying the facts of the case to the law, the Chamber held that international humanitarian law was applicable even though substantial clashes were not occurring at the time and place the crimes allegedly were committed. It considered it sufficient that the alleged crimes were "closely related to the hostilities occurring in other parts of the territories controlled by the parties to the conflict."[59]

The ICTY has subsequently established certain criteria based on which one can examine the existence of a *nexus* between alleged crimes and an armed conflict. According to its jurisprudence, "[t]he armed conflict need not have been causal to the commission of the crime, but the existence of an armed conflict must, at a minimum, have played a substantial part in the perpetrator's ability to commit it, his decision to commit it, the manner in which it was committed or the purpose for which it was committed."[60] To conclude that acts were closely related to an armed conflict, it is sufficient to establish that "the perpetrator acted in furtherance of or under the guise of the armed conflict."[61] Moreover, the ICTY

---

54 Ibid at para 67.

55 Ibid.

56 Ibid.

57 Ibid para 70.

58 See also Lubell and Derejko, "A Global Battlefield?" (n 18) at 69.

59 ICTY, *Tadić* (n 26) at para 70.

60 ICTY, *Prosecutor v Dragoljub Kunarac, Radomir Kovač, Zoran Vuković* (IT-96–23&23/2), Appeals Chamber Judgment (June 12, 2002) at para 58.

61 Ibid.

noted that the following criteria may be taken into account to establish a nexus: the fact that the perpetrator is a combatant, the fact that the victim is a non-combatant, the fact that the victim is a member of the opposing party, the fact that the act may be said to serve the ultimate goal of a military campaign, and the fact that the crime is committed as part of or in the context of the perpetrator's official duties.[62]

B. Non-International Armed Conflicts with Cross-Border Dimension

NIACs are sometimes not limited to the territory of a single state party, but have a cross-border dimension. For example, hostilities may affect the territory of a neighboring state. A state may also conduct military operations against non-state armed groups outside its own territory with or without the territorial state's consent.[63] This raises the question if and to what extent Common Article 3 accommodates the possibility of NIACs outside a High Contracting Party's own territory. Put differently, is the application of Common Article 3 limited to "internal armed conflicts" or does it encompass NIACs with a cross-border dimension?

The wording of Common Article 3 seems to suggest its limitation to armed conflicts occurring within the territory of one state only. In accordance with the ordinary meaning its reference to the "territory of *one* of the High Contracting Parties" (emphasis added) is understood in contradistinction to "two," "three," or "several" High Contracting Parties' territories. However, given that Common Article 3 is not only addressed to state armed forces but equally binds non-state armed groups, it can be considered "only logical"[64] that the provision requires a territorial link to a High Contracting Party. Common Article 3 can therefore be interpreted to refer to the territory not of "one," but "a" state party to the Conventions. Given the Conventions' universal ratification it is moreover argued that any armed conflict today takes place on the territory of a High Contracting Party and that the territorial reference within Common Article 3 has lost importance in practice.[65]

Common Article 3's context furthermore suggests that NIACs can be negatively defined as all armed conflicts not covered by Common Article 2—regardless of their geographical extent. In order to avoid a gap in protection, which could not be

---

62   Ibid at para 59.

63   These conflicts are frequently referred to as "spill-over armed conflicts," "multinational armed conflicts," and "transnational armed conflicts." However, the terms only classify such situations factually, not legally.

64   N Melzer, *Targeted Killing in International Law* (OUP 2008) at 258.

65   ICRC, "How is the Term 'Armed Conflict' Defined in International Humanitarian Law?," Opinion Paper (2008) 3. Nevertheless, the possibility of a non-international armed conflict occurring on the territory of a newly created state not (yet) party to the Conventions continues to exist.

explained by states' concerns about their sovereignty,[66] the Convention would thus comprehensively govern all types of armed conflict and classify them as either "international" (Common Article 2) or "non-international" (Common Article 3). The US Supreme Court has applied this contextual approach to interpreting Common Article 3 in its *Hamdan* decision when it held that that the term "armed conflict not of an international character" is used in contradistinction to a conflict between states.[67]

The object and purpose of international humanitarian law is used for arguments both in favor of and against a territorial limitation of Common Article 3. On the one side, it is argued that Common Article 3 should be applied only with territorial constraints in order to prevent a crowding-out of international human rights law, if the object and purpose of international humanitarian law is to set certain minimum standards where the more protective international human rights law regime cannot be applied due to the armed conflict.[68] On the other side, it is noted that the effects of armed conflicts which international humanitarian law seeks to govern do not necessarily stop at border crossings. As much as detainees should not be deprived of protection because Common Article 3's safeguards cease at the frontier,[69] "fighters" should arguably not be able to escape the reach of international humanitarian law by using third country "safe havens" unable or unwilling to respond to the threat they pose.[70]

Thus the wording, context, as well as the object and purpose of Common Article 3 are "rather ambiguous"[71] regarding its potential extraterritorial application. The resort to supplementary means of treaty interpretation, most importantly the preparatory work for the drafting of Common Article 3, is hence appropriate.[72]

---

66   Sassòli, "Transnational Armed Groups" (n 37) 9. See also D Jinks, "September 11 and the Laws of War" (2003) 28 Yale J of Int'l Law 1. Jinks notes that "Common Article 3 was revolutionary because it purported to regulate wholly internal matters as a matter of international humanitarian law" and argues that "[i]f the provision governs wholly internal conflicts, as the 'one state' interpretation recognizes, then the provision applies *a fortiori* to armed conflicts with international or transnational dimensions," at 41.

67   US Supreme Court, *Hamdan* (n 34) at 562.

68   R Geiß, "Armed Violence in Fragile States: Low Intensity Conflicts, Spillover Conflicts, and Sporadic Law Enforcement Operations by Third Parties" (2009) 91 Int'l Review of the Red Cross 127, at 138.

69   J Pejic, "Common Article 3" (n 21) at 15.

70   R Chesney, "Who May Be Killed? Anwar al-Awlaki as a Case Study in the International Legal Regulation of Lethal Force" in T McCormack, M Schmitt and L Arimatsu (eds), (2010) 13 *Yearbook of International Humanitarian Law* 3, at 37.

71   R Bartels, "Timelines, Borderlines and Conflicts. The Historical Evolution of the Legal Divide between International and Non-International Armed Conflicts" (2009) 91 Int'l Review of the Red Cross 35, at 60.

72   In light of the importance attached to the historic circumstances surrounding the adoption of Common Article 3 by some commentators, it is suitable to recall the subsidiary nature of supplementary means of treaty interpretation, such as Common Article 3's *travaux*

While the first drafts of a provision extending the Conventions' application to NIACs suggested by National Red Cross and Red Crescent Societies as well as governmental experts proposed language limiting its application to cases of armed conflict "within the borders of a State"[73] and "in any part of the home or colonial territory of a Contracting Party,"[74] the International Committee of the Red Cross' (ICRC) consolidated draft Conventions proposed different wording. The text, submitted to National Societies and states with a view to discussion and adoption at the XVIIth International Conference in Stockholm,[75] provided that,

> [i]n all cases of armed conflict which are not of an international character, especially cases of civil war, colonial conflicts, or wars of religion, which may occur *in the territory of one or more of the High Contracting Parties* [emphasis added], the implementing of the principles of the present Convention shall be obligatory on each of the adversaries.[76]

Unfortunately, it remains unclear why the ICRC chose to extend the provision's geographical scope from the territory of one state (possibly including its colonies) to the territory of one *or more* High Contracting Parties. When the XVIIth International Conference adopted the draft Conventions, the provision's reference to the territory of one *or more* of the High Contracting Parties remained unchanged.

However, the Diplomatic Conference convened in Geneva in 1949 quickly abandoned it. Following the first working party's first proposal,[77] none of the subsequent drafts reverted to the Stockholm draft's wording. That is, none of the draft provisions presented to the Plenary Assembly therefore still referred to the territory of one *or more* of the High Contracting Parties—including the second working party's proposal, which was finally adopted and today constitutes

---

préparatoires (Art. 32 VCLT). Cf. R Ash, "Square Pegs and Round Holes: Al-Qaeda Detainees and Common Article 3" (2007) 17 Indiana Int'l & Comp Law Review 269.

73   ICRC, "Report on the Work of the Preliminary Conference of National Red Cross Societies for the Study of the Conventions and of Various Problems Relative to the Red Cross" (Geneva, July 26-Aug 3, 1946) (ICRC 1947, Series I, No 3a) at 15.

74   ICRC, "Report on the Work of the Conference of Government Experts for the Study of the Conventions for the Protection of War Victims" (Geneva, April 14–26, 1947) (ICRC, 1947, Series I, No 5b) at 8.

75   ICRC, "Draft Revised or New Conventions for the Protection of War Victims established by the International Committee of the Red Cross with the assistance of Government Experts, National Red Cross Societies and Other Humanitarian Associations" ICRC (1948).

76   Ibid 5.

77   Swiss Federal Political Department, "Final Record of the Diplomatic Conference of Geneva of 1949" vol. II, Section B, Summary Record, Fifth Meeting of Special Committee of Joint Committee, at 46.

Common Article 3.[78] Given the lack of record as to why its geographical scope was first extended to "one or more" territories of High Contracting Parties and later limited, Common Article 3's preparatory work remains only of limited value in assessing its potential extraterritorial application.

However, state practice after World War II supports the application of Common Article 3 to "spill-over" armed conflicts. The ICRC as well argues that conflict parties remain at a minimum bound by Common Article 3 and customary international humanitarian law because "the spillover of a NIAC into adjacent territory cannot have the effect of absolving the parties of their IHL obligations simply because an international border has been crossed."[79] For example, hostilities between the Colombian armed forces and FARC (Fuerzas Armadas Revolucionarias de Colombia) members on Ecuadorian territory are widely considered covered by Common Article 3.[80] Similarly, US drone strikes on Pakistani territory are regarded as governed by Common Article 3 if they are conducted in relation to the NIAC occurring in Afghanistan. Furthermore, Common Article 3's application to "multinational" armed conflicts has increasingly found support. Recent examples, such as the International Security Assistance Force's (ISAF) support to the Afghan government, have shown that states seem to accept the application of at least Common Article 3 in cases in which they assist another state in its fight against non-state armed groups outside their national borders.[81]

IV. Conclusion

Most stakeholders and analysts of the drone debate agree that "drones ... are here to stay,"[82] and will likely be used within and outside conventional theaters of armed conflict to an even greater extent in the future. This is generally explained by their low cost, efficacy in targeting individuals in inaccessible areas and regions without military ground presence, as well as the averting of danger for a state's own personnel.[83]

---

78   Seventh Report drawn up by the Special Committee of the Joint Committee, Art 2, Para 4 ("Application of the Conventions to Armed Conflicts Not of An International Character") (July 16, 1949) ibid at 120 and Report drawn up by the Joint Committee and presented to the Plenary Assembly, ibid at 128.

79   ICRC, "Challenges Report" (n 23) at 9.

80   For a legal classification of "Operation Phoenix" (March 1, 2008), see F Szesnat and AR Bird, "Colombia" in Wilmshurst, *Classification of Conflicts* (n 17) 203, at 232.

81   See G Westerwelle, "Paving the Way for a Responsible Handover: Germany's Engagement in Afghanistan after the London Conference" German Parliament (Feb 10, 2010).

82   Heyns, "Report of Special Rapporteur" (n 41) at para 12.

83   See R Ratnesar, "Five Reasons Drones Are Here to Stay" *Bloomberg* (May 23, 2013); see also Chapter 10 by Marek Madej.

The legality of their use in cross-border counterterrorism operations is undisputedly determined by different branches of international law. While the legality of the resort to armed force is governed by the *jus ad bellum*, the lawfulness of targeted killings of individuals by drones is decided based on international humanitarian law in case of armed conflict and international human rights law and domestic law "outside of armed conflict." Given the differences in regulating the lawful use of force under international human rights and humanitarian law, it is crucial to determine whether or not a targeted killing by a drone takes place inside or outside of armed conflict.

Current state practice and *opinio iuris* do not support the US construction of a single, global NIAC in which international humanitarian law applies even outside of "hot battlefields." Criticism by (legal) practitioners and scholars is based on arguments founded not only in international humanitarian law, but also international human rights law and *jus ad bellum*.

An analysis of Common Article 3, including its wording, context, object, and purpose, shows that the law of NIACs may be applied extraterritorially despite its denomination, for example, to spill-over armed conflicts. State practice confirms that non-international armed conflicts may have certain cross-border dimensions. However, what distinguishes a NIAC with cross-border dimensions from a global NIAC is the occurrence of both required criteria (i.e. threshold of violence and degree of organization of conflict parties) in at least one territory. Present international humanitarian law thus seems to require a nexus to an existing NIAC and does not allow for the aggregation of both conditions across the globe. The reason for favoring a global non-international armed conflict seems fairly obvious: It would provide a legal basis for targeting individuals outside of "hot battlefields" even in case of an end of combat operations in conventional theaters of armed conflict, such as Afghanistan.[84]

The criteria according to which a nexus to an existing armed conflict can be established also remain debatable. Some make use of a territorial approach and note that international humanitarian law applies in "spill-over areas." They argue, for example, that the law of NIAC applies to drone strikes conducted in relation to the Afghan conflict only if carried out within the Afghan-Pakistani border region.[85] However, the delimitation of spill-over areas raises concerns about the degree of

---

84 In addition to arguing the existence of a global non-international armed conflict, the US has increasingly relied on a self-defense argument independently or conjunctively with its claims under international humanitarian law. Many commentators see this as an indication for a possible declared future end of the US' armed conflict with Al-Qaeda, the Taliban, and associated forces.

85 See S Breau, M Aronsson and R Joyce, "Drone Attacks, International Law, and the Recording of Civilian Casualties of Armed Conflict" Oxford Research Group Discussion Paper (June 2, 2011) 12, who argue that drone strikes outside the Federally Administered Tribal Areas (FATA) and the North-West Frontier Province (NWFP) are governed by the rules of law enforcement.

arbitrariness attached to "distance-based reasoning."[86] Substantive approaches, such as the nexus test, thus increasingly receive support. If one accepts the ICTY's nexus test to be appropriate in this regard, international humanitarian law would apply to all acts having a sufficient nexus to an existing NIAC—even if committed elsewhere and on the territory of a neutral state.

Yet, opinions diverge on the question of whether a "fighter" or civilian directly participating in hostilities having a nexus to an existing NIAC and present on the territory of a neutral state may be directly targeted. The ICRC has taken the position that a person does not carry a NIAC to the territory of a non-belligerent state. It argues that this would potentially expand the application of the rules on the conduct of hostilities to multiple states according to a person's movements and would, in effect, signify the recognition of the concept of a "global battlefield." The ICRC therefore argues in favor of applying the law enforcement model to the use of force against individuals on the territory of a non-belligerent state.[87]

Others assert that while international humanitarian law does not follow individuals, it pursues activities. If a "fighter" or civilian directly participating in hostilities continues to engage in a NIAC from a new location, he/she thus remains targetable.[88] While the distinction between an "individual"- vs "activities"-based approach arguably does not matter for civilians directly participating in hostilities, "fighters" in NIACs would hereby only be targetable based on their continued engagement in the armed conflict and not on their status alone.[89]

Most commentators agree that the spreading of armed conflict, that is, the "creep of war,"[90] to neutral states should be prevented. Disagreement exists, however, whether this is best done through a restrictive interpretation of the geographical scope of application of international humanitarian law or the applicable *jus ad bellum*. The future debate on the boundaries of the battlefield should thereby consider both the protective and permissive nature of international humanitarian law. While Common Article 3 based on its wording only contains safeguards for persons taking no active part in hostilities, the accompanying customary international humanitarian law regime arguably provides for a possibility to detain "fighters" without charge and to target them as long as they

---

86   Lubell and Derejko, "A Global Battlefield?" (n 18) at 82.

87   ICRC, "Challenges Report" (n 23) at 22.

88   Lubell, "War (?)" (n 17) at 449.

89   It has also been suggested that while international humanitarian law should apply globally, a distinction should be made between "zones of active hostilities," "peacetime zones," and "lawless zones" with enhanced substantive and procedural standards applying in the latter two. Accordingly, an individualized threat finding, a least-harmful means test as well as meaningful *ex ante* and *ex post* procedural safeguards should limit targeting and detention operations outside "zones of active hostilities," see Daskal, "Geography of the Battlefield" (n 42) at 1192.

90   Ibid at 1173.

are not *hors de combat*.[91] A consistent approach in defining the boundaries of the battlefield regarding persons located on neutral territory would require that the laws of war are equally applicable to the targeting and detention without charge on the one side, and procedural safeguards of detainees on the other side. After all, international humanitarian law's claim to represent a compromise between military necessity and humanity must also be considered and expressed in the definition of its scope of application.

The principle of reciprocity, imposing equal rights and obligations on all conflict parties, should furthermore be taken into account in further deliberations. Drones are usually operated and controlled by an entire crew consisting of a pilot, payload operator, and imagery intelligence analysts, who are at times physically dispersed across different countries. If the law of NIAC is applicable to drone strikes with a nexus to an armed conflict on neutral territory, it arguably also applies to attacks by organized non-state armed groups against control bases or launching stations located elsewhere in reverse.[92] Determining the boundaries of the battlefield is thus not only relevant for the location of the target, but also of the drone crew.

In view of predictions that an increasing number of states will acquire drones, international consensus on the geographical scope of application of international humanitarian law is of utmost importance. Next to vocal groups of scholars,[93] states should primarily state their legal position, for example, on whether (1) international humanitarian law requires an assessment of the severity and frequency of armed attacks occurring within defined geographical boundaries, (2) it is legitimate to aggregate armed attacks occurring in geographically diverse locations, and (3) non-international armed conflicts can exist without finite territorial boundaries.[94] The debate may ultimately not only clarify the applicable rules on the use of drones, but serve to inform the debate on the geographical scope of the laws of war more generally.

---

91 The detention of "terrorist suspects" remains relevant for this debate as individuals potentially continue to be captured and held in facilities outside "hot battlefields," including in secret locations, in relation to the purported armed conflict. See United Nations, "Joint Study on Global Practices in Relation to Secret Detention in the Context of Countering Terrorism" UN Doc A/HRC/13/42 (Feb 19, 2010) at para 120.

92 This would arguably also be the case where attacks are directed not at launching stations or similar objects, but members of drone crews. The status of drone operators depends on whether they are members of the regular armed forces or intelligence agencies, such as the CIA. However, it is widely accepted that the operation of a drone amounts to direct participation in hostilities leading to a loss of protection from direct attack even in cases where drone operators are not entitled to combatant status. See LR Blank, "After 'Top Gun': How Drone Strikes Impact the Law of War" (2012) 33 Univ of Penn J of Int'l Law 675, at 707.

93 Daskal, 'Geography of the Battlefield' (n 42) at 1169.

94 Emmerson, "Report Special Rapporteur" (n 15) at para 71.

Chapter 4

Lethal Force and Drones: The Human Rights Question

Gloria Gaggioli[1]

I. Introduction

Drones are being used—and are likely to be increasingly used—in peacetime policing. For the time being, the types of activities state authorities conduct with drones in peacetime are mainly related to surveillance. The likelihood that drones could be armed to use force domestically in order to maintain or restore public security, law, and order in the near future should, however, not be underestimated. This is even more so if one takes into account advances in technology and the fact that, today, drones come not only in the shape of large unmanned combat air vehicles (UCAV), but may also be miniature aerial—and sometimes weaponized—systems. In the multifaceted fight against terrorism, armed drones have moreover been widely used against persons extraterritorially, including outside armed conflict situations. Even in armed conflicts, drones may be used—like in peacetime—to conduct law enforcement activities (e.g. to deal with riots or other forms of civilian unrest).

The rules on the conduct of hostilities provided for in international humanitarian law (IHL) do not apply outside armed conflict situations or (in the context of an armed conflict) outside hostilities. The use of lethal force by means of drones in such cases is governed exclusively by international human rights law (HRL) and its rules and standards for the use of force (hereafter the HR law enforcement paradigm). It is thus not surprising that the international community often calls on states using armed drones to do so in a way that complies with their international legal obligations, not only under the Charter of the United Nations and under IHL, but also under HRL.[2]

The practical and legal consequences of this position remain, however, unexplored. While the legal and humanitarian issues posed by drone strikes

---

1   I would like to thank Jelena Pejić, Senior Legal Adviser at the ICRC, for the fascinating discussions we had on this topic and for her useful, thoughtful, and inspiring comments on this chapter.

2   See eg *Protection of Human Rights and Fundamental Freedoms while Countering Terrorism* UN Doc A/RES/68/178 (Jan 28, 2014) at paras 1 and 6 (s); *European Parliament Resolution on the Use of Armed Drones* 2014/2567(RSP) (Feb 25, 2014) at para F.

have been extensively examined under the lens of IHL, the issue of the potential conformity of drone strikes with the law enforcement paradigm has been overlooked. Beyond the issue of the extraterritorial application of HRL, can the use of force by drones (ever) respect the principles of absolute necessity and proportionality? How can an escalation of force procedure be applied in such situations? These are some of the key questions this chapter will address.

This chapter is divided into three parts. It first explores the various contexts in which the use of force by drones in law enforcement arises. Second, it recalls the legal requirements for the use of force derived from HRL, and more precisely the prohibition of arbitrary killing. Third, it analyzes whether the use of force by drones can comply with the HR law enforcement requirements, not only regarding the use of force as such, but also regarding obligations before and after the use of force. It also tackles briefly other indirect and less "visible" legal and policy issues that might emerge if drones are used to conduct law enforcement activities.

II. When Does the Human Rights Question Arise?

A. In Peacetime and Situations of Violence Not Reaching an Armed Conflict

As highlighted by Christof Heyns, UN Special Rapporteur on Extrajudicial, Summary or Arbitrary Executions: "The use of armed drones happen[s] in conflict and counterterrorism situations, but also increasingly in ordinary policing and law enforcement."[3]

In certain developed countries, drones are presently used for a number of purposes in peacetime.[4] They can perform highly advanced surveillance and can help to detect fires, collect data on suspected offenders or for relief personnel working in areas affected by natural disasters.[5] Drones can also do border control and security operations. For instance, they are used along the US–Mexico border to

---

3   *Human Rights Council Holds Panel on Remotely Piloted Aircraft or Armed Drones in Counterterrorism and Military Operations. Panel of the Human Rights Council on Remotely Piloted Aircraft or Armed Drones in Counterterrorism and Military Operations* (Sept 22, 2014) <http://www.unog.ch/80256EDD006B9C2E/%28httpNewsByYear_en%29/BCE56E D914A46D40C1257D5B0038393F?OpenDocument> accessed Sept 2014.

4   "Law Enforcement Agencies Using Drones List Map" *Governing the States and Localities* (Jan 16, 2014) <http://www.governing.com/gov-data/safety-justice/drones-state-local-law-enforcement-agencies-license-list.html> accessed Sept 2014.

5   Peter Maurer (Web Interview), *The Use of Armed Drones Must Comply with Laws* (May 10, 2013) <https://www.icrc.org/eng/resources/documents/interview/2013/05–10-drone-weapons-ihl.htm> accessed Sept 2014; Dan Roberts "FBI Admits Using Surveillance Drones over US Soil" *The Guardian* (June 19, 2013) <http://www.theguardian.com/world/2013/jun/19/fbi-drones-domestic-surveillance> accessed Sept 2014.

detect and track drug smugglers and human traffickers.[6] Drones can even do crowd-control. In May 2014, a South African company created a brand new unmanned aerial system which is able to shoot pepper spray and non-lethal paintballs to mark offenders. It can also employ strobe lights and on-board speakers to send verbal warnings.[7] It was reported that Turkey purchased this "riot drone."[8]

Media report that around 80 countries have some kind of drone capability, even if not many states resort yet to using drones in peacetime policing. As technology evolves and becomes more affordable, the day where a number of state authorities will have and use this technology for such purposes might come sooner rather than later.

It cannot be ruled out either that drones able to project lethal or potentially lethal force might be used in the future for conducting law enforcement activities in peacetime or internal disturbances and tensions.[9] This might not be overly surprising given the amount of force that states use in internal disturbances and tensions, for instance, in the fight against criminality to combat drug cartels.[10] The increasing use of combat weapons such as M-16 rifles, armored trucks, and grenade launchers in the context of peacetime law enforcement has been a subject of concern for some years now.[11] One case in point is the widely reported event in Ferguson, Missouri where demonstrators protesting the fatal police shooting of a teenager were confronted with heavily armed, militarized, police officers.[12] The acquisition by police units, such as SWAT units (Special Weapons and Tactics), of armed drones would seem to be just one further step in this worrying evolution where police officers are armed, equipped, and operating like soldiers.

---

6 Aliya Sternstein, "Obama Requests Drone Surge for U.S.-Mexico Border" *Defence One* (July 9, 2014) <http://www.defenseone.com/threats/2014/07/obama-requests-drone-surge-us-mexico-border/88303/> accessed Sept 2014.

7 "Riot Control Drone Armed with Paintballs and Pepper Spray Hits Market" *RT* (June 19, 2014) <http://rt.com/news/167168-riot-control-pepper-spray-drone/> accessed Sept 2014.

8 "Turkey Cracks Down: Skunk Riot Drone Will Fire Paint and Pepper Balls" *WorldTribune.com* (Ankara, June 27, 2014) <http://www.worldtribune.com/2014/06/27/turkey-buys-skunk-riot-drone-payload-including-paint-pepper-balls/> accessed Sept 2014.

9 Conor Friedersdorf, "Congress Should Ban Armed Drones Before Cops in Texas Deploy One" *The Atlantic* (May 24, 2012) <http://www.theatlantic.com/national/archive/2012/05/congress-should-ban-armed-drones-before-cops-in-texas-deploy-one/257616/> accessed Sept 2014.

10 Jane Perlez, "Chinese Plan to Kill Drug Lord With Drone Highlights Military Advances" *The New York Times* (Feb 20, 2013) <http://www.nytimes.com/2013/02/21/world/asia/chinese-plan-to-use-drone-highlights-military-advances.html?_r=0> accessed Sept 2014.

11 Clyde Haberman, "The Rise of the SWAT Team in American Policing" *The New York Times* (Sept 7, 2014) <http://www.nytimes.com/2014/09/08/us/the-rise-of-the-swat-team-in-american-policing.html?_r=0> accessed Sept 2014.

12 Ibid Alcindor Yamiche and Bello Marisol, "Police in Ferguson Ignite Debate about Military Tactics" *USA Today* (Aug 19, 2014) <http://www.usatoday.com/story/news/nation/2014/08/14/ferguson-militarized-police/14064675/> accessed Sept 2014.

As highlighted by the International Committee of the Red Cross (ICRC) President, Peter Maurer, it is clear that: "If and when drones are used in situations where there is no armed conflict, it is the relevant national law, and HRL with its standards on law enforcement, that apply, not international humanitarian law."[13]

B. In Extraterritorial Counterterrorism Operations and Armed Conflict Situations

HRL applies at all times.[14] Even if some rights can be derogated from "in time of public emergency which threatens the life of the nation,"[15] this is not the case for the prohibition of arbitrary deprivation of life, which is non-derogable and from which the limits for the use of force against individuals in HRL are derived.[16]

The human rights question related to the use of force by drones arises not only in peacetime or in situations of internal disturbances and tensions, but also in extraterritorial counterterrorism operations (which may or may not amount to an armed conflict depending on the situation) and, in certain cases, within situations of armed conflict as well.

1. Extraterritorial use of force in the territory of a non-belligerent state

The extraterritorial targeting of individuals by means of drones has been commonplace in the context of the "war on terror." The United States and United Kingdom, for instance, conducted a number of extraterritorial drone strikes against Al-Qaeda, the Taliban, and associated forces. In many cases, such attacks took place in the context of an acknowledged armed conflict (e.g. in Afghanistan or Iraq) and were governed by IHL. Drone strikes have also been carried out against these and other terrorist groups, including the Islamic State of Iraq and Syria (ISIS), in contexts where the group's legal classification is disputed (e.g. in Pakistan, Philippines, Somalia, Syria, Yemen). It is not the place here to attempt a classification of these various contexts at different points in time. The facts are often unclear and evolving so quickly that it is difficult to conduct an accurate legal analysis. It is sufficient to say that at least some drone strikes have allegedly

---

13   Maurer (n 5).

14   There is, however, a minority view according to which human rights law (HRL) does not apply to armed conflicts. See, Legality of the Threat or Use of Nuclear Weapons, Advisory Opinion, ICJ Reports 1996, at para. 24.

15   Article 4 of the International Covenant on Civil and Political Rights (ICCPR). See also Article 15 of the European Convention on Human Rights (ECHR); Article 27 of the American Convention on Human Rights (ACHR). However, the African Charter on Human and Peoples' Rights (ACHPR) does not include such a provision, therefore no derogation is possible. See *Commission Nationale des Droits de l'Homme et des Libertés v Chad* (ACommHPR, 1995) at para 21.

16   ICCPR, Article 4(2); ACHR, Article 27(2). In the ECHR however the right to life is non-derogable "except in respect of deaths resulting from lawful acts of war" (ECHR, Article 15(2)).

been conducted outside the territory of the parties to an ongoing armed conflict, that is, in non-belligerent states.[17]

There are different legal opinions as to whether the targeting of an individual in the territory of a non-belligerent state is governed by the conduct of hostilities under IHL or the law enforcement paradigm under HRL.[18] The answer to this question depends mainly on what the geographical scope of IHL is. (For a full discussion see Chapter 3 by Katja Schöberl.) Under one school of thought, "humanitarian law is not territorially delimited but governs the relations between the belligerents irrespective of geographical location."[19] A person considered as a legitimate target in relation to an ongoing armed conflict would thus be targetable under the conduct of hostilities paradigm wherever that person is located. Pursuant to other views,[20] which the ICRC shares,[21] IHL does not apply outside the territories of the parties to an armed conflict. The use of force in a non-belligerent state must therefore comply with the rules pertaining to law enforcement. This is also the view of the author of the present contribution.

Another equally controversial issue arises and is related, this time, to the question of the geographical scope of application of human rights law. A minority of states do not accept the extraterritorial application of human rights treaties.[22] Since 1995, this has been the position of the US, which emphasizes, notably, that

---

17　Stuart Casey-Maslen, "Pandora's Box? Drone Strikes Under Jus ad Bellum, Jus in Bello, and International Human Rights Law" [2012] 94, 886 IRRC at 616; Jennifer C Daskal, "The Geography of the Battlefield: A Framework for Detention and Targeting Outside the 'Hot' Conflict Zone" (2013) 161 Univ of Penn Law Review 1165, at 1188.

18　ICRC, *International Humanitarian Law and the Challenges of Contemporary Armed Conflicts* 31IC/11/5.1.2 (Oct 2011) at 22 (ICRC Challenges Report); Jelena Pejić, "Extraterritorial Targeting By Means of Armed Drones: Some Legal Implications" [2015] IRRC.

19　Nils Melzer, *Study on the Human Rights Implications of the Usage of Drones and Unmanned Robots in Warfare* (European Parliament, 2013), at 21, <http://www.europarl.europa.eu/delegations/en/studiesdownload.html?languageDocument=EN&file=92953> accessed Sept 2014. See also Michael N Schmitt, "Extraterritorial Lethal Targeting: Deconstructing the Logic of International Law" [2013] 52 Columbia J of Transnat'l Law, at 99.

20　Pejić (n 18). See also *European Parliament Resolution on the Use of Armed Drones* (n 2) at F; *Concluding observations on the fourth periodic report of the United States of America* UN Doc CCPR/C/USA/CO/4 (April 23, 2014) at para 9. See also Chapter 3 by Katja Schöberl in this volume.

21　ICRC Challenges Report (n 18) at 22; *ICRC Statement before the Human Rights Council* (Sept 22, 2014) <https://www.icrc.org/sites/default/files/document/file_list/icrc_statement_to_hrc_22_sept_2014_drones_eng.pdf> accessed Sept 2014.

22　For the US, see below (n 23). For the position of Israel, see eg Human Rights Committee (HRC), *Sixty-third session. Summary Record of the 1675th meeting: Consideration of the Initial Report of Israel* UN Doc CCPR/C/SR.1675 (July 21, 1998) at paras 21 and 27; HRC, *Addendum to the Second Periodic Report, Israel* UN Doc CCPR/C/ISR/2001/2 (Dec 4, 2001) at para. 8.

Article 2, paragraph 1, of the International Covenant on Civil and Political Rights provides that it applies "to all individuals within its territory *and* subject to its jurisdiction."[23]

This position has been widely criticized, and even within the US government. In 2010, Harold H. Koh, then Legal Advisor to the State Department, wrote an internal memorandum in which he contended that: (1) the strict 1995 interpretation is not compelled by either the language or the negotiating history of the Covenant; (2) this interpretation is actually in significant tension with the treaty's language, context, as well as object and purpose, along with interpretations of important US allies, the Human Rights Committee and the International Court of Justice (ICJ), and developments in related bodies of law; and (3) in fact, the Covenant does impose certain obligations on a State Party's extraterritorial conduct under certain circumstances.[24] Koh then distinguishes between, on one hand, obligations to respect Covenant rights (that he defines as "when a State is itself obligated not to violate those rights through its own actions or the actions of its agents") which would persist extraterritorially when a state exercises authority or effective control over the person or context at issue and, on the other hand, obligations to ensure Covenant rights (that he defines as "affirmatively acting to protect individuals abroad from harm by other states or entities") which would apply only when individuals are both within its territory and subject to its jurisdiction.[25] This rather original (though not very convincing) distinction is not made by the ICJ or human rights bodies, but was probably an attempt to find a middle ground that would be acceptable to the government. The Obama government maintained however the 1995 interpretation in 2014[26]—a matter of concern and regret for the Human Rights Committee (and for the international community more widely).[27]

Nonetheless, the minority view as expressed by Koh, and the substantial internal debate it created, might have indirectly led the US to adopt a more

---

23   *Statement of State Department Legal Adviser, Conrad Harper*, 53rd session, 1405th meeting of the HRC, UN Doc CCPR/C/SR 1405 (April 24, 1995) at para 20 (emphasis added); US Department of State, *Second and Third Periodic Report of the United States of America to the UN Committee on Human Rights Concerning the International Covenant on Civil and Political Rights*, Annex 1 (October 21, 2005) <http://www.state.gov/g/drl/rls/55504. htm> accessed Sept 2014. *United States Responses to Selected Recommendations of the Human Rights Committee* (Oct 10, 2007) at 1–2 <http://2001–2009.state.gov/documents/organization/100845.pdf> accessed Sept 2014. For legal writings on which the US Government rely, see eg Michael J Dennis, "Application of Human Rights Treaties Extraterritorially in Times of Armed Conflict and Military Occupation" [2005] 99 AJIL, 119–141.

24   Harold H Koh, Legal Advisor to the US State Department, "Memorandum Opinion on the Geographical Scope of the International Covenant on Civil and Political Rights" (Oct 19, 2010) at 4.

25   Ibid.

26   *Fourth Periodic Report of the United States of America to the Human Rights Committee* UN Doc CCPR/C/USA/4 and Corr. 1 (May 22, 2012) at paras 504–505.

27   Concluding Observations: USA (n 20) at para 4.

nuanced approach regarding the extraterritorial application of human rights treaties. The United States acknowledgment of the extraterritorial application of the UN Convention against Torture in November of 2014 is worthy of note.[28]

Inversely, human rights bodies and the ICJ have clearly recognized the extraterritorial application of human rights treaties at least in cases of occupation or detention.[29] Outside such situations, in particular when it comes to extraterritorial targeting (without control over the territory or person), the extraterritorial reach of HRL remains however a matter of ongoing legal debate. In the *Banković* case,[30] concerning the NATO bombing of Radio Television of Serbia in Belgrade, the European Court of Human Rights considered that the bombing's victims did not enter into the jurisdiction of the European allied states because this notion is essentially territorial. The Court subsequently softened its position[31] but it is unclear what the Court's decision would be if another "Banković-type" case arose. Other human rights bodies tend to accept the extraterritorial application of human rights treaties for a wider spectrum of situations.[32] The ICRC (like many commentators) held the view that, in any case, customary law prohibits the arbitrary deprivation of life without territorial limitation.[33] It is furthermore often overlooked that the jurisdiction should be established not only through the state

---

28   Committee against Torture, *Fifty-third session, Concluding observations on the third to fifth periodic reports of United States of America*, UN Doc CAT/C/SR 1276 and 1277 (Nov 20, 2014) at para 10. The wording used in the UN Convention against Torture (CAT) is however different from the wording used in the ICCPR. See, eg Article 2(1) of the UN CAT: "Each State Party shall take effective legislative, administrative, judicial or other measures to prevent acts of torture *in any territory under its jurisdiction* (emphasis added)"; see also Articles 5, 11, 12, 13 and 16.

29   *Legal Consequences of the Construction of a Wall in the Occupied Palestinian Territory* (n 14) at paras 107–113; General Comment no 31: *Nature of the General Legal Obligation Imposed on States Parties to the Covenant* [2004] UN Doc CCPR/C/21/Rev.1/Add.13, at para 10; *Al-Skeini and others v United Kingdom* App no 55721/07, ECHR (July 7, 2011) at paras 130–150; *Al-Jedda v United Kingdom* App no 27021/08, ECHR (July 7, 2011) at paras 74–86.

30   *Banković v Belgium and 16 other States* App no 52207/99, ECHR (Dec 12, 2001) at paras 54–82.

31   *Al-Skeini* (n 29) at para 142 (use of force on occupied territory); *Jaloud v Netherlands* App no 47708/08, ECHR (Nov 20, 2014) at para 152–153 (use of force at a checkpoint in Iraq).

32   *Burgos v Uruguay* [1981] UN Doc CCPR/C/13/D/52/1979, at paras 12.2–12.3; *Lilian Celiberti de Casariego v Uruguay* [1981] UN Doc CCPR/C/13/D/56/1979, at paras 10.2–10.3; General Comment no 31 (n 29) at para 10; *Equador v Colombia* Report no 112/10, IACHR (Oct 21, 2010) at para 99; *Disabled Peoples' International v USA*, Case no 9213, IACHR (Sept 22, 1987); *Salas v USA*, Case no 10.573, IACHR (Oct 14, 1993); *Armando Alejandre, Carlos Costa, Mario de la Peña et Pablo Morales v Cuba* Application no 11589, IACHR (Sept 29, 1999).

33   See ICRC Challenges Report (n 18) at 22. See also, Nils Melzer, *Targeted Killing in International Law* (OUP 2008) at 212.

using force extraterritorially, but also through the territorial state. The latter cannot evade its own HRL obligations by consenting to the intervention of a third state.

It is this author's contention that the geographical scope of application of IHL and HRL shall not be construed in a way that creates legal gaps. It is submitted that, from an international law perspective (and outside *jus ad bellum* considerations), any use of force by states must comply either with IHL or HRL.

2. Use of force in the territory of a belligerent state

The use of force by drones occurs mostly in armed conflicts on the territory of one of the belligerent parties, where the applicability of IHL is given.[34] Many examples exist, such as drone strikes taking place in Afghanistan, Gaza, Iraq, or Libya. When drone strikes are directed against persons or objects considered as legitimate military targets under IHL, the relevant legal framework is IHL, and more specifically the conduct of hostilities paradigm. The question (which is outside the scope of this chapter) then turns around whether the attack respects the IHL principles of notably distinction, proportionality, and precautions.

The question here is whether there are situations in armed conflicts in which the use of armed drones would not be governed by the conduct of hostilities paradigm. In many contemporary armed conflicts, particularly in occupied territories and in non-international armed conflicts (NIAC), armed forces are increasingly expected to conduct both combat operations against the adversary and law enforcement operations to maintain or restore public security, law, and order.[35] It is generally accepted that this latter type of operations is governed by applicable HRL rules and standards for the use of force, that is, the HR law enforcement paradigm.

The difficulty then lies in finding the dividing line between the conduct of hostilities and law enforcement paradigms. This is a matter of ongoing legal debate and has been addressed elsewhere.[36] What is clear is that even in armed

---

34 Ben Emmerson, *Report of the Special Rapporteur on the Promotion and Protection of Human Rights and Fundamental Freedoms while Countering Terrorism, Interim Report on the Use of Remotely Piloted Aircraft in Counter-Terrorism Operations* UN Doc A/68/389 (Sept 18, 2013) at para 59.

35 Gloria Gaggioli, *Report of the Expert Meeting on the Use of Force in Armed Conflicts: Interplay Between the Conduct of Hostilities and Law Enforcement Paradigms* (ICRC 2013) at 1 (ICRC Report on the Use of Force).

36 Ibid; Tristan Ferraro, *Expert Meeting Report on Occupation and Other Forms of Administration of Foreign Territory, Third Meeting of Experts: The Use of Force in Occupied Territory* (ICRC 2012); Nils Melzer, "Conceptual Distinction and Overlaps Between Law Enforcement and the Conduct of Hostilities" in Terry D Gill and Dieter Fleck (eds) *The Handbook of the International Law of Military Operations* (OUP 2010) at 33–49; Dieter Fleck, "Law Enforcement and the Conduct of Hostilities: Two Supplementing or Mutually Excluding Legal Paradigms?" in Andreas Fischer-Lescano et al (eds), *Frieden in Freiheit: Festschrift für Michael Bothe* (Nomos DIKE 2008) at 391–407; David Kretzmer, Aviad Ben-Yehuda and Meirav Furth, "Thou Shall Not Kill: The Use of Lethal Force in Non-International Armed Conflicts" [2014] 47/02 Israel Law Review 191–224; Ken

conflicts, there are situations—such as civilian unrest and other forms of civilian violence or criminal acts not amounting to direct participation in hostilities—that must be addressed under a law enforcement paradigm. For instance, when civilians riot in occupied territories or in a NIAC, force cannot go beyond what is authorized under a HR law enforcement paradigm. If fighters hide in the crowd, then they might be targetable under the conduct of hostilities paradigm and the two paradigms may apply in parallel. Also, when belligerents launch operations against common criminals, such as members of drug gangs or other forms of organized crime that are not a party to the armed conflict, such use of force is not governed by the conduct of hostilities, but by the HR law enforcement paradigm. Alleged criminals are civilians and remain protected against attacks under the IHL principle of distinction, unless and for such time as they directly participate in hostilities. These are just two obvious examples where the law enforcement paradigm kicks in. They are by no means meant to be exhaustive.

If drones were to be used in such situations, then they would have to comply with the law enforcement paradigm.

III. The Legal Requirements for the Use of Force under Human Rights Law

The law enforcement paradigm for the use of force is not detailed in human rights treaties. They only guarantee the prohibition of arbitrary killings, from which the set of rules and standards framing the use of force has been derived and further elaborated in soft law documents, such as the *Code of Conduct for Law Enforcement Officials* and the *Basic Principles on the Use of Force and Firearms by Law Enforcement Officials* and in human rights case law.[37] The *Basic Principles* expressly emphasize that "exceptional circumstances, such as internal political instability or any other public emergency may not be invoked to justify any departure from these basic principles."[38]

Watkin, "Use of Force during Occupation: Law Enforcement and Conduct of Hostilities" [2012] 94:885 IRRC 295–296.

37   UN Code of Conduct for Law Enforcement Officials [1979] (adopted by Resolution 34/169 of the UN General Assembly); UN Basic Principles on the Use of Force and Firearms by Law Enforcement Officials [1990] (adopted by the Eight UN Congress on the Prevention of Crime and the Treatment of Offenders and welcomed by Resolution 45/166 of the UN General Assembly. HR bodies often mention the *UN Code of Conduct*, as well as the *UN Basic Principles*, and they adopt generally the same standards when they have to assess the use of deadly force. See, for instance, Christof Heyns, *Report of the Special Rapporteur on Extrajudicial, Summary or Arbitrary Executions* [2014] UN Doc A/HRC/26/36, at para 43; *McCann and Others v United Kingdom* App no 18984/91, ECHR (Sept 27, 1995) at paras 138–140; *de Oliveira v Brazil* Case no 11.599, IACHR (Feb 24, 2000) at para 33; *Montero-Aranguren et al. v Venezuela,* Series C no 150, IACHR (July 5, 2006) at para 69.

38   Ibid, *UN Basic Principles*, at para 8.

A. Obligations Pertaining to the Actual Use of Force

1. Necessity

Under the law enforcement paradigm, force can be used only when "strictly" or "absolutely" necessary to pursue a legitimate objective.[39] In other words, the principle of necessity requires that force must be a last resort (*ultima ratio*). State officials must, as far as possible, apply non-violent means and may use force and firearms only if other means remain ineffective or without any promise of achieving the intended result.[40] If the use of force is unavoidable, only the smallest amount of force necessary may be applied. This implies that state officials must strive to arrest suspected criminals by using non-violent means insofar as possible ("capture-rather-than kill" approach); and, if force is unavoidable, there must be, as far as possible, a differentiated use of force (e.g. verbal warning, show of force, "less-than-lethal" force, lethal force).[41]

2. Proportionality

The kind and degree of force used must not only be necessary, but also strictly proportionate to the seriousness of the offence and the legitimate objective to be achieved.[42] State officials must also strive to minimize damage and injury to human life.[43] Concretely, the principle of proportionality requires a balancing between the risks posed by the individual *versus* the potential harm to this individual as well as to bystanders. In particular, if the individual is not posing a threat of death (or serious injury), the use of lethal (or potentially lethal) force would not be

---

39   Ibid, *UN Code of Conduct*, Art. 3; *UN Basic Principles*, principle 4; eg *McCann*, at para 149.

40   Ibid, *UN Code of Conduct* at Article 3, commentary para c). See also, eg *Montero-Aranguren* (n 37) at paras 67–68; *Zambrano Vélez et al v Equador,* Series C no 166, IACHR (July 4, 2007) at paras 83–84; *Mouvement Burkinabé des Droits de l'Homme et des Peuples v Burkina Faso,* Comm. 204/97, African Commission on Human and Peoples' Rights (2001) at para 43.

41   Ibid, *UN Basic Principles,* principles 4, 5, 9, 10; *UN Code of Conduct,* Article 3. For the case law, see, for example *Pedro Pablo Camargo v Colombia* ("Guerrero") [1982] UN Doc CCPR/C/15/D/45/1979, at para 13.2; *Alejandre* (n 32) at para 42; *Montero-Aranguren* (n 37) at para 75.

42   Ibid, *UN Basic Principles,* principle 5a; *UN Code of Conduct,* Art 3, commentary §b. For the case law on the proportionality principle: see *Neira Alegria v Peru,* Series C no 20, IACHR (Jan 19, 1995) at para 76. The case law does not always distinguish clearly between necessity and proportionality. See *"Guerrero"* (ibid) at para 13.3; *Alejandre* (n 32) at para 42; *Finca 'La Exacta' v Guatemala* Case 11.382, IACHR (Oct 21, 2002) at para 43; *Montero-Aranguren* (n 37) at paras 68 and 74; *Zambrano Vélez* (n 40) at paras 84–85; *Carandiru v Brazil* Case 11.291, IACHR (April 13, 2000) at para 91.

43   Ibid, *UN Basic Principles,* principle 5b; *Stewart v United Kingdom* App no 10044/82, ECommHR (July 10, 1984) at para 19; *Wolfram v RFA* App no 11257/84, ECommHR (Oct 6, 1986) at 213; See also *Montero-Aranguren* (n 37) at paras 68.

considered as proportional, even if the necessity requirement were to be fulfilled. This conclusion is not clear when one reads the provisions protecting the right to life in human rights treaties. Most of them only prohibit "arbitrary" deprivations of life, without providing more details.[44] This may give the false impression that it is possible to use lethal (or potentially lethal) force even if there is no threat to life or limb. The case law of human rights bodies clarifies though that the use of lethal (or potentially lethal) force can be envisaged only if there is a threat to life or limb.[45]

The *Basic Principles on the Use of Force and Firearms* further specify that firearms—and more generally, lethal or potentially lethal force[46]—may be permitted only if they pursue the following legitimate objectives: (1) self-defense or defense of others against the imminent threat of death or serious injury; (2) to prevent the perpetration of a particularly serious crime involving grave threat to life; and (3) to arrest a person presenting a danger of perpetrating such crimes and resisting the authority, or to prevent his or her escape.[47] The use of lethal (or potentially lethal) force while policing unlawful and violent assemblies and controlling persons in custody or detention is also limited to the previous situations.[48] Although a careful reading of these legitimate objectives raises a number of questions,[49] they help to

---

44   ICCPR, Art 6; IACHR, Art 4; AfCHPR, Art 4. As for the European Convention on Human Rights, it allows the use of lethal (or potentially lethal) force when it is absolutely necessary (a) in defense of any person from unlawful violence; (b) in order to effect a lawful arrest or to prevent escape of a person lawfully detained; (c) in action lawfully taken for the purpose of quelling a riot or insurrection. See ECHR, Art 2(2). Human rights bodies have interpreted the term "arbitrary" in a way that is consistent with the purposes mentioned in the ECHR. See, for instance, *Pedro Pablo Camargo* (n 41) at para 13.2.

45   Concerning Art 2(2)(b), see for example *Makaratzis v Greece* App no 50385/99, ECHR (December 20, 2004); *Natchova and Others v Bulgaria* App no 43577/98, 43579/98, ECHR (July 6, 2005); *Kakoulli v Turkey* App no 38595/97, ECHR (Nov 22, 2005). Regarding Art 2(2)(c), see for example *Stewart* (n 43); *Güleç v Turkey* App no 21593/93, ECHR (July 27, 1998). See also *de Oliveira v Brazil* Case (n 37) at para 33; *da Silva v Brazil* Case no 11.598, IACHR (Feb 24, 2000) at para 34; *Finca 'La Exacta' v Guatemala* (n 42) at paras 41–42.

46   Heyns (n 37) at para 71.

47   *UN Basic Principles* (n 37), principle 9. See also *UN Code of Conduct* (n 37), commentary to Art 3 (§a et §c).

48   Ibid, *UN Basic Principles*, principles 12–16.

49   In particular, it is not clear whether the temporal requirement of "imminence" attached to the threat of death that allows the use of force in self-defense and defense of others continues to exist in the pursuance of other legitimate objectives. Strikingly, the other legitimate objectives mentioned in the *Basic Principles* do not contain any express temporal requirement. It is submitted that the principle of "absolute necessity" inevitably implies a temporal requirement ("temporal necessity"). In this sense, see Melzer, *Targeted Killing* (n 33) at 228. While the very strict requirement of imminence, ie "a matter of seconds, not hours" (Heyns [n 37] at para 59), is intrinsic in the notion of self-defense or defense of others, the temporal requirement for the use of force to pursue other legitimate aims (such as arrest or prevention of escape) might however be less stringent. For example, in the case of a person suspected of having committed a series of murders, the text of the

ascertain that the use of potentially lethal force is legitimate only to protect human lives, or—in the case of self-defense at least—to avoid serious injury. In any event, *intentional* lethal use of force may only be made when strictly unavoidable in order to protect life. This is what Christof Heyns has called the "protect life principle"; that is, a life may be taken intentionally only to save another life.

3. Precaution

The principle of precaution is an additional principle of the law enforcement paradigm, although it does not appear explicitly in the treaty provisions pertaining to the right to life. It has been developed in human rights case law, through the theory of positive obligations.[50]

The European Court of Human Rights was the first to develop this principle in the famous *McCann v United Kingdom* case in 1995.[51] In this case, the Court held that when analyzing the right to life, it is appropriate to look not only at the actions of state officials using force, but also at the planning/organization and control of the operation.[52] States must take all feasible precautions to minimize recourse to deadly force; the aim being always to limit damage and injury, and respect and preserve human life.[53] The principle of precaution has been applied systematically by the European Court in subsequent cases,[54] including in situations of armed conflict.[55] It has also been taken up, albeit more briefly, by other human rights bodies and experts, such as the Inter-American Court of Human Rights[56] and the Special Rapporteur on Extrajudicial, Summary or Arbitrary Executions.[57] Other human rights bodies refer more vaguely to an obligation to "prevent" violations of the right to life.[58]

Basic Principles may be read as not prohibiting the use of firearms as a last resort to arrest such a person provided that this person is held to continuously represent a "grave threat to life." To avoid abuses, the likelihood that the person might commit again in the near future "particularly serious crimes involving grave threat to life" should be particularly high.

50   *General Comment* no 31 (n 29) para 8.

51   *McCann* (n 37) at paras 150 and 194; see also *Ergi v Turkey* App no 23818/94, ECHR (July 28, 1998) at para 79.

52   *McCann* (n 37) at paras 202–214.

53   *UN Basic Principles* (n 37), principle 5(b).

54   *McCann* (n 37); *Oğur v Turkey* App no 21594/93, ECHR (May 20, 1999) at paras 83–84; *Gül v Turkey* App no 22676/93, ECHR (Dec 14, 2000) at paras 84–86.

55   *Isayeva, Yusupova and Bazayeva v Russia* App no 57950/00, ECHR (Feb 24, 2005) at paras 188–201.

56   *Neira Alegría* (n 42) at para 62; *Montero-Aranguren* (n 37) at para 82; *Zambrano Vélez* (n 40) at para 89.

57   Heyns (n 37). See also Philip Alston, *Report of the Special Rapporteur on Extrajudicial, Summary or Arbitrary Executions* [2006] UN Doc E/CN.4/2006/53, at paras 53–54.

58   *General Comment no 6: Right to life (Article 6)* [1982] UN Doc HRI/GEN/1/Rev.8, at para 3; *Rickly Burrell v Jamaica* [1996] UN Doc CCPR/C/57/D/546/1993, para 9.5; *National Commission on Human Rights and Freedoms v Chad* (AfCommHPR, 1995) at para 22; *Montero-Aranguren* (n 37) at para 71.

B. Obligations Before and After the Actual Use of Force

HR treaties and case law indicate that, in addition to the aforementioned principles pertaining to the actual use of force in a specific operation, the HR law enforcement paradigm involves obligations before and after the actual use of force.[59] It is submitted that obligations can be framed as deriving from the principles of legality and accountability.

1. Legality

For the principles of necessity, proportionality, and precaution to be effectively applied, the principle of legality inherent in the right to life requires governments to set an appropriate domestic legal framework (including not only proper laws but also directives such as rules of engagement). This legal framework must restrict the use of force to the maximum extent possible and stipulate the limited circumstances in which state officials can use force in accordance with international law.[60] All treaties protecting the right to life state that this right must be "protected by law."[61] The law must be published and accessible to the public to be considered as a sufficient legal basis for the use of force.[62] In order to ensure that the law is translated into practice, governments must provide an adequate training to their state officials, including in alternatives to the use of force and firearms (such as non-violent methods of arrest and techniques).[63] Governments should also equip state officials with various types of weapons and ammunition, including alternative means to firearms (such as water cannons and other "less-lethal weapons"[64]). This allows a differentiated use of force and restrains the use of means capable of causing death or injury.[65] Governments should also equip state officials with self-defensive equipment (such as shields, helmets, bullet-proof vests, etc.) in order to decrease the need to use weapons of any kind.[66]

---

59   For the distinction between obligations pertaining to the actual use of force and those intervening before and after the actual use of force, see: *ICRC Report on the Use of Force* (n 35) at 43.

60   *UN Basic Principles* (n 37), principles 1 and 11.

61   ICCPR, Art 6(1); ECHR, Art 2(1); ACHR, Art 4(1) combined with Article 1 of the AfCHPR. See also *General Comment No. 6* (n 58) at para 3; *Makaratzis* (n 45) at paras 56 to 72; *Pedro Pablo Camargo* (n 41) para 13.3; *Montero-Aranguren* (n 37) at para 75; *Zambrano Vélez* (n 40) at para 86.

62   *Natchova* (n 45) at para 102; Heyns (n 37) at 10; Melzer, *Targeted Killing* (n 33) at 114.

63   *UN Basic Principles* (n 3/) principles 18–21. For the case law, see for example, *Hamiyet Kaplan and Others v Turkey* App no 36749/97, ECHR (Sept 13, 2005) at para 51–55; *Rickly Burrell* (n 58) at para 9.5; *Montero-Aranguren* (n 37) at paras 77–78; *Zambrano Vélez* (n 40) para 87.

64   *UN Basic Principles* (n 37) principle 3.

65   *UN Basic Principles* (n 37) principle 2. For the case law, see for example *Güleç* (n 45); *Hamiyet Kaplan* (n 63).

66   Ibid.

2. Accountability

The accountability principle requires that after a particular operation where the use of force has resulted in death or injury, state officials must report the incident promptly to their superiors.[67] An effective investigation must be conducted each time a person has been killed (outside death penalty cases) or at least each time there is a credible allegation of a violation of the right to life.[68]

Human rights bodies have further clarified that, to be effective, the investigation must be led expeditiously and with due diligence by an independent and impartial body. The next-of-kin must be given an opportunity to participate in the investigation process and all possible steps must be taken in order to gather evidence, including hearing witness testimony, conducting ballistic examinations, medico-legal examinations, etc.[69] This obligation to conduct effective investigations into killings has not only been applied by human rights bodies in peacetime, but also in armed conflicts.[70]

The criteria to consider an investigation as effective seem to be very demanding. Human rights bodies tend to recognize however that the nature and degree of the investigation may vary in different circumstances;[71] for instance, the notion of due diligence in peacetime requires gathering evidence immediately after the facts, while this may be impossible in armed conflicts. What "due diligence" is will therefore vary considerably depending on the circumstances. The involvement of the next-of-kin in the procedure might also be difficult depending on the context and given the amount of classified information at stake.[72] While in peacetime and in developed countries an autopsy is required to consider the investigation into a killing effective, this might be impossible in many situations of conflict.

In summary, the use of force under the HR law enforcement paradigm can be analyzed in five stages:

---

67   *UN Basic Principles* (n 37), principles 6 and 22; *UN Code of Conduct* (n 37), commentary (c) to Art 3.

68   See eg, among many others, Alston (n 57) at paras 35–36; *McCann* (n 37) at para 161; *Mapiripán Massacre v Colombia*, Series C no 134, IACtHR (Sept 15, 2005) at paras 216–241; *Montero-Aranguren* (n 37) at para 75; *Commission Nationale des Droits de l'Homme et des Libertés v Chad* (n 15) at para 22.

69   For steps to be taken in order to gather evidence see *UN Manual on the Effective Prevention and Investigation of Extra-Legal, Arbitrary and Summary Executions*, ST/CSDHA/12,1991, III. Model Protocol for a Legal Investigation of Extra-Legal, Arbitrary and Summary Executions (*Minnesota Protocol*).

70   *Isayeva v Russia* App no 57950/00, ECHR (Feb 24, 2005); *Al-Skeini* (n 29) at para 161–177; Alston (n 57) at para 36. For additional case law and practice see *ICRC Report on the Use of Force* (n 35) at 50–51.

71   Alston (n 57) at para 36; *Isayeva* (ibid) at para 210; *Mapiripán Massacre* (n 68) para 238.

72   See *The Public Commission to Examine the Maritime Incident of 31 May 2010* (Jan 2010) Part two, at 145–146, available at <http://www.turkel-committee.com/index-eng.html> accessed Dec 2014.

1. Was the use of force made in accordance with a domestic legal framework that is suitable for securing the prohibition of arbitrary killings in accordance with international law? Were state officials adequately trained, armed and equipped to minimize the use of lethal (or potentially lethal) force? (legality)
2. When an operation was planned, did the state seek to minimize the possibility of recourse to lethal force as well as death and injury at the initial organizational stage? (precaution)
3. When force was used by state officials, was it absolutely necessary to use this kind and degree of force at the time to achieve a legitimate objective? (necessity)
4. Was the anticipated harm caused by the law enforcement official to the suspected offender and to bystanders proportionate in comparison to the seriousness of the threat posed by the suspected offender and the legitimate objective to be achieved? (proportionality)
5. In case of death or serious injury, did the State conduct an effective investigation after the facts? (accountability)

IV. Confronting the Reality of Armed Drones with Human Rights Law

A. Respecting the Principles of Necessity, Proportionality, and Precautions?

Given the progresses in technology and the different ways in which armed drones can be used, the answer to the question whether armed drones can ever respect the principles of necessity, proportionality, and precautions should be nuanced. A distinction needs to be made between: (1) "combat drones"—this category comprises large UCAV, such as those that are currently used for extraterritorial counterterrorism operations but also miniature aerial vehicles (MAV) that may be weaponized; and (2) "law enforcement drones" that are able to give warnings, shoot pepper spray, and that might also in the near future be armed with lethal weapons.

1. Combat drones for law enforcement?
The first time a large UCAV was allegedly been used outside the scope of the battlefield was in 2002 when the Central Intelligence Agency (CIA) killed six purported Al-Qaeda members in Yemen.[73] Since then, it has been argued that more

---

73 See Casey-Maslen (n 17) at 616; David Kretzmer, "Targeted Killing of Suspected Terrorists: Extrajudicial Executions of Legitimate Means of Defence" [2005] 16:2 EJIL 171, at 171–172. At that time, it was quite clear that there were no hostilities between the government of Yemen and Al Qaeda and that the US did not intervene in a pre-existing armed conflict on Yemeni soil. It is also clear that a mere drone strike does not give rise to a NIAC. It is unclear, however, whether the government of Yemen consented in advance to that attack. If that was not the case, the drone attack would have violated *jus ad bellum*

drone strikes against alleged terrorists in countries that are not at war or against persons who were not legitimate targets under IHL have been conducted. Until now, drone strikes have occurred in non-Western countries, but persons suspected of terrorism are increasingly recruited worldwide, including in Western countries by ISIS for instance.[74] It cannot be ruled out that in the future drone strikes could be envisaged in Europe for instance if the territorial country consents to it or is otherwise unable or unwilling to remove the alleged threat posed by these persons. Contemplating this idea is disturbing, but why? Would the HR law enforcement paradigm prohibit such strikes under all circumstances?

Under the principle of *necessity*, combat drones may be used only if *absolutely necessary* at the time to use this kind and degree of force to reach a particular legitimate objective, which implies the application as far as possible of a graduated use of force. Because a drone strike is, by definition, an *intentionally lethal* (rather than potentially lethal) use of force,[75] it must be proven in a particular case that not only non-violent, but also potentially lethal, means are ineffective or without any promise of achieving the intended result, and that the use of intentionally lethal force is *strictly unavoidable*. Combat drones are usually not equipped to effect warnings or to apply less-than-lethal or potentially lethal force. This, in and of itself, gives rise to serious concerns as to the ability of these weapons to be used in a way that respects the law enforcement principle of absolute necessity. True, under certain circumstances, an escalation of force procedure might simply prove impossible and, thus, directly using potentially lethal or even intentionally lethal force may be allowed. A graduation of force must nevertheless be attempted as far as possible.

Drone strikes—as intentional lethal force—must be strictly unavoidable and aim at protecting life exclusively in order to be considered as *proportionate*.[76] It is tempting to consider persons suspected of being leaders of dangerous criminal/ terrorist groups as necessarily constituting a threat to life. But, under the law enforcement paradigm, the threat posed by a specific individual does not depend

---

requirements and give rise to an international armed conflict. Al-Qaeda would however not be considered as a party to that armed conflict and IHL could not justify the targeting of a member of Al-Qaeda. An analysis under HRL would therefore still be required. In subsequent attacks on Yemeni soil, the government consented to US interventions. Some contend that, in recent years, a separate NIAC emerged between the government of Yemen and Al-Qaeda in the Arabian Peninsula (AQAP) and that the US has intervened in this conflict in support of the government of Yemen and is therefore a party to that conflict. See, e.g. Benjamin R. Farley, "Targeting Anwar Al-Aulaqi: A Case Study in U.S. Drone Strikes and Targeted Killing," (2012) 2 American Univ National Security Law Brief 57, at 71–73.

74   "European jihadists: It ain't half hot here, mum. Why and how Westerners go to fight in Syria and Iraq" *The Economist* (Aug 30, 2014) <http://www.economist.com/news/middle-east-and-africa/21614226-why-and-how-westerners-go-fight-syria-and-iraq-it-aint-half-hot-here-mum> accessed Sept 2014.

75   See Heyns (n 37) at para 71; Melzer (n 33) at 424.

76   *UN Basic Principles* (n 37) principle 9.

on his/her membership in a particular group.[77] It depends on his/her actual and individual conduct. It is also questionable whether the mere fact of plotting, planning, or financing a criminal group without actually being the one killing at the end of the chain is sufficient to constitute a threat to life under HRL. For instance, if a person hires a hit man in peacetime, the police would unquestionably be authorized, under certain circumstances, to use force against the hit man as a last resort to prevent the murder. But it would seem completely inappropriate for the police to use the same force against the person having hired the hit man (unless this person tries to violently evade arrest).

The person against whom force may be used must furthermore represent an imminent, or at least continuous, threat to life. This derives from the temporal requirement inherent in the principle of necessity.[78] The mere fact that a person has committed a violent crime can in no way justify the use of potentially lethal (let alone intentionally lethal) force. Past conduct can at best inform the assessment of the likelihood and seriousness of the threat a person represents. For instance, if a person is suspected of having committed a murder 20 years ago, this would definitely not be sufficient to consider using lethal or potentially lethal force in order to arrest that person. The lethal or potentially lethal use of force cannot be punitive, but only preventive.[79] In practice, the justification often used to resort to drone strikes against a person is the broad notion of self-defense and defense of others. It is unclear whether the invocation of self-defense in such contexts derives from *jus ad bellum* or law enforcement considerations or whether it is seen as a separate basis for the use of force.[80] The legal basis and contours of self-defense remain shrouded in doubts and deserve to be clarified. This exercise goes far beyond the purpose of this chapter. It suffices here to say that, under HRL, self-defense requires an "imminent threat." The notion of imminence in this context is—and must be understood as being—extremely strict: "a matter of seconds not hours."[81] If a person is put on a targeting list long before the strike, it is doubtful that the criterion of imminence is fulfilled.[82] It is also unlikely in practical terms that an extraterritorial drone strike can be ordered and conducted at a very last minute to prevent an "imminent threat of death," although this is not altogether impossible.

The principle of *proportionality* also requires that the harm caused by the state official to the suspected offender and to bystanders is proportionate in comparison

---

77 Melzer (n 33) at 425.

78 See (n 49).

79 Melzer (n 33) at 239.

80 Pejić (n 18) at 7; Kenneth Anderson, "Targeted Killing and Drone Warfare: How We Came to Debate Whether There is a 'Legal Geography of War'" in Peter Berkowitz (ed), *Future Challenges in National Security and Law* (Hoover Institution, Stanford University 2011) at 8.

81 Heyns (n 37) at para 59.

82 *Human Rights Council Panel on Armed Drones* (n 3): Shahzad Akbar.

to the seriousness of the threat posed by the suspected offender and the legitimate objective to be achieved. It is a strict proportionality test and, in contradistinction with IHL, this test would not be fulfilled if there is more than minimal "collateral damage." True, the unintentional killing of a bystander would not necessarily lead to a violation of the right to life: HR bodies have accepted limited and unforeseen casualties among bystanders in rare cases.[83] This exception remains nonetheless extremely narrow. The objective of any law enforcement operation must be to minimize death and injury. Given the firepower of combat drones, they might cause high casualties among bystanders depending on where and how they are used. In populated areas, for instance, large UCAV will almost inevitably lead to unintended death that would most probably be considered as unauthorized under the strict HR proportionality test. If, conversely, the suspected offenders posing a threat to life are isolated, in the desert for instance, the drone strike would not cause any unintended death and the proportionality test (under that limb) might be respected. As the technology evolves, drones with more limited firepower, such as MAV, are being developed and can potentially improve respect for the HR principle of proportionality.

Under the principle of *precautions*, a drone strike must also be analyzed taking into account the planning and control phase of the operation. In that phase too, states must minimize the possibility of recourse to lethal force as well as minimize the risk of death and injury. The biggest problems for the conformity of drone strikes with law enforcement requirements lie here. If drone strikes are planned with the premeditated aim of killing the suspected offenders and if the possibility of arresting the person or applying a graduation of force is not even envisaged, then the operation will be difficult to reconcile with the principle of precautions.[84] The only situation which could justify not envisaging an arrest under the law enforcement paradigm would be the total absence of control over the area where the person is by both the territorial state and the state sending the drone and the material impossibility to arrest that person before the threat to life posed by this person materializes. If that is not the case, it would go against the fundamentals of law enforcement not to plan the operation with the objective to arrest the person by means of a differentiated use of force. A drone strike should not be envisaged as a law enforcement operation per se, but rather be contemplated, in the worst-case scenarios, as the *ultima ratio* at the very end of an operation.

To conclude, one can say with Nils Melzer that:

> In the final analysis, the resort by States to targeted killing [drone strikes in our case] as a method of law enforcement is extremely problematic. On the one hand, it cannot be ruled out that, in extreme circumstances, the prevention of

---

83   *Andronicou and Constantinou v Cyprus* App no 25052/94, ECHR (Sept 10, 1997) at para 194; *Kerimova and Others v Russia* App no 17170/04, ECHR (May 3, 2011) at para. 246.

84   Alston (n 57) at para 33; Melzer (n 33) at 425.

an unlawful attack against human life may exceptionally require, and justify, the intentional killing of a perpetrator. On the other hand, it cannot be ignored that the method of targeted killing involves great risks, and may easily lead to situations which are diametrically opposed to the principles and values underlying the normative paradigm of law enforcement.[85]

2. Toward "law-enforcement drones"?

New types of drones are developed and commercialized to address situations typical of law enforcement such as riots or the fight against criminality. Some of these "law enforcement" drones, in contradistinction with combat drones, can employ strobe lights and on-board speakers to send verbal warnings, shoot pepper spray, solid plastic balls, and non-lethal paintballs to mark offenders or transport an 80,000-volt Taser dart to zap criminals.[86] Although these drones cannot yet use lethal weapons, even so-called less-than-lethal crowd-control weapons can kill. For instance, given the high number of solid plastic balls that can be shot per second or high voltage of the Tasers used, these means should be considered at least as potentially lethal weapons.[87]

In the case of law enforcement drones, the application of the principles of necessity, proportionality, and precautions gives different results compared to combat drones. Regarding the principle of *necessity*, the possibility of doing a graduation of force seems to exist and to be already built into some of these law enforcement drones. Regarding the principle of *proportionality*, it also seems that these types of drones are developed to respond to imminent threats rather than to kill preselected "targets" depending on their membership. The ability to target precisely the alleged offender is supposed to exist, thanks notably to video and audio surveillance systems allowing the drone operator to see precisely what is happening on the ground, although an on-site assessment of the situation should always be preferred.[88] In terms of *precautions*, law enforcement drones seem to be developed notably to minimize death and injury.

There is thus no inherent flaw in law enforcement drones that would necessarily lead to the conclusion that they violate law enforcement principles. Depending on

---

85   Ibid For an even stronger position against "combat drones" under law enforcement, see Alston (n 57) at para 33; Emmerson, *Interim Report* (n 34) at para 24; Heyns (n 37) at para 136.

86   "CUPID drone to 'shock the world' with 80,000 volt stun gun" *RT Question More* (March 8, 2014) <http://rt.com/usa/drone-taser-gun-security-650/> accessed Sept 2014.

87   See, for instance, the website of one of the companies commercializing "riot control" drones <http://www.desert-wolf.com/dw/products/unmanned-aerial-systems/skunk-riot-control-copter.html> accessed Sept 2014.

88   Eric Brumfield, "Armed Drones for Law Enforcement: Why it Might be Time to Re-Examine the Current Use of Force Standard" (2014) 46:3 McGeorge Law Review 543 at 570.

the situation, however, the use of such means might either be lawful or unlawful. If a Taser drone is used to arrest an offender despite the possibility of a non-violent arrest, the principle of necessity would be violated. If rubber bullets are fired by a law enforcement drone in a disproportionate manner at a rioting crowd, then the principle of proportionality would be violated. The principle of precaution would not be respected if law enforcement operations are planned mostly or exclusively with drones; even in cases where they might not allow minimizing death and injury. The fact that drones are cheaper or permit keeping law enforcement officials safe does not constitute a sufficient excuse.

Law enforcement drones therefore seem *prima facie* capable of respecting principles of necessity, proportionality, and precautions. This does not mean that they do not pose any legal and policy issues (see below).

B. Can Domestic Frameworks Regulating Armed Drones and Investigative Measures Render them Appropriate Means for Law Enforcement?

1. Regulating drones

The legality requirement relates in general terms to the law enforcement framework rather than the use of specific means such as drones. Given however the peculiarities of drones in terms of opening up options for the use of force going beyond usual weapons, it could be argued that strong legal and policy frameworks governing the use of drones extraterritorially and domestically should be developed.

For instance, the US has endeavored to establish a domestic framework that governs the extraterritorial use of lethal force—by drones notably—against persons considered as terrorists.[89] Criteria for drone strikes outside the Afghanistan theater against Al-Qaeda and its associated forces have been developed. Such strikes can be conducted only when the United States has no ability to capture or when the risk to US troops and bystanders is too high; when there is a continuing and imminent threat to the American people; when no other government is able or willing to effectively stop the terrorist threat; and when there is near-certainty that no civilians would be killed or injured. Interestingly, some of these criteria are common to the notion of

---

89   This has been done through the *Authorization for Use of Military Force Against Terrorists* (AUMF), which has been passed by Congress in 2001, a few days after the 9/11 attacks. It has been complemented in 2013 by a *Presidential Policy Guidance* that formalizes the standards and procedures for reviewing and approving operations to capture or employ lethal force against terrorist targets "beyond the Afghan theatre," in countries such as Yemen or Somalia that the US qualifies as being 'outside areas of active hostilities, see Barack Obama, "Remarks by the President at the National Defense University," Washington, DC (May 23, 2013) <http://www.whitehouse.gov/the-press-office/2013/05/23/remarks-president-national-defense-university> accessed Sept 2014. See also Stephen W Preston, "The Framework under US Law for Current Military Operations," US Senate, Committee on Foreign Relations (May 21, 2014).

self-defense under HRL.[90] A number of commentators criticized these criteria as being overly vague.[91] The lack of transparency regarding the implementation of this policy is also problematic.[92] The reason for this secretive approach is theoretically obvious: If this counterterrorism strategy is meant to work, then the way targets are selected, information is collected, etc. has to remain classified; but it raises legitimate concerns as to possible arbitrariness of those strikes.[93] To increase oversight, some have suggested establishing a special court or an independent oversight board in the executive branch to evaluate and authorize in advance extraterritorial drone strikes (see the final section of this book for a discussion on this idea.)[94] Others recommend a less formal "nonpartisan independent commission" to review lethal UCAV policies.[95] These new mechanisms might bring useful solutions, but they certainly do not solve the issue overall. Every precaution should be taken when attempting to regulate the use of drones to avoid the possible perverse effect of legitimizing combat drones outside the territory of the parties to an armed conflict. Combat drones are not appropriate tools for law enforcement, be it domestically or abroad. This should be the starting point of any legal and policy framework aiming at regulating the use of combat drones outside hostilities.

Regarding the use of law enforcement drones, it seems wise to suggest to strictly regulate in advance at the domestic level the use of such technology and to check the conformity of such means with international and domestic law enforcement requirements.[96] Although there is no equivalent to Article 36 of Additional

---

90   For the US however the international legal basis of such strikes against "Al Qaeda and its associated forces" is to be found both under self-defense (as interpreted by the US) as well as under IHL even "outside areas of active hostilities," and not under HRL. See Gen John P Abizaid and Rosa Brooks (co-chairs of the Task Force), Stimson Report, *Recommendations and Report of the Task Force on US Drones Policy* (Stimson 2014) at 34.

91   See, for instance, "Killing Americans Abroad: Is the Obama Administration Justified?" *Al Jazeera America* (June 24, 2014) <http://america.aljazeera.com/watch/shows/inside-story/articles/2014/6/24/drones-memo-releasewastheobamaadministrationju stified.html> accessed Sept 2014.

92   Stimson Report (n 90) at 32; Philip Alston, *Study on Targeted Killings, Addendum to Report of the UN Special Rapporteur on Extrajudicial, Summary or Arbitrary Executions*, UN Doc A/HRC/14/24/Add.6 (May 28, 2010) at paras 87–92; Emmerson, *Interim Report* (n 34) at para 41; Ben Emmerson, *Report of the Special Rapporteur on the Promotion and Protection of Human Rights and Fundamental Freedoms While Countering Terrorism* UN Doc A/HRC/25/59 (March 10, 2014) at para 36. See also *Concluding Observations: USA* (n 20) at para 9.

93   Stimson Report (n 90) at 32 and 36.

94   Of note, without proposing a particular oversight body, the Human Rights Committee recommends an "independent supervision and oversight of the specific implementation of regulations governing the use of drone strikes." See *Concluding Observations: USA* (n 20) at para 9.

95   Stimson Report (n 90) at 43.

96   See a proposal in this sense by Eric Brumfield (n 88).

Protocol I for law enforcement means under HRL (a provision requiring states to assess a new weapon's ability to conform with all legal obligations), states should nevertheless carefully assess if (and under which conditions) new law enforcement means can respect the HR law enforcement paradigm before allowing their use.

2. Accountability

The lack of transparency regarding extraterritorial drone strikes raises serious concerns under the accountability principle when those strikes are conducted outside the scope of the battlefield.[97] An investigation must be conducted each time drone strikes are conducted under the law enforcement paradigm[98] even if most of the information related to those strikes remains classified at least for a certain time.

For instance, an oversight body, such as a specific civilian or military tribunal, could review the strikes *ex post* without disclosing to the wider public the information on which it bases its analysis. Some regional human rights bodies have considered that military tribunals inherently lack the requisite independence and impartiality when investigating military personnel suspected of serious human rights violations, such as the killing of civilians in military operations.[99] There are ways, nevertheless, to ensure that the military investigative body is legally and practically separated from those who allegedly committed a violation and, therefore, independent and impartial. In a similar way, the Stimson Report has suggested, for instance, the establishment in the US of a non-partisan independent commission of experts to review particular past drone strikes outside of hot battlefields.[100] Although this body would not be judicial in nature and, for this and other reasons, would not be considered as sufficient under the HR law enforcement requirements, this proposal points in the right direction. This same report also proposes that, in order to improve transparency and accountability to the wider public, drone strikes should generally be acknowledged by the United States after the fact.[101] It also suggests the public release of a first report on "the approximate number and general location of targeted UAV [unmanned aerial vehicles] strikes; the number of individuals known to have been killed and their organizational affiliations; the number and identities of any civilians known to be killed, and the approximate number of strikes carried out by the military versus the CIA" and of a second report detailing the "legal basis under domestic and international law for the United States conducting targeted killings."[102]

---

97   See (n 92). It is true that under IHL, there is no obligation to conduct an investigation except notably when there is an allegation of a war crime. This lower standard for investigations however does not apply outside hostilities (and allegedly outside the geographical scope of IHL) because the governing paradigm is law enforcement.

98   *Concluding Observations: USA* (n 20) at para 9.

99   *ICRC Report on the Use of Force* (n 35) at 56.

100   Stimson Report (n 90) at 43.

101   Ibid at 14 and 42.

102   Ibid.

These suggestions are inspiring. However, even if increasing transparency and providing for accountability mechanisms would improve the situation, this cannot alter the conclusion that combat drones are not appropriate law enforcement means and should be relied on only in proper hostilities.

Regarding law enforcement drones, they should in no way escape from regular accountability mechanisms, whether they are used at home or abroad to maintain or restore public security, law, and order. As any other means, they may in a particular situation be used in a way that is contrary to law enforcement principles. Accountability mechanisms are there to prevent such occurrences. Although this might not appear particularly problematic at first sight, accountability could be hampered by the possible difficulty (and sometimes) impossibility to merely attribute the use of force as discussed below.

C. What Are the Other Legal and Policy Issues Raised by Armed Drones?

In addition to law enforcement rules and principles, there are other legal and policy issues that should be taken into account when assessing armed drones.

(1) *Attribution*: An inherent feature of all armed drones is that they are unmanned vehicles. It may thus be extremely difficult to know who is operating the drone. Is it a state, and if yes, which state? Which entity within the state (e.g. Department of Defense or Intelligence)? Is it a non-state actor? Who is the individual who ordered or operated the strike? How many persons were involved? This leads to particularly difficult challenges in terms of attribution and thus also accountability.[103]

(2) *Proliferation*: Given the relative inexpensiveness[104] of armed drones and the fast evolutions in technology, the risk of uncontrolled proliferation is high.[105] Armed drones may be purchased not only by states, but also by organized non-state armed groups. In September 2014, CNN stated that Hezbollah used a combat drone.[106] The Hamas proudly claims it can now produce armed drones.[107] ISIS uploaded a video to YouTube that showed aerial views of a Syrian military base allegedly destroyed by a drone.[108] A South African company sold 25 "riot drones" to a mining company.[109] Individuals can acquire small drones equipped with

---

103   Ibid at 33.

104   Brumfield (n 88). Cf. Stimson Report (n 90) at 22.

105   Ibid at 21.

106   Peter Bergen and Emily Schneider, "Hezbollah armed drone? Militants' new weapon" *CNN* (Sept 22, 2014) <http://edition.cnn.com/2014/09/22/opinion/bergen-schneider-armed-drone-hezbollah/> accessed Sept 2014.

107   Ibid.

108   Ibid.

109   "Riot Control Drone ... " (n 7).

Tasers in order to incapacitate intruders.[110] These evolutions do require reflection, and the possible adoption of appropriate laws and policies to prevent the harmful proliferation of armed drones.[111]

(3) *Violence increase*: The use of "combat drones" has been criticized as being a factor of instability that may escalate rather than suppress violence.[112] Communities affected by armed drones, because of the resentment such means may provoke, might become in turn more violent. Similarly, a world where the police are far from the communities and maintain law and order through law enforcement drones may increase violence rather than play a deterrent role.

(4) *Impact on human rights other than the right to life*: The use of lethal force by drones might challenge the presumption of innocence and the right to a fair trial as well as the right to an effective remedy for both the direct victims and their next-of-kin.[113] Law enforcement drones can affect the right to privacy or freedom of assembly and of expression because of the fear they may stir up in the public.[114]

(5) *Ethical and psychological impact*: HR experts have noted that the use of armed drones creates an atmosphere of fear in the affected communities, which live with the continuous uncertainty as to whether, when, and where a strike will be performed.[115] In a web-interview, the ICRC President acknowledged this potential psychological impact of drones and stated that it is a matter of concern for his organization.[116] This atmosphere of fear could also exist if law enforcement drones were increasingly used domestically or abroad. Recent studies also demonstrated that drone operators are particularly vulnerable to post-traumatic stress as a result of prolonged observation of their "targets" and the damage caused by the strike.[117]

V. Conclusion

The use of drones has grown considerably since the early 2000s. Armed drones are used in contexts of conflicts governed by IHL, and they are also used to target alleged terrorists extraterritorially and outside of the battlefield where the

---

110   "CUPID drone ... " (n 86).

111   Stimson Report (n 90) at 27.

112   Ibid at 30.

113   Ibid at 36; Melzer (n 33) at 426.

114   "Riot Control Drone ... " (n 7).

115   *Human Rights Council, Panel on Armed Drones* (n 3).

116   Maurer (n 5).

117   Elisabeth Bumiller, "Air Force Drone Operators Report High Levels of Stress" *New York Times* (Dec 18, 2011) available at <http://www.nytimes.com/2011/12/19/world/asia/air-force-drone-operators-show-high-levels-of-stress.html> accessed Sept 2014.

governing framework is HRL. The latter framework would also apply if armed drones were used domestically in peacetime and in situations of violence not reaching the threshold of an armed conflict in order to maintain peace, security, and order. A number of companies are already developing drones for law enforcement purposes, and as a result some law enforcement agencies are eager to get this new technology which is cheaper and implies lesser risks for their staff.

The HR law enforcement paradigm is strict. There are obligations for the actual use of force (necessity, proportionality, and precautions) and obligations that intervene before and after the actual use of force (legality and accountability). Lethal (or potentially lethal) force may be used only as a last resort and only the smallest amount of force necessary may be applied. The kind and degree of force used must not only be necessary, but also strictly proportionate to the seriousness of the offence and the legitimate objective to be achieved. Intentional lethal force may be used only when strictly unavoidable in order to protect life. All precautions must be taken at the planning and control phase of an operation in order to avoid, as much as possible, the use of force and to minimize death and injury. Well before the use of force, states must put in place an appropriate legal and administrative framework in order to secure the right to life. State officials must be adequately trained and provided with various weapons, ammunitions, and equipment allowing for a differentiated use of force. Each use of force by state officials must be investigated.

Large UCAV of the kind that are used in armed conflicts and counterterrorism operations are not appropriate means for law enforcement: they cannot employ any graduation of force; they comport a high risk of causing deaths among bystanders; and they tend to be used against individuals because of their membership in a criminal group rather than because these persons pose an imminent threat to life. The circumstances under which the use of such means might be justified under law enforcement are thus extremely rare. Consequently, combat drones should therefore not be used outside armed conflict situations and hostilities.

Instead, there are new types of armed drones that are being developed to maintain or restore public security, law, and order. These law enforcement drones can give warnings and are armed with less-than-lethal weapons such as Tasers or rubber bullets. In contradistinction with combat drones, they can effect an escalation of force, the type and degree of force they apply is less violent, and they are meant to avoid the use of potentially lethal force. If used in an adequate manner, these drones are presumably able to respect the HR law enforcement principles.

Combat drones and law enforcement drones are thus different and lead to different conclusions when analyzing their compliance with the HR law enforcement paradigm. They have however common features and, to that extent, raise similar issues in terms of attribution, accountability, proliferation, impact on increase in violence, and ethical and psychological impacts. Thus it is recommended that informed legal and policy decisions on the use of armed drones in law enforcement take into account all these various dimensions.

SECTION II
Through the Lens of Morality—
Axiological Validity

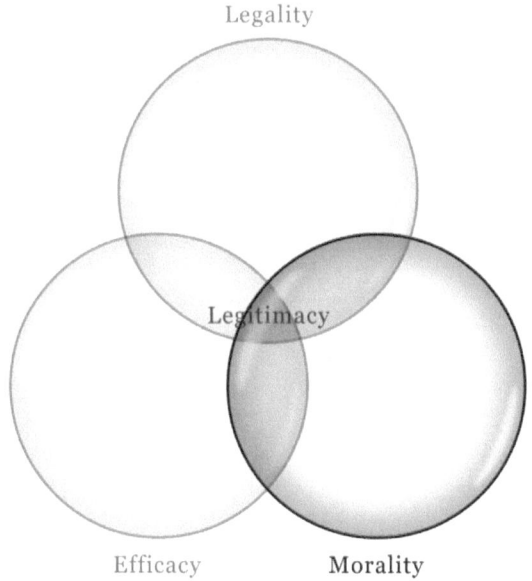

Chapter 5

Old Ideas in New Skins: The Sixteenth Century Debate on Artillery

Alexis Keller

I. Introduction: On the Emergence of Military Technology in Sixteenth Century Europe

Normative ethics traditionally witnesses a clash between three major theories: *deontology*, an approach based on Kant's belief that an action is morally justified if it can be universalized and is based on a sense of duty or respect for a principle; *consequentialism*—of which utilitarianism is a version—which asserts that an action is right if it produces the best consequences for the individuals concerned; and *virtue ethics* which, unlike the two earlier theories that assess actions on the basis of their rationality, judges the moral character of the person performing the action.

This framing can be used to good purpose in the debate on drones. The deontological position can affirm that under the proper conditions the use of drones is not only permitted, but also morally necessary. This, for example, is the opinion of the philosopher Bradley Jay Strawser. He believes that if an agent performs an action that is indeed morally justified, although it may imply certain risks, there is a moral obligation to protect this person whenever possible, unless "there exists a countervailing good that outweighs the protection of the agent."[1] The author admits that the obligation to protect the soldier may be in conflict with the obligation to protect civilians (the dilemma before those who authorized high-altitude bombing in Serbia).[2] He therefore adds that it is necessary to protect the pilot "up until protecting the warrior impedes his/her ability to act behave justly in combat"[3]—for example by not allowing this person to apply the principle of distinction. In that

---

1 Bradley Jay Strawser, "Moral Predators: The Duty to Employ Uninhabited Aerial Vehicles" (2010) 9, 4 J of Military Ethics 342, at 343.

2 For a legal discussion on the NATO bombing of Radio Television of Serbia in Belgrade as it would relate to the human rights of those on the ground, see the European Court of Human Rights decision *Banković v Belgium and 16 other States* App no 52207/99, ECHR (Dec 12, 2001) at paras 54–82. Although it ultimately ruled in 2001 that the bombing's victims did not enter into the jurisdiction of the European states because protection of human rights is essentially territorial, the Court's position has unstiffened since then (see Chapter 4, section II).

3 Strawser (n 1) at 348.

case, how can the point of equilibrium be determined? To what extent should the warrior compromise their own safety to save other lives? To what extent should the pilot be exposed to danger? There is no answer to these questions.

Very often, the ethical debate leads to the crucial question: just how many civilian deaths are drones responsible for? (An in-depth discussion of civilian casualties can be found in Chapter 7 by Avery Plaw and Carlos R. Colon.) Arguments regarding the enhanced ability of drones to narrow their targeting and greatly reduce civilian deaths are typically consequentialist. This has produced widely disputed figures which, according to reports, could be placed anywhere between less than one percent and 90 percent.[4] There is no means of verifying these figures because strikes are often conducted in areas lying beyond the control of a country's central government making it difficult to count the casualties or even distinguish between combatants and civilians. The dead are generally buried immediately and data regarding casualties may be manipulated.[5] Each group's assessment of casualties fits in with its core argument: local groups overestimate the number of victims while western powers underestimate them and others know that the truth lies somewhere between the two. So there is little point in brandishing the figures of civilian casualties as they fail to prove that drones are the least deadly means of carrying out these assaults.

One of the most serious objections that could be raised against the use of drones is that criteria related to the use of armed force (*jus ad bellum*) are blurred to such an extent that it becomes difficult to apply the criterion of the last resort (*ultima ratio*).[6] It is nevertheless difficult to prove this point: in fact, we do not know if the use of drones *really* leads to war. What we can prove on the other hand is that they provide us with multiple tactical advantages that do not have the effect of instigating a war per se. It is therefore important not to confuse the moral relevance of drones as a whole with their use by the CIA or other intelligence agencies for targeted killings.

There is also a third approach to the ethical debate on drones, which consists of reflecting on the "virtues" that we desire to see among soldiers (*virtue ethics*). The moral malaise currently observed in the public debate on drones and in some

---

4   Of course, these numbers also vary from inside the territory of an armed conflict to outside. See Chapter 7 generally, and Tables 7.1 and 7.2 specifically. For claims at the higher end of the spectrum see eg Daniel Byman, director of research and a senior fellow in the Center for Middle East Policy at Brookings, "Do Targeted Killings Work?" *Brookings Institute* (July 14, 2009) available at <http://www.brookings.edu/research/opinions/2009/07/14-targeted-killings-byman> accessed Dec 2014.

5   This methodological problem of counting the dead is well documented, see eg Human Rights Clinic at Columbia Law School, *Counting Drone Strike Deaths* (Oct 2012) available at <http://web.law.columbia.edu/human-rights-institute/counterterrorism/drone-strikes/counting-drone-strike-deaths> accessed Dec 2014. See also Chapter 11, sections III-V by S Barela.

6   See D Brunstetter and M Braun, "The Implications of Drones on the Just War Tradition" (2011) 25, 3 Ethics and Int'l Affairs 346 ff.

military circles is based not so much on the opposition to drones as a matter of principle (*deontology*) or on the grounds of their effectiveness (*consequentialism*) as on a particular perception of the soldier colored by what is sometimes described as a romantic view of war that has existed for ages.

Killing from a safe distance has always incited indignation. It should be noted that when the Second Council of the Lateran banned the use of crossbows (except against infidels) in 1139, it did so in the name of cruelty—although it was really because they were equally effective against Christians. The crossbow was certainly not a new weapon: it was no more than a portable version of the ballista transformed by the Romans into a military weapon in the 4th century. It was, however, treated as an "inferior" weapon and those wielding crossbows were considered bandits, if not assassins.[7] When the French knights were decimated by Welsh archers in the Battle of Crécy in 1346, they were shocked "that a crude treacherous weapon should be used against them, a weapon that allows a cowardly warrior under cover to hurl a projectile indiscriminately against a valiant [knight] who, watchful and ready, naively waits to engage an equal in single combat."[8] As a matter of fact, medieval knights professed the highest disdain for bows and arrows which were considered weapons fit for cowards having been designed to avoid the hand-to-hand combat that was deemed noble. The latter struck from afar and were associated with feigned flight, a tactic commonly used in the Steppes.[9]

Viewed from this angle, there is nothing new about the moral indignation aroused by the use of drones today. It is a continuation of the ancient belief that the physical risk involved in hand-to-hand combat is the only yardstick against which the courage of a warrior can be judged. Present-day debates on drone pilots, their so-called cowardice, their lack of honor and the morality of their action are no more than a repetition of ancient quarrels that have persisted since the invention of firearms.[10] After their invention in the 14th century, it took them two centuries to win their spurs. They were banned in Egypt under the Mameluks and in Japan under the Samurai as they were not considered suitable weapons for the ruling classes. They were also denigrated in Europe. Writers like Ariosto, Cervantes, Shakespeare and Milton—among several others—referred to them as the devil's spawn.

---

7   See M Van Creveld, *The Transformation of War* (The Free Press 1991) at chap 3.

8   R Caillois, *Bellone ou la pente de la guerre* (Flammarion [1963] 2012) at 69 (my translation).

9   It is worth pointing out that the bow was treated quite differently in areas under Muslim domination as also among the Persians, Indians and Chinese. For these peoples, the bow was the preferred weapon of heroes and archery. In addition to being used to train young men for hunting and fighting wars, it also played a role in initiatory rites. Considered as a weapon of royalty, the bow was often perceived as a bridge between the sky and the earth. In this connection, see the informative observations of Franco Cardini, *La culture de la guerre* (Gallimard 1992) at 60 ff.

10   From the viewpoint of both effectiveness and output, care should be taken not to treat firearms as a single block: heavy mounted artillery cannot be equated with individual small arms.

Nevertheless, with the beginning of the Italian Wars in 1494, firearms gained ground and artillery, which became the principal weapon at the technological level, played a decisive role in the conceptual transformation that had set in. The harmful impact of this "military revolution," to borrow Geoffrey Parker's expression, was attributed to the fact that it not only questioned the ethics of chivalry but also the classical Roman model.[11] Thus it broke away from two equally outmoded systems of antithetical values and proposed a new approach where the notion of progress and the prestige of being associated with scientific disciplines like geometry and mathematics created a new image of the soldier and led to the recasting of the *jus in bello*. This new military perspective rejected the aristocratic and moralistic conception of war based on the principles of chivalry as well as the ideological and cultural perception of the humanists and replaced them with a new scientific and technical approach to war. Thus, even though sixteenth century literature is replete with classical and medieval references containing a vast repertoire of enduring examples, their meaning and function have changed over time. The "ancient" model of the soldier can no longer be considered inimitable; instead he becomes the counter-example for the "modern" soldier. We find that some humanists exhibit a greater condescendence toward medieval knights and ancient heroes. In 1530, the historian Mambriano Roseo distanced himself from ancient and chivalric models. He made no mention of Roman foot soldiers or Teutonic knights but only spoke of the musketeers and valiant soldiers he observed during the siege of Florence.[12]

The novelty of a weapon is frequently a significant reason for its rejection: whether effective or not, it always threatens to overturn traditional notions about the methods of warfare and, essentially, its true nature. This explains why so-called technical weapons are generally invented in times of rapid technological advances. Artillery is not an exception to this rule. Just like drones today, it emerged at a time when there were technological upheavals and technical innovations. From being confined within ramparts and forts in its early stages, it moved to the battle field in 1510. It made a mark because of its speed, its lightness and its efficacy. Several technical innovations (manufacture of gunpowder, drilling and calibration of guns, rationalization of calibers) improved the precision of shooting, reduced slack periods and increased the death rate. In Ravenna, the French cavalry was decimated in 1512 by the cross-fire of enemy guns. Their rejection and condemnation by contemporary chroniclers reflect the other side of the successful efforts made by Italians to adopt these new weapons and familiarize themselves with them.

In my contribution, I will present the different arguments used in this intellectual debate, which began with the advent of artillery. I will try to show how the emergence of this new military technology at the dawn of the modern era gave rise to a new code of military values and conduct in the place of the traditional chivalric

---

11   Geoffrey Parker, *The Military Revolution: Military Innovation and the Rise of the West, 1500–1800* (Cambridge Univ Press 1988).

12   M Roseo, *L'assedio et impresa di Firenze cun tutte le cose successe …* (Apresso Girolamo Cartolari 1530) in *Guerre in ottava rima* (Panini 1989) Vol IV, at 117–128.

code. I will stress in particular the aesthetic and ludic dimensions of the new soldier that appeared in response to the need to democratize military recruitment and extend it to new "engineers." I will finally propose in my conclusion that going beyond radically different historical contexts, the discourse on these "accursed and abominable contraptions," as Ariosto describes them in his famous poem *Roland furieux*, is strangely reminiscent of the present-day discourse on drones.

In this study, we have given priority to military treatises, mainly published in Italy,[13] because they have the advantage of dealing with war and soldiers at the same time—some even written for and by soldiers themselves. We believe that no other literary genre illustrates better the reconciliation between letters and arms and between the theory and practice of war. It is nevertheless necessary to distinguish between the two sources of this vast body of writing. The first consists of exhaustive treatises covering all branches of military art and addressing a diversified public of amateurs and professionals. *The Art of War* (1521) by Machiavelli is its most successful expression. The second source tends toward specialization and makes for more difficult reading. It consists for the most part of treatises of a technical nature, which are sometimes manuals and sometimes scattered observations on their work or on the processes employed in different arts. They make a decisive contribution that lies between scientific knowledge and technical and practical know-how, and they have conclusively contributed to the birth of cooperation between scholars and technologists.[14] This body of literature created by artists, engineers and highly skilled craftsmen includes, notably, the writings of Brunelleschi, Piero della Franscesca, Leonardo da Vinci, Paolo Lomazzo, two treatises by Dürer on geometry and fortifications (1525 and 1527), *Pirotechnia* by Biringuccio (1540), Niccolò Tartaglia's work on ballistics (1537), Besson's *Teatro di macchine* (1569), Guidobaldo del Monte's *Mechanicorum libri* (1577) and Lorini's work on fortifications (1597). It is in these works by skilled artisans or engineers that we see the first reflections on artillery and the new concept of the modern soldier.[15]

II. Rejecting the Ancients

In 1520, an intellectual assault was launched against artillery by two powerful socio-cultural groups: the aristocrats and the humanists. The arguments advanced by the former, listed by Ariosto, are well known. We must remember, however,

---

13 After 1494 Italy became Europe's major battlefield.

14 Regarding this topic, see Paolo Rossi's seminal work, *Les philosophes et les machines, 1400–1700* (PUF 1996).

15 This enormous output—which is not confined to this short list—corresponds to a renewal of interest in mathematical and technical writings going back to classical antiquity. It is important to note that the first printed edition of Euclid's works appeared in Venice in 1482 and initiated the first formulations on the new concept of science and scientific progress.

that the aristocrats were against firearms not only because they were considered "ignoble" and killed from a distance, but also because they required professional training and a technical background which were not at all compatible with the aristocratic lifestyle. Finally, they were costly. Hence, artillery was the preferred weapon of national monarchies as opposed to a feudal aristocracy which did not have the financial means to acquire a battery of heavy guns.[16]

Humanists too contributed to this criticism by proposing a reinterpretation of ancient, particularly Roman, military science. Less clear-cut than the aristocrats, humanistic thinking revealed a conflict between respect for classical writings and the demands of an age when the art of war was undergoing a transformation as a result of technological development. A major difficulty faced by this group was adapting Roman theories to a war in which firearms played an increasingly important part. In this respect, it is significant to note that the ones who displayed the greatest distrust toward the use of the new weapons were the most ardent supporters of war "conducted in the antique style," *more antiquo*—led by Machiavelli, their most illustrious representative. In the third book of *The Art of War*, Machiavelli assigns a negligible tactical role to artillery because it appears only at the beginning of the battle and is withdrawn after firing a single salvo. What discredits this new weaponry in the eyes of the Florentine writer is not its immorality, but its inefficacy, of which he is firmly convinced. What is the use of firing, he asks, when all that the infantrymen have to do is to bend down to avoid the cannonballs passing over their heads? But Machiavelli was not an amateur in military matters. As a matter of fact, he was personally in charge of the Florentine militia, wrote several small books on this subject, recruited militia members, selected officers and took care of military supplies during the siege of Pisa in 1509. But the Roman model remained the ultimate reference for him.[17]

The debate on the imperfection of firearms united the humanists into a coherent opposition. Two strategies were employed. The first consisted of ignoring or denigrating artillery, underestimating its efficacy and emphasizing its limitations. The second strategy consisted of "appropriating" the invention either by claiming that the Romans were already aware of it or by depreciating its novelty and putting it down as a part of ongoing technical development.

The engineer Roberto Valturio thus claimed in 1472 that he was convinced that his contemporaries had not invented anything new, neither at the technical nor at the intellectual level, and that firearms had existed even in the remote past.[18] This opinion was shared by Enea Silvio Piccolomini (Pope Pius II) who reported in

---

16   See M van Creveld, *Technology and War: From 2000 BC to the Present* (The Free Press 1991) at 88–90.

17   See F Gilbert, *Machiavelli, the Renaissance of the Art of War in Makers of Modern Strategy* (EM Earle 1952).

18   According to him, the bombard was invented by Archimedes: "Archimedis, ut putatur, inventu, eo tempore quo Marcellus Syracusas obsidebat," R Valturio, *Trattato circa le cose militari* (Verona, 1472) at 5.

his *Commentarii rerum memorabilium quae temporibus suis contigerunt* (1462) a conversation with Frederico Montefeltro on the historical importance of the Trojan War and if the soldiers of antiquity were similarly armed to those of his own time. Pius II answered this question in the affirmative, quoting as authority Homer and Virgil. The Pope claimed that both military art and armaments had reached such a high degree of perfection during antiquity that the moderns were incapable of either copying or attaining the same perfection as their predecessors and certainly not of excelling them. As a result, the discovery or invention of artillery was treated simply as an illusory effect of ignorance and oversight.

Even though this view supporting the ancients is not very surprising coming as it does from a humanist Pope, it did not have unanimous backing. Thus in his famous *Treatise on Civil and Military Architecture* (1492), the engineer Francesco di Giorgio Martini, not without irony, mentions those who claimed that the Greeks and Romans invented all the instruments of war and even went so far as to insist that they invented the earliest cannon and the bombard—an assault with artillery fire.[19] He insisted that it was a wrong hypothesis. Like the Renaissance architect Baldassare Peruzzi of Sienna,[20] he underlined that the moderns had developed a weapon that was far superior to the war machines built by the ancients.

"Antiquity must make way for the present century" stated Mario Savorgnano in 1595 in the book on artillery of his military treatise.[21] Artillery and military architecture were at the forefront of the new ideas and techniques put forth by the moderns. The firm conviction that they were distinct from the ancients and that they had also overtaken them signified progress. The invention of the bombard was considered on a par with the invention of the compass, of printing and even the discovery of America. The cannon became a "great discovery" and bore testimony to the ability of human beings to excel themselves. The distrust of firearms—particularly artillery—could be attributed to nostalgia and naïve sentimentality. "Should the cannon be rejected? Must we go back to the battering-ram?" asked R. Hitchcock, the sixteenth century English writer.[22] Similarly, the Italian engineer and architect Guilo Brancaccio, in Book VI of his *Commentaries* on the harquebus, ponders on the sophistication of cannons. Are they not,

> an obvious proof that the human mind is extremely refined and that there is nothing among the things we need that we cannot make in the most exquisite manner and, what is even more important, they were not made then, and above all that we have weapons that are much superior to theirs [the ancients], so much

---

19   F Di Giorgio Martini, *Trattati di architettura ingegneria ed arte militare* (Maltese 1967).

20   B Peruzzi, *Trattato di architettura militare* (Gonelli 1982) at 114 ff.

21   M Savorgnano, *Arte militare terrestre e maritima secondo la ragione et uso de piu valorosi capitani antichi e moderni ...* (Venezia, 1595) Vol IV, at 257 (my translation).

22   R Hitchcock, *A General Proportion and Order of Provision to Victuall a Garrison of One Thousand Souldiours* (1591) at 30.

> so that I am surprised that after having invented the harquebus and artillery, we
> have not yet become masters of the world once again.[23]

So for many sixteenth century writers, Roman antiquity was no longer an unattainable model but a simple benchmark for judging their progress. Modern wars, they pointed out, were otherwise more difficult than the ones fought by the ancients.

Criticism of the military exploits of the ancients by most historians and military historians from 1550 onwards should be viewed from this perspective. They remind us that Caesar was certainly a great soldier but that he was also a tyrant, that Hannibal was lustful, that Alexander was a heavy drinker and that Fabius Cunctator was lucky to find himself fighting wars in which his hesitant nature proved to be an advantage. The Romans were able to accomplish so many military exploits only because the circumstances were in their favor. But the situation in the sixteenth century was totally different. The existence of numerous citadels and forts was more conducive to sieges than to battles. The change in the nature of the terrain led to a change in technological requirements. If Italian, French and Spanish soldiers did not seem to measure up to Roman soldiers, it was not because the lack of wars gave them fewer opportunities to display their valor, but mainly because they did not have enough weapons.

Those who upheld the superiority of modern weapons used one main argument to support their claim that technical inventions in general, and inventions in the realm of military technology in particular, were a living demonstration of the progress made by humanity. In his *Methodus ad facilem historiarum cognitionem* (1566), Jean Bodin affirmed that the knowledge acquired by the moderns was definitely more impressive than that of the ancients. It was an age when thinkers displayed an unprecedented fecundity of imagination. He wrote,

> We would willingly admit that they [the ancients] discovered numerous sciences
> useful to mankind beginning with the movement of celestial bodies; they studied
> the regular path of astral bodies, the admirable trajectories of stars and planets;
> intrigued by the mysteries of nature, they studied it carefully and provided an
> explanation for numerous problems. But they also left numerous problems
> unsolved that we are now passing on to posterity completely solved. *And, if we
> look at them closely, there is no doubt that our discoveries are equal and often
> better than those of the ancients.* Nature's secrets have been unveiled and useful
> medicines have been discovered. I will not speak of the method of determining
> longitude ... nor of catapults and ancient war machines which, as compared to
> ours, *were children's toys.*[24]

---

23   GC Brancaccio, *Della nova disciplina militare ...* (Venezia, Apresso Aldo 1582) Vol I, Book VI, at 116 (my translation).

24   J Bodin, *Methodus ad facilem historiarum cognitionem*, in Pierre Mesnard *Œuvres philosophiques de Jean Bodin* (PUF 1951) at 227–228 (my translation and emphasis).

Basically, by affirming the inadequacy of the ancient model in the realm of military technology, the supporters of these new firearms were able to invalidate the confrontation between the two worlds and avoid any value judgment on the new technology. Guichardin—who was undoubtedly thinking of Machiavelli—makes fun of those who persist in admiring the ancients, such as the old philosophers who constantly blame the present and praise the past. All that we have of antiquity are a few selected pieces. Hence it is not possible to compare entities that are so different in nature. The flow of history moves incessantly and past experiences are not of much use in the current military situation. They should be seen for what they are: a valuable lesson, but obsolete nonetheless.

III. The Skills of the New Warrior

The humanistic tradition which claims that the art of war requires a high level of knowledge and thinking evolved in a specific direction after 1530: the knowledge of military matters became a science and the soldier an expert. This semantic shift is fairly well documented in the terminological passage of the word "art" to "science" and "technology."

The strengthening of the links between war and technology in early modern Europe is an indication of the growing complexity of wars and their dependence on skilled personnel, especially technical professionals, who can be likened to chess players, and no longer simple foot soldiers and cavalrymen. From 1520 onwards, the idea that war implied a total destruction of the enemy and its resources was progressively abandoned. War is now seen as a result of rigorous technical planning that covers roads, bridges, fortresses, supply depots, replenishment centers, precision instruments and mechanisms. This explains the growing importance of gunners struggling to get themselves recognized as full-fledged soldiers, a designation reserved essentially for combatants.

There is no more ambiguity about the need to confer a new status on the modern soldier in military literature as in *New Science* (1537), Tartaglia's treatise on ballistics, or in *Pirotechnia* (1540) written by the engineer Biringuccio, whose neologistic title denotes the emergence of a new technology.[25] Later, in 1624, the architect Francesco Tensini too claimed a special status for military architecture in his book *On the Subject of Fortification*. "It is undoubtedly a science due to its basic principles and its formal perfection that it owes to mathematics, which is a demonstrative science, but it is an art because of its purpose: fortifying and defending a site on the basis of fixed and well-defined rules."[26]

Gradually, military treatises stopped dealing with war as an entity: they became more specialized and addressed themselves to a more restricted readership. This

---

25   V Biringuccio, *Pirotechnia* (Venezia, 1540) most specifically Vol X.

26   F Tensini, *Della Fortificazione*, in *Arte militare da varii autori* (L Carter 1840) (my translation).

had an effect on their readability and accessibility. They were no longer read by everybody, but only by those who could. This led to a growing chasm between the theoretical approach and the practical approach. Technology ruptured the unity of knowledge. The different categories of soldiers found it more and more difficult to communicate with each other as evidenced by the quarrel between Ruscelli and Tartaglia.[27] Ruscelli's treatise is a simple manual of practical instructions to be followed on the battle field or tricks of the trade. Besides, the book is explicitly meant for trainee gunners. Tartaglia, on the other hand, is a theoretician without any field experience as he has admitted in the preface of his *New Science*. He sweeps aside the criticism of practitioners and upholds the strictly speculative nature of his approach.

Quite logically, the increasingly technical nature of artillery made it more difficult to deal with ethical issues. As military treatises got more specialized, they were not able to communicate ideas and values effectively. Enthusiasm for scientific and technological progress obscured the problem of purpose. Generally speaking, the new engineers did not consider themselves accountable for the consequences of their discoveries. This explains why they were embarrassed by their own writings and verbally condemned war even though their object was to teach readers how to use weapons more efficiently. Whether they are demonstrating that a cannon can reach its maximum range when the angle is 45 degrees or whether they are codifying dueling rules, the authors of these treatises display an unlikely inconsistency since their stated intentions are incompatible with the ends they pursue. The editorial inconsistencies in Tartaglia's treatise are an apt illustration. Thus he asked himself in 1540 whether it was necessary to publish and divulge discoveries that would only increase the sufferings of mankind.[28]

To get rid of the moral stigma attached to these new ignoble weapons, it was necessary to talk of their beauty as well as their effectiveness. Thus, Biringuccio ended his book *Pirotechnia* on a joyous note, even though the evident objective was to explain the most efficacious techniques of war.

> After having talked of the useful, powerful and ingenious effects of numerous [types of] lethal and harmful fire on all living things, I did not want this piece of writing to end on a tragic note, and I would still like to talk to you of fires that are lit on festive and pleasurable occasions and [point out] that people, contrary to the previous lot, do not flee from them but seek them.[29]

---

27 G Ruscelli, *Precetti della militia moderna tanto per mare quanto per terra* ... (Venezia, 1568).

28 This is what made him destroy his manuscript the first time. It was a spectacular gesture, but the mathematician nevertheless rewrote his work even though he had to work under pressure due to lack of time and could have used the looming Turkish threat as convincing excuse.

29 Biringuccio (n 25) at Chapter 10 (my translation).

The engineer plays on the multiple facets of pyrotechnics, an art capable of producing both fireworks and cannon. Seen from this angle, the new military techniques are capable of bringing both happiness and misfortune to mankind. They can be beautiful and lethal. But Biringuccio never questions the ends: it is only a matter of circumstances and context. What is more important is to ensure that the instructions for use are precise and to guarantee the result.

Changes in the nature of artillery brought in their wake changes in the language used to describe it. Though there were many more restrictions on writing, numerous new possibilities opened up in the areas of drawing and calculation. After having accepted the authority of the ancients for so long, the very foundations of the complementarity between letters and arms were undermined. Writers started making greater use of illustrations, diagrams, charts and maps. The lack of adequate words to describe this new firearm and its use is very evident in sixteenth century military treatises. Machiavelli was among the first to reintroduce charts in 1520 in his treatise *The Art of War*. Despite his prejudices, approximations and selective memory, the Florentine used drawings and illustrations to supplement literary explanations. Similarly, in the preface to his 1595 treatise, Mario Savorgnano claimed he was dissatisfied with the means available to him. So he decided to introduce etchings as well as the comparative maps of the battles of Ravenna/ Cannes, Trebia/Marignan, Xantippus/Cerignola. Illustrations ceased to be mere supplements as they were able to express ideas that could not be conveyed through writing. As a result, drawing became an indispensable element in the training of artillery-captains. Modern officers and soldiers were obliged to learn drawing just as they were expected to be competent engineers.

Most military theorists admitted that the era of the old fashion soldier ended after 1540. His place was taken by the engineer armed with his set-square and time tables. The alliance between arms and letters fizzled out. The professionalization of soldiers and specialization in the art of warfare brought about the democratization of armed forces. The development of artillery, drawing and architecture led to the emergence of new categories of military personnel: gunners, bombers, engineers and pyrotechnic gunners who were experts in fortification. Most of these men came from modest social backgrounds, but they rose in the hierarchy because of their professional competence. Thus Girolamo Cattaneo, who wrote a famous treatise on fortifications in 1584, wanted his work to be equally useful to persons of "average intelligence." This book was expressly meant for professionals. The knowledge of Latin was not necessary: all that was needed was a certain amount of professional experience and the rudiments of mathematics. Selection was no longer based on the candidate's linguistic skills or cultural background: technical competence and the ability to calculate would henceforth be more important.[30] While Cattaneo points out the importance of a soldier's ability to think and think clearly, Niccolò Tartaglia goes much further. In his *Questions and Diverse Inventions* (1546), he declares that there must be collaboration between theory and

---

30   G Cattaneo, *Dell'arte militare* ... (Brescia, appresso Pietro Marchetti 1584).

practice, knowledge and experience.[31] As a mathematician he finds it necessary to test the abstract validity of his theories by comparing them with the empirical observations of professionals.

These ideas are also reflected in the works of some philosophers and representatives of official mindset. Between 1530 and 1580, they discussed several common topics in their writings: the processes used by artisans, engineers and technicians were fairly important for the progress of knowledge; these processes ought to be appreciated for their cultural value and cultured men—that is men of letters—should give up their traditional disdain for operations and practical work and stop regarding knowledge as consisting purely of rhetoric and contemplation. They should devote themselves to the study of the arts and technology. These same ideas were expressed with great lucidity by the philosopher Juan Luis Vives, a friend of Erasmus and Thomas More and a tutor in the royal court of England besides being a man of immense culture and refinement. In his comprehensive treatise *De disciplinis* of 1531, Vives invites European scholars to pay serious attention to technical problems related to the construction of machines, agriculture, weaving, navigation and war in the part entitled *De tradendis disciplinis*.[32] The cultured man, he asserts, should not be ashamed of going to workshops and farms, talking to craftsmen and asking them questions while paying close attention to the details of their work. Moreover, in *De causis corruptarum atrium*, he added that the knowledge of nature is not in the hands of philosophers and dialecticians but in the hands of peasants and craftsmen who understand technical problems better than the great philosophers.[33]

The polemics against book-learning thus endorsed this new kind of knowledge: practical knowledge. Helped by the development of mathematics, this new knowledge was within the reach of common people. In this sense, it was more democratic than the humanities whose conditions of study made them an elitist discipline. It gave an intellectual aura to have practical hands-on experience, which it lacked earlier, and brought the battle field closer to the scholar's study. As weapons and technology improved, these new highly qualified soldiers became indispensable. In addition, they provided craftsmen and those trained in other mechanical professions a real opportunity to improve their status.[34] In keeping with the times, in 1604 in his treatise *About the Soldier*, Giovanni Altoni advised

---

31   N Tartaglia, *Quesiti et inventioni diverse* (Venezia, Baparini 1546).

32   Juan Luis Vives, [1531] *De Disciplinis / Savoir et enseigner : I. De causis corruptarum artium / Les causes de la corruption des arts. II. De tradendis disciplinis / La transmission des savoirs* (Tristan Vigliano tr, Première edition 2013).

33   Juan Luis Vives, *De causis corruptarum artium* (Basilea, 1555): [*melius agricolae et fabri norunt quam ipsi tanti philosophi*] at 410.

34   As far back as in 1590, the Republic of Lucques required its gunners to pass rigorous examinations in mathematics and ballistics. See *Examine da farsi al bombardieri quale desidera essere approvato per gli offitii sopra la muraglia* [Examination for bombardiers wanting to be posted on ramparts], text from Lucques (G Arrighi 1969) at 17–19.

officers to get acquainted with new military techniques if they did not want their prestige and authority to suffer due to the "superiority of engineers."[35]

Consequently, the engineer was better equipped than the scholar to win battles, capture fortifications, organize parade grounds and set up military camps. All through the sixteenth century, the primacy given to technical and scientific disciplines completely overturned the traditional approach to war and its perception. What mattered most, more than the ethical principles of combat or the purpose of a conflict, was winning the battle with the use of knowledge based on mathematics, geometry and technology.

IV. Transforming the Reality of War

In the sixteenth century, even as soldiers gradually became engineers firmly entrenched behind their competence and technical jargon, in the civilian world they became objects of a truly aesthetic cult. They were demilitarized in many ways, especially from the sartorial point of view, or the manner of dress.

This phenomenon is illustrated by three works: *Praise* by Paolo Giovo (1546), *The Idea of the Temple of Painting* by Giovanni Paolo Lomazzo (1590) and Cesare Vecellio's work on the collection of costumes, *About Ancient and Modern Dress in Various Parts of the World* (1590). The study of these three books gives a civilian dimension to the military figure. In addition to local and historical differences, Giovio tries to bring out the timeless essence of the military physiognomy or appearance and Lomazzo the iconographic archetype of the soldier while Vecellio highlights the continuity in various styles of dressing. However, they all deal with the military personage among many others. The soldier is submerged in the midst of the civilian population attesting to his integration in society and, generally speaking, in late sixteenth century culture.

The trend of collecting weapons is also a manifestation of the non-military approach to arms. This meant elevating the status of the firearm from a proud instrument to that of a useless ornament. Brantôme describes the famous arms collection of Strozzi, an Italian captain in the service of the French king. Though unlucky on the battlefield, he was considered by his contemporaries as a very cultured person and was known for his valuable collection of weapons.[36] This collection followed a diachronic logic that allowed the viewer to compare modern and ancient weapons and follow their evolution as well as a synchronic logic bringing together European and exotic weapons. It not only contained classical defensive and offensive weapons but also different kinds of accessories and military devices. It was supposed to be exhaustive and boasted of rare pieces. The weapon was not only used, but it was also compared with others, studied and served as

---

35   G Altoni, *Il Soldato* (Firenze, 1604).

36   P de Brantôme, *Œuvres complètes de Pierre Bourdelle, seigneur de Brantôme* ... (édition Lalanne 1864–1882) Vol II, at 215 ff.

an object of contemplation. The room housing the arms collection thus became a reading-room and public arsenals became a compulsory stop in tourist itineraries.[37]

This timeless dimension of war and soldier became even more blatantly visible in the performing arts. The success of games, festivals and military bands in the sixteenth century bears witness to the demilitarization of public opinion. Thus, military parade performed an important social function in Florence because, instead of inciting wars, they helped pacify public opinion. As clearly explained by Salvadori in 1616, the aim of these military exhibitions was to entertain a peace-loving people, lull the multitude in "the idleness of peace" and certainly not to incite them to war. They were meant to be a diversion rather than a dissimulation. The vocabulary belonged to the realms of illusion, imitation and entertainment. War thus became a smoke-screen, a play performed by actors who were mimed by dancers and applauded by the spectators.[38]

If war was increasingly aestheticized and presented as a choreographed event in the sixteenth century, this was done undoubtedly to tone down the effects of reality. It should not be forgotten that firearms—and particularly artillery—had transformed the battlefield. The negative aspects of military life are numerous beginning with the torment and sufferings of soldiers. As the Italian writer and historiographer Luigi da Porto observed in 1528,

> It was a miserable sight seeing the countryside strewn with so many dead lying on the ground in the most diverse positions, or buried in ditches under enormous horses; others still had their feet caught in the stirrups while the rest of the body was dragged along the ground by the horse; some were pierced by lances, others by swords and still others shattered to pieces by the fury of the artillery, as dead as they could be. One could hear the heart-rending lamentations of the dying which sounded so terrible and pitiable at nightfall that no heart could be so inhuman as to withhold tears on hearing them, and nobody could utter words that were harsh enough to condemn the cruelty of the merciless rulers whose greed to enlarge their already vast realms even the slightest bit was the cause of so many murders, so many fires and so many evil deeds. Although they

---

37   In the sixteenth century, the Venetian arsenal became a stopping place in the tour of Italy. The early tourist guides even included it among the *Mirabilia Urbis* like the famous churches. It should be noted that this tradition continued well after the sixteenth century. Pretty women, pilgrimage centers and powerful cannon figured among the curiosities that interested European tourists.

38   A Salvadori, *De la guerra di bellezza, festa a cavallo fatta in Firenze* (Pignoni 1616). Johan Huizingua and Roger Caillois have emphasized the metaphorical dimension of war. Huizingua's ideas on the play element in war (also present in civilization, religion and poetry) complement the belief that play is a civilizing factor while Caillois insists on the affinity between war and celebration considering both as a waste, a suspension of moral standards, collective exaltation, an eclipse or weakening of physical sensitivity and the instinct for preservation. See John Huizinga, *Homo Ludens. Essai sur la fonction sociale du jeu* (Gallimard 1951); Roger Caillois, *Les jeux et les hommes* (Gallimard 1967).

were born like all other men, not poor but completely naked, and were initially content to be fed a little milk and with the swaddling clothes they were wrapped in, their rapacity quickly grew beyond these riches and, soon, even the largest kingdoms were not big enough for them.[39]

Death in battle was not however the most terrifying peril: the conditions accompanying such deaths made them more painful and artillery in particular played a fearsome role. It inflicted terrible wounds that mutilated soldiers and crippled them.[40] When a soldier did return home, but with one or two missing limbs, he was incapable of working; in such cases, if the family was not in a position to feed an unproductive mouth, he had to seek public assistance or was reduced to begging. Such a situation was far removed from the "joyful" war glorified in military festivals and celebrations in Florence, Pisa and Venice. What one saw in areas ravaged by war was utter desolation. The chroniclers of this age describe a long saga of misery. Thus the English ambassador sent to Bologna to attend the coronation of Charles V in February 1530 significantly noted in his diary,

> The country [Italy] is in a pitiful state; in many places, one sees neither men nor horses, and the towns are in ruin and laid waste. Between Vercelli and Pavia, over a distance of fifty miles, an expanse of the most fertile wheat fields and vineyards that can be imagined has become so desolate that we saw no man or woman working in the fields. ... Vigevano has been totally destroyed, Pavia has suffered the same fate and rouses pity: children weep in the streets as they beg for bread and they are dying of hunger. The entire population of these districts and many others in Italy—as the Pope confirmed to us—is dead or has moved to other places because of wars, famine and the plague. There is no hope that Italy will be able to recover for many years.[41]

The successful portrayal of some conflicts faithfully translates the miseries brought by war. In the *Massacre of the Innocents* painted by Pieter Bruegel the Elder in 1563, presently housed in a museum in Vienna, the harsh and desolate winter landscape in which the painter has located the episode depicts a savagery that is as old as mankind. There seems to be a total absence of sound because everything is covered by snow as in the most commonplace sequences in cinema. Evil thus appears in all its banality, something that man can never get used to. Jacques

---

39   Luigi da Porto, *Lettere storiche dal 1509 al 1528* (B Bressan 1857) at 312 (my translation).

40   It is true that in the sixteenth century many wounds, even those that were not serious, proved to be mortal because they were not properly tended to or were accompanied by complications, infections and, in the army where horses were omnipresent, tetanus.

41   "Accounts of the English ambassadors dispatched to Bologna for the coronation of Charles V, September 1529" in *State Papers (Henri VIII)* (1830–1852) at 226 (my translation).

Callot's etchings, including the disturbing series of prints depicting the horrors of war [*Misères de la guerre*] bring back to life scenes from antiquity that were later repeated in the works of Goya. There is an attempt to hide war behind a mask, to dress it up somehow but to no avail. Always, and everywhere in Europe, one sees the same impoverished people begging for help, the same weeping women, the same terrified children. The depiction of war has its limitations.

V. The Legal Implications of the Use of Artillery

Though in different ways, sixteenth century Europe witnessed the decline of chivalric ethics and the code of honor associated with it. The latter had not really been restrictive and after 1530 it seemed to be devoid of any theoretical or social foundation. The increasing use of firearms made it inoperable and replaced it with a new code of values, new and varied means of expression and new conceptual and formal tools provided by science. War became a "science" or an "independent art" whose methodology and, particularly, efficacy had to be developed because, since fighting was inevitable, it was better to do it as efficiently as possible. And the gunner and the bombardier were a part of this new military scheme.

This transformation in the art of war was accompanied by an important change in the law of nations and the development of artillery gave rise to an extensive debate on the right of war. The massacres in the New World and the growing fear aroused by the development of firearms on the European continent gave birth to a new vision of *jus gentium*—or law of nations—intended to control political violence based on the natural rights to life and property and on respect for the traditional practices of mankind as a whole.[42] These rights had to take the place of medieval laws of war that were more concerned with the formalities related to the official notification of hostilities and the treatment of prisoners who could be held to ransom. The canons of chivalric ethics did not suffice anymore; it was necessary to control and regulate, as soon as possible, issues like piracy, reprisals, self-defense, the right to go to war and the rights of civilians and prisoners in accordance with the principles of this *jus gentium* that all states were bound to respect.

The theologian and jurist Francisco de Vitoria (1486–1546) played a central role in this regard. Undoubtedly, when he published *De Indis* [*On the American Indians*] in 1539 followed by *De Juri Belli* [*On the Laws of War*] in the same year, he was not really concerned with the massive development of a new military technology. He was more interested in presenting a critical exposé of all the titles claimed by the Spaniards to legitimize their domination over territories that, on their arrival, were already populated and governed by native chiefs.[43] However,

---

42   For a parallel discussion of this historical transition see Chapter 6 by Steven J. Barela.

43   This led him to analyze in detail the status of these "barbarians" and the power [potestas] they exercised in the public and private domains. See Francisco de Vitoria,

Vitoria's reflections on *jus in bello* are remarkable because they clearly take into account the technological transformation of the art of war. The advent of new weapons allowed the author of *De Indis* to make a clear distinction for the first time between *jus ad bellum* and *jus in bello* and formulate rigorous limitations for the latter, most of which are still valid. Natural law proscribes the massacre of innocent people. It affirms that some categories of the population should be spared. It also justifies the obligation to protect non-combatants (women, children, farmers, travelers and those belonging to religious orders).

Vitoria had a thorough knowledge of the new military technology and the theoretical issues related to it. Accounts of Ottoman and Portuguese conquests in North Africa in which artillery was used extensively were well known all over Europe[44] and Vitoria was perfectly aware of the ravages wrought by artillery. He disapproved in private of the terrifying behavior by Spanish soldiers on the battlefield.[45] It was not a coincidence that he devoted the first few pages of the third part of *De jure belli* to the issue of the "Massacre of Innocents" in terms that leave little doubt that he was fully aware of the terrible consequences of these new ignoble weapons. Vitoria expounded:

> First, it is never lawful in itself to intentionally kill innocent persons. This is proved, in the first place, by Exodus 23:7, where it is said "the innocent and righteous slay thou not." Second, the foundation of the just war is the injury inflicted upon one by the enemy, as shown above; but an innocent person has done you no harm. ... Third, within the commonwealth it is not permissible to punish the innocent for the crimes of the evil and, therefore, it is not permissible to kill innocent members of the enemy population for the injury done by the wicked among them. Fourth, the war would otherwise become just on both sides and it is clear that the innocent would also have the right to defend themselves. ... It follows that even in wars against the Turks we may not kill children, who are obviously innocent, nor women, who are to be presumed innocent, at least as far as the war is concerned (unless, that is, it can be proved of a particular woman that she was implicated in guilt). It follows also that one may not lawfully kill travelers or visitors, who happen to be in the enemy's territory, who are presumed innocent. And the same is true of clergy and monks, unless there

---

De Indis in *Vitoria: Political Writings*, Anthony Pagden and Jeremy Lawrance (eds) (Cambridge Univ Press 1991) at 239–252.

44   See John F Guilmartin, "Ideology and Conflict: the Wars of the Ottoman Empire, 1453—1606" (1988) XVIII J of Interdisciplinary History 721, at 721–747; John Vogt, "Saint Barbara's Legion: Portuguese Artillery in the Struggle for Morocco, 1415–1578" in *Military Affairs* (Society for Military History 1977) at 176–182.

45   In a letter addressed to his friend Miguel de Arcos, he talks of the Spanish conquest in very emotional terms. His "blood freezes" on reading his own accounts. He feels that the conquest was a real butchery and that the Indians were "most certainly innocent in this war." Vitoria, "Letter to Miguel de Arcos" in (n 43) at 331–333.

is evidence to the contrary or they are found actually fighting in the war. I think there can be no doubt about this.[46]

Vitoria saw the introduction of artillery as stimulating a new type of constraint. Hence it is arguably the introduction of this new technology that pushed him toward the critical shift not just in moral, but also in legal, understandings of acceptable killings. And even though he qualifies his statement a few lines later by authorizing violence in exceptional cases and in some justified cases of siege (just war), he adds immediately,

> Nevertheless, we must remember the point made a moment ago that care must be taken to ensure that the evil effects of the war do not outweigh the possible benefits sought by waging it. If the storming of a fortress or a town garrisoned by the enemy, but full of innocent inhabitants, is not of great importance for eventual victory in the war, *it does not seem to me permissible to kill a large number of innocent people by indiscriminate bombardment in order to defeat a small number of enemy combatants.* Finally, it is never lawful to kill innocent people, even accidentally and unintentionally, except when it advances a just war which cannot be won in any other way.[47]

In contributing to a reassessment of *jus in bello*, Vitoria also added to the more clarified interpretations of what would constitute a legitimate anticipatory attack. In other words, he advanced the idea that it would be illegal and immoral to kill before a wrong had been committed—more precisely, a verifiable material harm. As he explained:

> Against this, one may ask whether it is lawful to kill people who are innocent, but may yet pose a threat in the future. … It is perhaps possible to put forth a defence of this kind for killing innocent people in such cases, but I nevertheless believe that it is utterly wrong. It is never right to commit evil, even to avoid greater evils. It is quite unacceptable that a person should be killed for a sin he is yet to commit.[48]

The significant development on this question of preventive war, emerging here with Vitoria, will be discussed in the following chapter on "imminence."

VI. Conclusion: Old Ideas in New Skins

Almost five centuries after the appearance of artillery, we are once again debating the utility, efficacy, legality and morality of a recently invented military technology:

---

46   Vitoria, "De jure belli" in (n 43) at 315.
47   Ibid at 315–316.
48   Ibid at 316.

armed drones. As we have tried to suggest, there is nothing entirely new about the current criticism of this new technology. It is based on an old line of thinking that first emerged in Europe in the sixteenth century when the cannons and bombards were introduced. Like their predecessors during the Renaissance, some critics try to distance themselves emotionally from the unmanned aircraft and its targets that pose a real ethical problem for them. They stress the lack of moral accountability and, in a society addicted to video games, the difficulty of making a distinction between the real and the virtual leading to the risk of the dehumanization of the adversary.[49] Other opponents denounce the trend to treat war as a spectacle or a form of entertainment. They condemn the visuals of airstrikes on social media describing them as "drone-pornography."[50] At the other extreme, the victims of targeted attacks use the same arguments as the French knights of the sixteenth century. They accuse the Israelis and Americans of being "cowardly," of refusing to fight like "real men."[51] They reiterate the old idea that hand-to-hand combat and physical risk are the only standards worthy of judging the virtue of a soldier.

The so-called Medal of Honor scandal in the United States should be viewed from this perspective. In 2012, the Pentagon proposed to create a medal to honor soldiers involved in drone attacks and cyber warfare which, in terms of prestige, would be higher than the Purple Heart (bestowed on persons wounded or killed in combat). Veterans' associations immediately protested against the move on the grounds that "there is a fundamental difference between those who fight remotely, or via computer, and those fighting against an enemy who is trying to kill them."[52] Responding sympathetically to the argument, the Pentagon cancelled the new medal for drone and cyber warriors in 2013.

However, war no longer resembles a duel and it should not be considered as one. Acknowledging that the classical perception of war persists even today and that it does not really apply to present-day conflicts does not mean that the use of drones is unjust because it encroaches on the right of self-defence. As sixteenth century theologians-cum-jurists observed, a new military technique should not be condemned per se but for the manner in which it is used. Vitoria is very clear on this point and he accordingly modifies the *jus in bello*. It should be noted that he introduced two exceptions to the ban on the killing of non-combatants. The first exception is based on the criterion of intentionality. From this viewpoint, the killing of civilians—which should be reduced as much as possible—may be considered

---

49   ML Cummings, "Automation and Accountability in Decision Support System Interface Design" (2006) 32, 1 The J of Technology Studies 23, at 25–26.

50   R Rosenbaum, "Ban Drone-Porn War Crimes" in *Slate Magazine* (Aug 31, 2010) available at <www.slate.com/articles/life /the_spectator/2010/08> accessed Dec 2014.

51   Interview with Peter W Singer, "Robotique et adversaires irréguliers : du bricolage à la bataille des narrations" (mars 2010) Hors-série n° 10, Défense et Sécurité Internationale 77.

52   James Koutz, national commander of the American Legion, "Pentagon Calls off New Medal for Drone, Cyber Warriors" AFP (April 15, 2013) available at <http://www.securityweek.com/pentagon-calls-new-medal-drone-cyber-warriors>.

as the undesired and unfortunate effect of a just goal or war. The second exception is based on the principle of proportionality. Hence the undesirable effect—loss of civilian population—should be proportionate to the objective. It is therefore necessary to reduce these losses as much as possible, even if it poses a greater risk to the combatants.

Despite the sophistication of his analyses, Vitoria's theory of *jus in bello* has left many moral issues unsolved. To what extent is the criterion of non-intentionality psychologically plausible? In an age when computer simulation is widely used to create different military scenarios, how can we ignore consequent civilian losses? Is it possible to plan or deliberate a military action without mixing up ends and means? From the perspective of virtue ethics, the emergence of drones provides us with the opportunity to reflect anew on these problems in the light of the debate on firearms and artillery that arose in Europe in the sixteenth century.

Chapter 6

The Question of "Imminence": A Historical View on Anticipatory Attacks

Steven J. Barela

I. Introduction

Revealing the centrality of temporal constraints on addressing threats with drone killing, President Obama's former top counterterrorism advisor first explained in 2011, "the question [of targeting] turns principally on how you define 'imminence.'"[1] A Department of Justice White Paper providing details on this significant question was leaked to the press in February 2013. The authors of the memo concluded that "an 'imminent' threat of violent attack against the United States does *not* require the United States to have clear evidence that a specific attack on US persons and interests will take place in the *immediate* future."[2] Thus the current administration has plainly widened this term beyond its traditional meaning by discarding the "immediacy" standard.

Of course, this follows the tack taken by the previous president. While the Bush administration insisted after 9/11 that the US "must *adapt* the concept of imminent threat to the capabilities and objectives of today's adversaries"[3], which has come

---

1 John O Brennan, Assistant to the President for Homeland Security and Counterterrorism, "Strengthening Our Security by Adhering to Our Values and Laws," Speech at Harvard Law School (Sept 16, 2011), available at <http://www.whitehouse.gov/the-press-office/2011/09/16/remarks-john-o-brennan-strengthening-our-security-adhering-our-values-an> accessed April 2014.

2 US Department of Justice, Office of Legal Counsel (OLC), "Lawfulness of a Lethal Operation Directed against a US Citizen Who Is a Senior Operational Leader of Al-Qa'ida or an Associated Force," White Paper, draft copy, Office of Legal Counsel, (Nov 8, 2011) (my emphasis) at 7, available at <http://msnbcmedia.msn.com/i/msnbc/sections/news/020413_DOJ_White_Paper.pdf> accessed April 2014. Even though the original OLC Memo has now been released via court order (US Court of Appeals for the Second Circuit, *The New York Times Company et al v United States*, Case No 13–422-cv (April 21, 2014): David Barron, Acting Assistant Attorney General, "Re: Applicability of Federal Criminal Laws and the Constitution to Contemplated Lethal Operations Against Shaykh Anwar al-Aulaqi" (July 16, 2010)) redactions have removed any reference to the definition of "imminence."

3 The White House, "The National Security Strategy of the United States of America," (September 2002) (my emphasis) at 15, available at <http://www.globalsecurity.org/military/library/policy/national/nss-020920.pdf> accessed April 2014.

to be known as the Bush Doctrine, those in charge of the of the current program of drone killing have asserted that "the traditional conception of what constitutes an 'imminent' attack should be *broadened* in light of the modern-day capabilities, techniques, and technological innovations of terrorist organizations."[4] The coincidence in argumentation for modifying this timeworn norm is unmistakable.

At the time of the attacks the United States Department of Defense (DoD) Dictionary recognized two distinct forms of anticipatory military action—just as we will see in this chapter. Their definitions shared with the general public informed the debate on the subject and reflected the historical understandings behind the classifications. However, a deeply regrettable expunging of these terms has occurred in the DoD Dictionary under the Obama administration while it has been operating its drone program based on interpretations of law and morals that were long ago rejected. This adaptation is becoming entrenched and thus warrants continued academic attention as other governments conform to this reading of the appropriate standards based on so-called operational realities.[5]

In this chapter I will trace the historical debate on *imminence* to demonstrate how this standard acts as the final constraint for legitimate anticipatory defense. I will particularly focus in on an instructive bifurcation in the just war tradition that occurred in the seventeenth century on this very question. This division in the development of the moral doctrine, and the ideas advanced that helped give birth to international law, are particularly illuminating since this chapter treats an inherent place of overlap between legality and morality. By fleshing out the reasoning underpinning past considerations three points will become clear: (1) why the branch embracing more objective standards took root and burgeoned; (2) that there is no middle ground between the two positions; and (3) why there is a need for rejecting anew the "just feare" standard used today for triggering the use of killing across borders.

II. Law and Morals

As we have seen in the first section of this volume, there are three principal bodies of international law that come into play when speaking about the use of lethal force aimed at an individual found across an international border. In the current literature it has often been the case that the debate is couched in terms of humanitarian vs human rights law. However, as drone attacks have taken place outside of declared combat zones and across new international borders with consent being denied publically by the local government,[6] we have seen that there is valid and compelling reason for

---

4   Brennan (n 1) (my emphasis).

5   Daniel Bethlehem, "Note and Comment: Self-Defense against an Imminent or Actual Armed Attack by Nonstate Actors" (2012) 106 AJIL 769, at 773.

6   For an example of public denials see Ben Emmerson, UN Special Rapporteur on Counter-Terrorism and Human Rights "Statement of the Special Rapporteur Following

analyzing the third body of law: *jus ad bellum*. Thus, this chapter will build off of the work of Christian J. Tams and James G. Devaney in Chapter 1 to address the questions raised there regarding anticipatory attacks from a historical perspective.

As noted in Chapter 1, the relevant UN Charter provisions and International Court of Justice (ICJ) judgments do not provide a conclusive answer to the proper interpretation of the use of force and thus further investigation is necessary to understand the present constraints.[7] Although this prohibition has been referred to as the "cornerstone" of the Charter regime, the reading of this foundational provision has not remained fixed. In particular, the use of force against individuals alleged to be members of groups using terrorist tactics has been increasingly recognized by states as a right that can be exercised unilaterally.[8]

Yet the argument put forward here is that real and recognizable temporal limitation must be found within the existing framework otherwise we place ourselves outside of a "system of mutual forbearances and compromise which is the base of both legal and moral obligation" as expounded by H.L.A. Hart.[9] This consummate legal positivist found that every moral and legal code includes a prohibition against free, unconstrained violence, and eloquently continued with language clearly referencing Thomas Hobbes,

> Social life with its rules requiring such forbearances is irksome at times; but it is at any rate less nasty, less brutish, and less short than unrestrained aggression for beings thus approximately equal.[10]

Thus when interpreting Article 2(4) of the UN Charter that prohibits "the threat or use of force against the territorial integrity or political independence of any state," it is necessary to remember that a reciprocal restraint must exist, and do our best to delineate its minimum requirements. In addition, in a self-administered system it cannot be the case that the exception of self-defense found in Article 51

Meetings in Pakistan," United Nations Human Rights, Office for the High Commissioner of Human Rights, Islamabad, Pakistan (March 14, 2013), available at <http://www.ohchr.org/ en/NewsEvents/Pages/DisplayNews.aspx?NewsID=13146&LangID=E> accessed April 2014. Cf. Reports that Pakistan has secretly given consent for drone strikes, Greg Miller and Bob Woodward, "Secret memos reveal explicit nature of US, Pakistan agreement on drones" *Washington Post* (October 24, 2013) <http://www.washingtonpost.com/world/ national-security/top-pakistani-leaders-secretly-backed-cia-dronecampaign-secret- documents-show/2013/10/23/15e6b0d8–3beb-11e3-b6a9-da62c264f40e_story.html> accessed April 2014. As it is a decisive legal question whether the ruling government has given consent for this use of force within its territory, and due to the immense difficulty of verifying such consent, this question will be set aside here.

7 See eg Andrea Bianchi, "The International Regulation of the Use of Force: The Politics of Interpretive Method" (2009) 22 Leiden J of Int'l Law 651.

8 CJ Tams, "The Use of Force against Terrorists" (2009) 20 EJIL 359.

9 HLA Hart, *The Concept of Law* (OUP 1961, 1994) at 195.

10 Ibid.

allows for a wholly subjective and unverifiable trigger to military action. Much of the historical dispute over anticipatory attacks principally turned on this precise subjective/objective distinction.

Of course, it must be recognized that Hart argued in 1961 that the inequality of states renders the necessity of a prohibition against free violence as unsuitable to the international realm. However, there is good reason to doubt that this reasoning continues to be applicable. The fact that the most powerful governments now regularly invoke self-defense in protecting themselves against non-state actors suggests that such disparity is irrelevant because individuals can cause significant harm on a level that can be construed as parity. Additionally, the proliferation of weapons of mass destruction among states, and the fear that they could end up in the hands of violent persons, again changes this calculus of great inequality in the international sphere.

It is argued here that the same interpretation of the law that construes violence from individuals as sufficient justification for an appeal to self-defense by states (UN Security Council Resolutions 1368 and 1373)[11] logically also requires an understanding that the difference in strength among states and individuals is no longer wholly relevant in the international sphere. That is to say, threats from individuals and weaker states have become equally perilous and the individual analogy (often eschewed) is considered to be valid and applicable to the discussions of this chapter. While further historical reasoning for preserving this connection between domestic law and war will be presented below, it is also valuable to remember that legitimacy is of great concern for sound counterterrorism, and common citizens can easily understand such a correlation. Therefore this chapter should be recognized as a work treating both legality and morality as we investigate the constraints on anticipatory force since there is a veritable point of overlap between the two embodied in the requirement of a ban on unconstrained violence.[12]

III. Tracing the Historical Debate

Since this need for mutual forbearances in and among human societies is of such importance it is possible to find similar debates in our shared history. Thus to provide shape to the constraints on the use of force that cannot be resolved

---

11   UN Security Council, UN Doc S/RES/1368 (Sept 12, 2001) and UN Doc S/RES/1373 (Sept 28, 2001). Immediately identifying this as a significant problem was Antonio Cassese, "Terrorism Is Also Disrupting Some Crucial Legal Categories of International Law" (2001) 12 EJIL 993. Nonetheless, this novel interpretation of individuals and non-state groups triggering self-defense carries with it further implications that are emphasized here in this chapter.

12   For further investigation into this overlap see Steven J Barela, *International Law, New Diplomacy and Counterterrorism: An Interdisciplinary Study of Legitimacy* (Routledge 2014) at 49–53 and 148–152.

through the examination of contemporary jurisprudence and precedent it is useful to use a historical view on the question of anticipatory defense as an insight into customary law. This brief survey will focus in on a particularly telling moment at the beginning of the seventeenth century.[13] During this period some advocated that "just feare" was sufficient for triggering a war (an argument which sowed no progeny) while others insisted that the criteria for launching an attack must be more objective and verifiable (argumentation that helped found international law).

Before delving into this revealing juncture, it is beneficial to first explore the origins of proactive defense in Roman times and within the classic just war doctrine. In the well-known speech, *Pro Milone*, Cicero defended an extended version of anticipatory defense on behalf of a friend who was on trial for the killing of a fellow Roman. Cicero's argument was that justification for self-defense extended to the fear of future hostility. In recognizable terms (a filiation that can be traced back to Thucydides and forward to Hobbes' translated phrase "necessity of nature")[14] Cicero spoke of conflict by way of arms as being governed by "law not of the statute-book, but of nature."[15] Emphasizing the inherent character of defensive measures, Cicero went on to explain that it was "a law which we possess not by instruction, tradition, or reading, but which we have caught, imbibed, and sucked in at Nature's own breast."[16]

In this speech Cicero also articulated what has largely become conventional wisdom expressed in the proverb, *inter arma enim silent leges*. As he explained, "that should our lives have fallen into any snare, into the violence and the weapons of robbers or foes, every method of winning a way to safety would be morally justifiable. *When arms speak, the laws are silent*; they bid none to await their word."[17]

However, two elements must be remembered here. Firstly, Cicero argued explicitly that this idea could not be found within the realm of legality ("the laws are silent"), and that his broad reading of morality would allow for any and all measures employed ("every method of winning a way to safety would be morally justifiable"). Secondly, his defense of Milo was unsuccessful. His claim was found to be at the very periphery of decency, if not beyond it. Even in the Roman legal code, "the only 'fear' which could be pleaded in extenuation of an individual's act was an immediate and obvious one."[18]

---

13 Illuminated skillfully by James Turner Johnson, *Ideology, Reason and the Limitation of War: Religious and Secular Concepts 1200–1740* (Princeton Univ Press 1975) at 81–203.

14 Thucydides, *The Peloponnesian War*, Book V, Para 105, Thomas Hobbes (trans) (Univ of Chicago Press 1989) at 368.

15 Cicero, "Pro Milone" in NH Watts (trans), *The speeches / Cicero, Pro Milone; In Pisonem; Pro Scauro; Pro Fonteio; Pro Rabirio postumo; Pro Marcello; Pro Ligario; Pro rege Deiotaro* (Harvard Univ Press 1931) at 17.

16 Ibid.

17 Ibid.

18 Richard Tuck, *The Rights of War and Peace* (OUP 1999) at 21.

ı

In contrast to Roman law, one essential characteristic of the moral doctrine of just war at its nascent stage was that of creating a tight connection between internal law and warfare. Saint Augustine of Hippo made a direct correlation between the individual analogy in local statutes and the interaction of political bodies. It is this fusion of local legal rules and equitable treatment among communities in armed conflict that typifies the classic just war doctrine. One scholar has asserted that "the linkage between the theologians and the lawyers ... persisted throughout the Middle Ages, and is the most distinctive feature of the theology of war throughout the period."[19] In this insightful finding we can see the roots of the just war doctrine's later development into international law.

Of particular importance for our discussion is the powerful moral principle "double effect," devised in the classical period by Thomas Aquinas. This idea is central to whether there could ever be a moral justification for killing and is also pertinent for anticipatory defense. Thomas shrewdly drew an important distinction between carrying out a violent action with the intent to kill and a resulting death while defending one's own life:

> Nothing hinders one act from having two effects, only one of which is intended, while the other is beside the intention. Now moral acts take their species according to what is intended, and not according to what is beside the intention, since this is accidental. ... Accordingly the act of self-defense may have two effects, one is the saving of one's life, the other is the slaying of the aggressor.[20]

Not only was there an important moral distinction between the two acts, but this appreciation also provided an extremely useful guideline for determining under what circumstances self-defense could be plausibly invoked in advance. One interpretation of this principle that helps clarify its pertinence is that, "[t]he use of force had to be directed against the attack, not the attacker."[21] Thus an attack must have materialized, otherwise there is, in fact, nothing to repel. In other words, if anticipatory killing reaches too far into the future, the lethal act becomes the singular intention.

The first documented thinkers on *jus ad bellum* sketched out a basic requirement concerning an anticipatory attack. That is, just war must be in response to a specific wrongful act that has taken place. While there were three primary requirements for a just war in the classic doctrine—right authority, just cause, and right intention— implied all the way through is the stipulation that "just war be of a defensive or retributive nature only; offensive wars and wars of preemptive retribution are

---

19   Ibid at 57.

20   Thomas Aquinas (1265–1274) *Summa Theologiæ*, II-II: Q 64, Art 7 (internal citations omitted).

21   George P Fletcher and Jens David Ohlin, *Defending Humanity: When Force is Justified and Why* (OUP 2008) at 27.

not permitted."[22] This distinction remained vital throughout the just war tradition and continues to hold significance in our understanding today: viz offensive vs defensive war.

A. A Telling Bifurcation over Fear as a Trigger

Sixteenth-century Europe, as discussed in Chapter 5 by Alexis Keller, was struggling with the introduction of the new weaponry of artillery as blood spilled in the religious wars of the Protestant Reformation that lasted until the Peace of Westphalia in 1648 over 130 years later. At the same time, Europeans were confronting the ramifications of the 1492 "discovery" of the New World, inhabited by a people previously unknown to them. The carnage and bloodshed that this confrontation of peoples unleashed in a struggle for dominance and riches has been well documented. In the crucible of these violent clashes there was a proliferation of serious thought on the justness of war and conflict: some of it novel and progressive; and some of it self-absorbed and expedient.

This tumultuous historical moment spawned a pivotal development. It was during this time period in Europe that the classic doctrine of just war was secularized and became what is now recognized as the discernible roots of contemporary international law. Moreover, during this era of violent conflict some devout religious thinkers morphed the traditional just war doctrine into a justification for hostilities on spiritual grounds with contrary conclusions on anticipatory defense. Thought during this time period has been classified as a division between humanist and scholastic traditions, as well as being described as a bifurcation of the just war theory between a holy war doctrine and a secular doctrine. Regardless of which school of thought one adheres to, the division over what constitutes a justifiable anticipatory trigger for war is enlightening.

1. The first branch: holy war

One difference between the two sides of this bifurcation has been described as the rhetorical versus the philosophical.[23] The former was an approach taken by rhetoricians defending their population's interests before its own political bodies, and thus can be understood as representing a relatively narrow self-interest. The wider philosophical view looked beyond just one community to formulate a precept that could be acceptable to all populations and governments concerned with the justifications of war. The first drew extensively on the texts and rhetorical writings of the Romans who were openly skeptical of philosophy (often looking to

---

22 Johnson, *Ideology, Reason* ... (n 13) at 46.

23 See Tuck (n 18) at 16–77. Tuck suggests that the proper labels to denote the strands of thought on issues of war and peace through the sixteenth century is with the designations of "humanist" and "scholastic" traditions. However, to avoid including multiple terms for the bifurcating ideas discussed, these will be set aside here.

Cicero), and the second tradition was constructed from earlier Christian literature, along with the writings of the Greek philosophers.

One of the countries in which the just war doctrine was utilized to advocate for holy war (via the rhetorical method) was England. This is certainly not to say that this thinking did not exist in other parts of the continent. Rather, our attention is drawn to the English affair with holy war because it was brought into sharp relief by the tense rivalry with the staunch Catholicism and maritime dominance of Spain.[24] One specific characteristic of this doctrine of holy war was a shift in emphasis away from the traditional just war concept of a limitation on the Christian's right to make war to a focus on the permissions to go to war. This moved the attention to the positive right of defense and away from a negative right to be free from aggression: an important modification that impacts final conclusions.

Alberico Gentili—an Italian jurist who eventually settled in Oxford, England, as a professor of civil law—exemplifies particularly well the rhetorical/humanist tradition on this question. Additionally, his "influence is clearly discernable in the writings of Francis Bacon,"[25] whose days at Oxford coincided with Gentili's, and who has largely come to personify this standard based on fear, as we will see below. While not necessarily a proponent of holy war, Gentili's views on so-called defensive attacks reflect the thinking by some in England at the time. In his major work of 1588, *De jure belli*, Gentili treated many of the issues that were under debate concerning hostilities between states, and among them was that of an anticipatory attack. Gentili had few scruples about the idea of launching an assault before an injury was received. According to him, the ill intentions and rising power of another community certainly justified striking first in order to protect oneself. As he expressed it,

> no one ought to expose himself to danger. No one ought to wait to be struck, unless he is a fool. One ought to provide not only against an offence which is being committed, but also against one which may possibly be committed. Force must be repelled and kept aloof by force. Therefore one should not wait for it to come.[26]

Of note here is the clever rhetorical transition made from Thomas Aquinas in his own citation of Roman jurists: "[I]t is lawful to repel force by force."[27] It is the addition of the simple phrase "and kept aloof by force" that might at first blush seem logical and a minor alteration. In effect, however, this allowance drastically

---

24   Johnson, *Ideology, Reason* ... (n 13) at 81–149. For an explicit invocation of this terminology see Francis Bacon, *An Advertisement Touching an Holy Warre* Laurence Lampert (ed) (Waveland Press Inc 2000).

25   Cian O'Driscoll, *Renegotiation of the Just War Tradition and the Right to War in the Twenty-First Century* (Palgrave Macmillian 2008) at 31.

26   Alberico Gentili, *De Jure Belli*, Book II, in C Phillipson, (ed) JC Rolfe (trans) (OUP 1933) at 62, as cited by Tuck (n 18) at 18.

27   Thomas Aquinas (n 20).

changes the entire meaning of the phrase and irretrievably widens the possibilities of employing force.

Among the commonalities found in the other figures in England who advocated for the use of religion as a *casus belli* was on the point of anticipatory war. That is to say, those who promoted the idea of holy war also found it justifiable to forestall the evil intentions of another by initiating war rather than waiting for an imminent first strike. One of the earliest calls for this type of hostile action against Catholic Spain came from Stephen Gosson in his 1598 sermon at Paul's Cross Church in a parish where many high-ranking government officials worshiped. This "Trumpet for Warre" oration gained such popularity that publication ensued, and because there was no ongoing armed conflict with Spain at that time, "this sermon must be understood as a call to preemptive defensive war."[28]

This discourse anticipated the position of a more well-known English dignitary in the early seventeenth century, Sir Francis Bacon. While not a religious zealot, one of Bacon's writings in particular dealt directly with the subject at hand: "Considerations Touching a Warre with Spaine."[29] In this piece Bacon wrote judiciously, and his position at the court of Queen Elizabeth and then at that of King James gave him the experience necessary to craft his argument prudently. However, he was unmistakably favorable to holy war in this study document prepared for the King in 1624 while he was serving as a special royal adviser. He articulates the idea that armed struggle could be waged in anticipation of a wrong, and could be based solely on perceived malicious intentions.

To Bacon's mind, the protection of his religion was grounds enough to initiate a battle against the nation that had professed itself generally to be the protector of the Catholic world. Most importantly, he saw this battle as self-defense. The memorable, and often repeated, expression upon which he based this reasoning was that of a "just feare." Bacon mentions three examples of actions warranting justified trepidation: an upsetting of the balance of power; threats from states whose very nature makes them dangerous (rogue or hostile); and states seen to be committing acts perceived as aggressive. Of course, all measures are subjective in nature and as a consequence cannot be seen as truly limiting criteria. Bacon explained the specific circumstance facing England as such:

> [W]herein two things are to be proved, the one that a just feare (without an actuall invasion or offence) is a sufficient ground of Warre, and in the nature of a true defensive; the other that we have towards Spaine cause of just feare, ... not out of umbrages, light jealousness, apprehensions a farre off, but out of clear foresight of imminent danger.[30]

---

28 Johnson, *Ideology, Reason ...* (n 13) at 102.

29 Francis Bacon, "Considerations Touching a Warre with Spaine" [1624], in *Certaine miscellany works of the right honourable Francis Lord Verulam, Viscount St Alban* (London, 1629).

30 Ibid at 8.

We see here that Bacon certainly believed there to a justifiable reason to fear Spain, and that this was enough to trigger a war to be initiated (defensively) by England.

It is interesting to note the cunning rhetorical flourish employed at the excerpt closing. By invoking the language of an "imminent danger," there would be an understandable, and perhaps unsuspicious, draw to the logic presented. However, the word "foresight," no matter how clear, directly contradicts the imminence of which he speaks. It is nonsensical for something to be just about to happen at any moment, yet at some unspecific point in the future.

What is significant is that Bacon was indeed attempting to suggest criteria for a justifiable cause for hostilities within the just war doctrine for the international realm. If it could be discerned that another country that proclaimed a different faith was prepared for battle (regardless of whether this preparation was intended for its own defense or aimed at one specific country), then the initiation of armed conflict was said to be a just "response." Yet this argument employing the prevalent code of the time ran into a philosophical dead end.

Although the final result is quite similar, Bacon's views should be understood as distinct from Thomas Hobbes' work (who was, in fact, a clerk for Bacon at one point).[31] Hobbes' familiar theory of a *state of nature* was a methodical endeavor meant to create a coherent science of domestic politics. He conceptualized the manner in which omnipresent fear drove individuals into a single pact of association and submission, founding a society with a strong Leviathan at its helm: the definition of the modern state. The *state of nature* was only later used to describe the international sphere, and thus did not posit any moral or legal rules for limiting war. Without the presence of a power for all states to fear, each is left to provide for their own protection with whatever methods are seen fit. There are no constraining rules. It is rather the case that Hobbes' view is much more reminiscent of Cicero: both law and morals are irrelevant for questions of security.

2. The second branch: toward international law

The other side of this bifurcation during the transition to the modern era was the philosophical tradition that trekked toward a secular doctrine. During this same critical time period when the Christian church was splintering into warring factions and Europeans were trying discern their obligations to other peoples they were confronting in the New World, there was a particularly significant historical development. At this fateful moment of violent confrontation on multiple continents, there was the advancement of the just war doctrine into what is now recognized today as the origins of international law.[32]

---

31 Noel Malcolm, *Aspects of Hobbes* (OUP 2002) at 6–7.

32 JB Scott, *The Spanish Origin of International Law: Francisco de Vitoria and His Law of Nations* (Clarendon Press 1934), and B Hamilton, *Political Thought in Sixteenth-Century Spain* (Clarendon Press 1963).

Since one of the pressing concerns of the time was to expand the law of discovery and occupation to a breadth never before contemplated, it is not surprising that the most prominent scholars who first advanced a more objectivist just war doctrine were found in the Iberian peninsula: the Spanish Dominicans. In contrast to the other side of the bifurcation, the philosophical approach required a concern beyond the preservation of the commonwealth. The intention was rather to determine the universal rights of states by what could be termed a just war from differing perspectives. This branch believed in limiting anticipatory war much like their predecessors of the just war tradition—with one important advancement.

This is certainly not to say that Spain was the only location in which these ideas were to become developed and propagated. The geographical juxtaposition presented here might give the impression that the English busied themselves with justifications for religious war while the Spanish were contemplating foundations for an international order. This was not the case, and such a conclusion should not be drawn. Just as England provides a particular setting in which some of the most fervent and clear supporters of anticipatory war can be found, Spain similarly serves to highlight the first philosophers who further developed the just war doctrine and took it in the direction of an international law. Proof that conclusions based on geography would be erroneous is the simple fact that these Spanish origins of the international legal order were a mode of thought that produced an eminent lineage in other countries. This legal philosophy was ripe and continued to mature within the whole of the European continent over the following centuries.

Some of the clearest statements from the Spanish Dominicans on the question of the justness of advance military action came in the sixteenth century from Francisco de Vitoria, who generally wrote out his thoughts on war in the characteristic scholastic repertory of commentary on Thomas Aquinas and other Christian and medieval philosophers. Using a terminology and reasoning that reflected the just war tradition he inherited, Vitoria wrote that there is but "a single and only just cause for commencing a war, namely, a wrong received."[33] In this statement we find a vital distinction that was made by this group of Spanish philosophers—one that explains a great deal of the subject at hand.

During this development of the just war doctrine, the interpretation of the traditional claim of harm became focused on *material injury*.[34] Such a distinction makes the trigger for war more objective because it is demonstrable. Very importantly, this clarification also set in motion movement toward a legal concept of defense in international law with the standard for self-preservation moving toward verifiable evidence. This development is central to the definition of *imminence*.

Illuminating this point on anticipatory action, Vitoria espoused a position that falls in line with the concept of temporal proximity as characterized by most international lawyers today. He asserted that "defense can only be resorted to at the very moment of the danger, or, as the jurists say, *in continenti*, and so when

---

33 Cited in ibid *The Spanish Origin ...* at 208–209.
34 Johnson, *Ideology, Reason ...* (n 13) at 170–171.

the necessity of defense has passed there is an end to the lawfulness of war."[35] As further evidence (also presented in Chapter 5 by Alexis Keller), in a treatise delivered at Salamanca in 1539, Vitoria explored "whether it is lawful to kill people who are innocent, but may yet pose a threat in the future."[36] After citing an answer in the affirmative, Vitoria unequivocally proclaimed:

> It is quite unacceptable that a person should be killed for sin he has yet to commit. In the first place, there are many other measures for preventing future harm from such people, such as captivity, exile, etc. It is not lawful to execute one of our fellow members of the commonwealth for future sins, and therefore it cannot be lawful with foreign subjects either; I have no doubts on this score.[37]

Accordingly, we find a more clarified moral and legal position on anticipatory action emerging in the sixteenth century.[38]

The next step in the progression of ideas can be found in the work of Dutch jurist Hugo Grotius and his celebrated *De Jure Belli ac Pacis*. The methodology that he employed is an important element of his work that concerns us here, because there is an attention to what can be measured with demonstrable externals. This purveyor of international law shifted the tradition by replacing "right authority" with sovereignty, only sparsely treated the most subjective criteria of "right intention," and regulated "just causes" with what can be discerned by an objective observer.[39] Since the line between legal/illegal and just/unjust war pivots on this very distinction of objectivity vs subjectivity, the tack Grotius took foretells his stance.

Although Grotius offered up some of the most far-reaching sets of rights for a state to make war, he believed that there were only very limited conditions in which a state might strike an adversary before a harm had been done. Most importantly, in his work on the rights of war and peace, he outlined the necessity of an impending peril being both "immediate" and "certain" for a nation to take up arms. He wrote:

V. War in defence of life is permissible only when the danger is immediate and certain, not when it is merely assumed.

---

35   Vitoria, *The Spanish Origin ...* (n 32) at 203–204. However, Vitoria's position should not be taken as unwavering since he also believed that "one who has been contumeliously assaulted can immediately strike back, even if the assaulter was not proposing to make a further attack" (at 204), and that one's own self-respect can justify her from refusing to flee if by flight she might avoid the danger but compromise reputation (at 202).

36   Francisco de Vitoria, "De jure belli" in Antony Pagden and Jeremy Lawrance (eds) *Political Writings* (Cambridge Univ Press 1991) at 316.

37   Ibid.

38   Another Spanish Dominican contemporary, Luis de Molina, took a very similar stance that has been described as, "[w]ars in pursuit of glory, or pre-emptive strikes, were utterly forbidden," Tuck (n 18) at 52.

39   Johnson, *Ideology, Reason ...* (n 13) at 213–214.

> The danger, again, must be immediate and imminent in point of time. I admit, to be sure, that if the assailant seizes weapons in such a way that his intent to kill is manifest the crime can be forestalled; for in morals as in material things a point is not to be found which does not have a certain breadth. But those who accept fear of any sort as justifying anticipatory slaying are themselves greatly deceived, and deceive others.[40]

This explicit formulation surely coincides quite well with a contemporary understanding of the limits of anticipatory military action in that it focuses attention on the certitude that a coming attack is just about to befall a nation.

Grotius followed this assertion by entering into an extensive discourse over the concept of fear, offering numerous references in support of his view that this emotionally charged concept is an insufficient trigger for action. In rejecting fear as a just cause, Grotius cited both Greek and Roman authors who support his exclusion of advance strikes without immediacy from a legal framework. Grotius began by quoting the Roman Consul Gellius who explained that the normal human circumstance must not be understood as equal to those who find themselves forced into violent armed combat:

> When a gladiator is equipped for fighting, the alternatives offered by combat are these, either to kill, if he shall have made the first decisive stroke, or to fall, if he shall have failed. But the life of men generally is not hedged about by a necessity so unfair and so relentless that you are obliged to strike the first blow, and may suffer if you shall have failed to be first to strike.[41]

This extract is of particular note because the battle with terrorists is often depicted as though it is a fight to the death, and only complete eradication of the enemy will do.

Interestingly, Cicero was also cited querying, "[w]ho has ever established this principle, or to whom without the gravest danger to all men can it be granted, that he shall have the right to kill a man by whom he says he fears that he himself later may be killed?"[42] Grotius went on to cite Thucydides who declared, "[t]he future is still uncertain, and no one, influenced by that thought, should arouse enmities which are not future, but certain."[43]

These quotations of Cicero and Thucydides clearly support Grotius' conclusion that anticipatory defense must only be used when a threat is immediate and not supposed. Yet we can perhaps deduce that Grotius, as versed as he was in both

---

40 Hugo Grotius, *De Jure Belli ac Pacis,* Book II, Chpt. 1, Sec. V in James Brown Scott (ed), *The Classics of International Law* (Clarendon Press 1925) at 173 (original emphasis).

41 Ibid at 174.

42 Ibid.

43 Ibid.

traditions, was engaging in selective citation since Cicero has been identified as a central author constructing the intellectual roots of the other side of this bifurcation. At the same time, Thucydides was one of the Greek historians preferred by this same group, at least in part owing to the arguments presented in The Peloponnesian War for anticipatory attacks when discussing the case of the Mytilenaeans in the face of growing Athenian power.[44] Nevertheless, Grotius found the pieces of Cicero's and Thucydides' work that supported his own conclusions.

Grotius closed by drawing attention to the fact that if an attack has not yet been launched there are other less violent means of protecting oneself. Namely, if a strike is not proximate in time, then one must look to other methods of defense since it cannot be deemed legitimate to resort to killing when the future remains unwritten and will offer unforeseen means to resolution. Grotius proclaimed:

> Further, if a man is not planning an immediate attack, but it has been ascertained that he has formed a plot, or is preparing an ambuscade, or that he is putting poison in our way, or that he is making ready a false accusation and false evidence, and is corrupting the judicial procedure, I maintain that he cannot lawfully be killed, either if the danger can in any other way be avoided, or if it is not altogether certain that the danger cannot be otherwise avoided. Generally, in fact, the delay that will intervene affords opportunity to apply many remedies, to take advantage of many accidental occurrences; as the proverb runs, "There's many a slip 'twixt cup and lip."[45]

These pertinent and cogent arguments published in 1625 are of particular interest because they appear as an almost direct rebuttal of Francis Bacon's ideas put forward just one year earlier. Thus, one can clearly distinguish the bifurcation under discussion within the just war tradition when Grotius and Bacon are viewed side by side. Yet, beyond the strength of reasoning bolstering the case forwarded by Grotius are the ensuing historical developments. While the subjective rhetorical arguments embodied in the "just feare" standard ran aground shortly after they were advanced, the philosophical approach searching for objective and verifiable standards took root and matured into the international law we know today. Such a significant historical advancement on the question of "imminence" should not be overlooked.

B. Is There a Middle Ground?

Not every scholar has read this history in precisely the same manner, nor are all in agreement about the specific classifications for those who have weighed in on this question.[46] In general, however, it is agreed that Francis Bacon represents

---

44   Thucydides (n 14) Book III, Para 8–14, at 160–164.

45   Grotius (n 40) at 174–175.

46   See eg O'Driscoll (n 25) at 27–50.

an extreme in this debate with his view on the admissibility of "just feare." As for where Hugo Grotius stood on anticipatory attacks, the citations put forward here and his extended discussion of the dangers and follies of fear as a measure demonstrate the firmness of his stance. Yet an extremely relevant question in the literature, as well as in the discussion over current policy, is whether it is possible to actually find a reasonable and fixed standard that lies between the two positions of: (1) *imminence* and (2) *just fear*.

In the contemporary literature the *imminence* requirement is often represented by US Secretary of State Daniel Webster's words from the oft-cited 1842 *Caroline* affair suggesting that a lawful anticipatory attack occurs when the necessity of self-defense is "instant, overwhelming, leaving no choice of means and no moment of deliberation."[47] An expanded historical view would not only include Hugo Grotius, but also the seventeenth-century German legal philosopher Samuel von Pufendorf. He too spoke of the need for temporal constraint, explicitly excluding suspicion or fear as a just cause for military action. With a potent closing language that reminds us of the *Caroline* standard for its stark simplicity and ease of commitment to memory, Pufendorf asserted:

> [t]o render the Defence of our selves entirely Innocent, it is commonly thought a necessary Condition, that as to the time, the Danger be just upon us, or as it were in the very point of seizing us: And that no Suspicion or Fear, whilst yet uncertain is sufficient to justifie our assaulting another . . . yet before I can actually assault another under colour of my own Defence, I must have tokens and Arguments amounting to a Moral Certainty, that he entertains a Grudge against me, and hath a full design of doing me a Mischief, so that *unless I prevent him, I shall immediately feel his Stroke.*[48]

Thus we can place Grotius and Pufendorf alongside Webster, and juxtaposed to Bacon.

However, it has been argued that a defendable policy exists between the two positions. Michael Walzer is one of the best-known contemporary authorities on the just war tradition who has attempted to put forward a set of criteria meant to occupy such a space. Walzer is said to have staked out a position reminiscent

---

47   The Caroline Incident (exchange of diplomatic notes between Great Britain and the United States [1842]), reprinted in (1906) 2 Dig of Int'l Law 409, at 412. For examples of its discussion in the international law literature see Timothy Kearley, "Raising the Caroline" (199) 17 Wis Intl Law Journal 325; Maria Benvenuta Occelli, "Sinking the Caroline: Why the Caroline Doctrine's Restrictions on Self-Defense Should Not Be Regarded as Customary International Law" (2003) 4 San Diego Intl LJ 467; James A Green, "Docking the Caroline: Understanding the Relevance of the Formula in Contemporary Customary International Law Concerning Self-Defense" (2006) 14 Cardozo J Intl & Comp Law 429.

48   Samuel Pufendorf, *Of the Law of Nature and Nations*, Book II, chpt V, sec 6, (L Lichfield 1703) (my emphasis) at 145.

of eighteenth-century Swiss jurist Emmerich de Vattel: "Walzer's views on the moment at which a threat ought to generate a right to self-defense are almost perfectly aligned with Vattel's thoughts on anticipatory war."[49] Consequently, we will analyze these two authors together to assess this so-called middle ground.

Walzer found it necessary to reject Bacon's "just feare" standard much like was done in his time period, as it bore no intellectual fruit. Walzer ties the idea to the dangerous proposition of launching preventive war to maintain an illusory balance of power since "increments and losses of power are a constant feature of international politics."[50] At the same time, he also suggests that the "legalist paradigm" is too constraining and must be broadened for moral judgment. Since both legal and moral codes always include a prohibition against unconstrained violence (without which every other rule is rendered meaningless),[51] it is odd that Walzer insists upon this revision, which explicitly introduces subjectivity. Yet while he suggests that the stipulations should not be arbitrary, he continues: "men and women may well respond fearfully to a genuine threat, and their *subjective* experience is not an unimportant part of the argument for anticipation."[52] This begs an important question: Can unverifiable suspicions be anything other than arbitrary?

In an attempt to clear this high hurdle, Walzer put forward his idea of a "sufficient threat," an unavoidably vague phrase by his own admission. It is meant to capture: (1) a "manifest *intent* to injure"; (2) active preparation which "makes that *intent* a positive danger"; and (3) a general situation in which waiting (or anything other than fighting) will "greatly magnify the *risk*."[53] It does not take a linguist or a lawyer to readily discern that two out of three of these criteria offer an opening for dishonest actors to exploit because they are based upon perception. Moreover, they extend an avenue for even those acting in good faith to be deluded into war by a misinterpretation of intentions and events. Further underscoring the difficulty with this standard is the fact that Walzer admitted that "[i]t is not possible to put together a list" of acts that are sufficiently serious and that he "cannot specify a time span."[54] Thus what Walzer suggested are plainly not verifiable and objective constraints.

Even more telling are the historical examples that Walzer used to bolster the validity of his proposal. The first case is one that had been originally discussed by Vattel: The War of the Spanish Succession at the turn of the eighteenth century. While Vattel defended the preventive military intervention of the Grand Alliance to prevent a dynastic unification of France with Spain due to the "given proofs of imperious pride and insatiable ambition" from Louis XIV, he also concluded five

---

49   O'Driscoll (n 25) at 40.

50   Michael Walzer, *Just and Unjust Wars* (3rd ed, Basic Books 1977) at 77.

51   See section II above.

52   Walzer (n 50) (my emphasis) at 78.

53   Ibid (my emphasis) at 81.

54   Ibid at 80 and 81.

decades later that "it has since appeared that their policy was too suspicious."[55] Very interestingly, Walzer lamented the fact that Vattel did not revise his criteria in light of the changed historical facts.[56]

The second example that Walzer used is the Six Day War in June 1967 between Israel and Egypt, contending that "[t]he Israeli first strike is, I think, a clear case of legitimate anticipation."[57] Using the available accounts at the time of publication in 1977, Walzer made a compelling case for the anxiety and strain that Israel was under with a formidable buildup of Egyptian forces deployed onto the Sinai Peninsula on its border. Walzer asserted that the mobilization of its own forces in defense was unsustainable and that the action opened Israel to attack at any time. Hence this was "an almost classic example of just fear."[58]

Nevertheless, this was his reading of the records accessible at that time. Since then, however, declassified documentary evidence sheds new light on the events and what was known only to the decision-makers at that critical moment of choice.[59] It is certainly not within the scope of this chapter to delve into the historical details that would seem to beg for a revision of the criteria put forward by Walzer (just as he requested of Vattel), nor is it the intention to open a debate on such a volatile issue. Instead, put forward here are the relevant conclusions of one respected historian raising reasonable doubt:

> [i]t appears that scholars in the past have overvalued the military threat posed by the Egyptian build-up in the Sinai. New evidence from the archives suggests a limited remilitarization of the Sinai, amounting to nothing even remotely resembling an invasion force. ... there is no reason to treat the June 1967 War as the standard example of a preemptive war.[60]

The fact that we can be faced at a later date with new evidence that may cause us to question previous conclusions is nothing new. However, here we discuss armed hostilities. Both Walzer and Vattel placed great attention on the gravity of the calamities ushered in by war and consequently called for the utmost of prudence

---

55   Emmerich de Vattel, *The Law of Nations or the Principles of Natural Law in Four Books*, in J Chitty (trans) (1758) Book III, chpt 3, §44 at 311.

56   Walzer (n 50) at 79: "Having drawn his criteria so closely to his case, however, Vattel concludes on a sobering note: 'it has since appeared ... .' That is wisdom after the fact, of course, but still wisdom, and one would expect some effort to restate the criteria in its light."

57   Ibid at 85.

58   Ibid at 84.

59   Roland Popp, "Stumbling Decidedly into the Six-Day War" (Spring 2006) 60 Middle East Journal 281; cf. the older but well-researched and pro-Israel account of Michael Oren, *June 1967 and the Making of the Modern Middle East* (OUP 2002); or for another more balanced and recent Israeli historian, see Tom Segev, *1967: Israel, the War, and the Year That Transformed the Middle East*, Jessica Cohen (trans) (Metropolitan Books 2007).

60   Ibid Popp at 307.

when deciding upon its launching so as to avoid unjustified death and suffering. Yet both of them attempted to push the limits of anticipatory attack beyond the temporal constraint of *imminence*, while still insisting on the requirement of necessity.

Vattel opened this ambiguity. His first mention of the allowance to anticipate injury was followed directly by a warning that this must not occur "upon vague and uncertain suspicions" since a state must avoid "the imputation of becoming herself an unjust aggressor."[61] Adding to this confusion, he proclaimed that war must be a reaction to injury: "Then, *and not until then*, that nation has a right to repel the aggressor."[62] Yet Vattel immediately followed this statement with an assertion that would undermine the expressed firmness of his stance: "Further, she has a right to prevent the intended injury, when she sees herself threatened with it."[63]

It is my conclusion that the position of this respected jurist and well-known moralist is muddled and ambiguous on the question of anticipatory war. My reading is that each author seems to be arguing both sides of this debate, at once laying down strict boundaries followed by unclear exceptions. We have seen that there are great difficulties with the examples they use since the "facts" have not remained stable over time. However, it is submitted that one must blame the ground they attempt to traverse using the idiosyncratic notions of intent and risk. In other words, there is no middle ground between *imminence* and *just fear*; pushing beyond the former crosses a precipice that inevitably falls into alignment with the latter.

IV. Current Practice

A. Scrubbing the DoD Dictionary

The bifurcation we have seen vividly displayed in the seventeenth-century debate between Francis Bacon and Hugo Grotius is one that has continued to shape our understanding of the word "imminence" and our definitions of anticipatory attacks. One clear way to demonstrate this influence is through the dictionary created by the United States DoD since the two relevant categories of anticipatory attacks are indeed understood as separate and distinct within it. Specifically, in the DoD Dictionary through October 2009 there was a prominent difference between the terms "preemptive attack" and "preventive war." Although we have seen different applications of these specific terms in this chapter, the ideas in their separate definitions are in perfect orientation with what has been discussed. The 1994–2009 editions of the dictionary read:

---

61   Vattel (n 55) at 169.
62   Ibid at 304 (my emphasis).
63   Ibid.

- **preemptive attack**—An attack initiated on the basis of *incontrovertible* evidence that an enemy attack is *imminent*.[64]

- **preventive war**—A war initiated in the *belief* that military conflict, while not imminent, is inevitable, and that to delay would involve greater *risk*.[65]

In light of the analysis in this chapter, this distinction is significant since there is a moral and legal chasm between the two ideas. As can also be seen, there is a marked variation in tone found in these dictionary definitions. The language used to define "preemptive attack" is commanding, leaving no room for doubt about the incontestable high level of proof required—just like the *imminence* standard. While on the other hand, "preventive war" is clearly based on subjectivity allowing for varied individual interpretations—precisely as we have seen with *just fear*.

Of great interest here is the fact that the DoD Dictionary has now been disturbingly modified—twice. Like all dictionaries, this one from the DoD is regularly revised and updated, with the most recent version found online and its revision date clearly indicated. While the divergence between these two terms had been unmistakeably laid out in the versions traced at least as far back as 1994,[66] the edition amended through April 2010 no longer included the term "preventive war."[67]

As noted in the introduction to this chapter, the previous administration had already been notoriously involved in stretching the definition of "imminence" in the wake of 9/11 with what came to be known as the Bush Doctrine.[68] But under the Obama administration we find a much less vociferous action taken by his DoD, yet this action makes conversation on this critical question much more difficult by removing an important authority.[69] In its expelling of the term that clarifies the type of anticipatory attack based upon belief or future risk assessment, it appears

---

64   Department of Defense Dictionary of Military and Associated Terms, Joint Publication 1–02, April 12, 2001 (as amended through Oct 31, 2009), entry for 'preemptive attack,' (my emphasis) at 424, available at <http://jitc.fhu.disa.mil/jitc_dri/pdfs/jp1_02.pdf> accessed Jan 2015.

65   Ibid (my emphasis) at 428.

66   Ibid at <http://www.dod.mil/pubs/foi/joint_staff/jointStaff_jointOperations/913.pdf> accessed Jan 2015.

67   Ibid at <http://books.google.ch/books?id=Ap_En_k7r9AC&q=preemptive+attack#v=snippet&q=preemptive%20attack&f=false> accessed Jan 2015.

68   See eg R Kolb, "Self-Defence and Preventive War at the Beginning of the Millenium" (2004) 59 Zeitschrift für öffentliches Recht 111; M Leffler and J Legro (eds) *To Lead the World: American Strategy after the Bush Doctrine* (OUP 2008); M Gurtov and P Van Ness (eds) *Confronting the Bush Doctrine: Critical Views from the Asia-Pacific* (Routledge 2004); E Kolodziej and R Kanet, (eds) *From Superpower to Besieged Global Power: Restoring World Order after the Failure of the Bush Doctrine* (Univ of Georgia Press 2008).

69   For another work that cites the previous existence of entries for "preventive war" and "preemptive attack" see R Lawrence "The Preventive/Preemptive War Doctrine Cannot Justify the Iraq War" (Winter 2004) 33(1) Denver J of Intl Law and Pol 16, at 16–17.

as if the DoD no longer believes this type of anticipatory attack to exist. This is undoubtedly not the case, as we will see more clearly below; that is, attacks based on future intentions have become part and parcel of drone targeting.

Furthermore, the DoD Dictionary was not through with its purge in 2010. It further revised the wordlist by removing the second relevant term, "preemptive attack," in its March 2013 version.[70] These amendments, which can be described as a scrubbing of materials in light of the discussion in this chapter, are to be lamented. What is more, this disappearance from the US government's official security lexicon can also be established with a search in the current DoD Web site's "Dictionary of Military Terms."[71] Nonetheless, even if the historical record on this important question cannot be similarly scrubbed, this feckless effort is unwelcome.

B. "Operational Realities"

As for current practice and readings of the applicable standards, it has been argued that, behind the scenes, governments are starting to coalesce around an expanded interpretation of the term under discussion. For example, the Obama administration stated in 2011: "We are finding increasing recognition in the international community that a more flexible understanding of 'imminence' may be appropriate when dealing with terrorist groups."[72] Likewise, one barrister has suggested the same, citing UK government documents and "detailed discussions over recent years with foreign ministry, defense ministry, and military legal advisers from a number of states who have operational experience in these matters."[73] Most pertinently, he argues that "[t]here is little intersection between the academic debate and the operational realities" of lethal action directed against terrorist groups because "the consequences of inaction ... frequently trump a doctrinal debate that has yet to produce a clear set of principles that *effectively* address the specific operational circumstances faced by states."[74] Yet this approach is, once again, an attempt to push the legal and moral debate off the table. Much like Cicero and Hobbes,[75] the argument is that acute security concerns must trump legality and morality to give way to a narrowly understood concept of efficacy.

---

70   DoD Dictionary (n 64) at <http://www.lawfareblog.com/wp-content/uploads/2013/04/jp1_02.pdf> accessed Jan 2015.
71   DoD Dictionary of Military Terms, Managed by the Joint Education and Doctrine Division, J-7, Joint Staff, available at <http://www.dtic.mil/doctrine/dod_dictionary/?zoom_query=preventive+war&zoom_sort=0&zoom_per_page=10&zoom_and=1> accessed Jan 2015.
72   Brennan (n 1).
73   Bethlehem (n 5) at 773.
74   Ibid (my emphasis).
75   See notes 14–16 and accompanying text.

However, many of the tactical benefits remain buried in secret data unavailable to the public,[76] while much of the relevant information remains entirely inaccessible to all of those outside the extremist group itself due to the difficulties of determining membership and rank.[77] Even more concerning, the strategic efficacy of such attacks has not been demonstrated or conclusively evidenced through analysis.[78] If there were to be credible studies drawing systematic conclusions on the effectiveness of such targeting, the considerations could assuredly shift. Yet this is not where we find ourselves today.

Nevertheless, even concrete answers to the complex and varied questions of efficacy would not entirely exclude the tenets of morality and legality from assessments of the legitimacy of these operations. The "operational realities" argument being forwarded by a growing number of governments should thus be confronted head on for the absence of evidence regarding efficacy, along with its myopic view on legitimacy.

There is no doubt that in today's world, with an expanded attention upon security threats posed by individuals and violent groups, a government is frequently faced with alarming intelligence that begs for action. And in a democracy, the perpetual election cycle makes an administration particularly susceptible to pressure for manifest deeds. However, the essential question before us for armed drones is whether that action can and should rise to the level of lethal force across a new international border based on unverifiable concerns.

Attorney General Eric Holder made a speech in March 2012 in which he argued the case for necessity (in perfect coincidence with the leaked White Paper cited above since both were meant to address the circumstance of a targeted and killed US citizen, Anwar al-Awlaki). In it, the novel definition of *imminence* used by the Obama administration is explained:

> The evaluation of whether an individual presents an "imminent threat" incorporates considerations of the relevant window of opportunity to act, the possible harm that missing the window would cause to civilians, and the likelihood of heading off future disastrous attacks against the United States.[79]

---

76   See Chapter 10 by Marek Madej and my Chapter 11.

77   Extrapolating the numbers from Guantánamo makes this point when we calculate that 780 have been detained and 68, or 8.7 percent, are to be held until the end of hostilities: see Human Rights First, "Guantánamo by the Numbers" (last updated Jan 15, 2015) available at <http://www.humanrightsfirst.org/resource/guantanamo-numbers> accessed Jan 2015.

78   See my Chapter 11 in this volume on the difficulties raised by the question of strategic efficacy; and eg Audrey Kurth Cronin, *How Terrorism Ends: Understanding the Decline and Demise of Terrorist Campaigns* (Princeton Univ Press 2009); and for an empirical study see Jenna Jordan, "When Heads Roll: Assessing the Effectiveness of Leadership Decapitation" (2009) 18 Security Studies 719.

79   Eric Holder, Attorney General, Speech at Northwestern University School of Law (March 5, 2012) <http://www.justice.gov/iso/opa/ag/speeches/2012/ag-speech-1203051.

While there are various avenues to pursue in dissecting this three-part test, the subjectivity is plain and thus where this leads is obvious. I will instead mention only a couple of points. The first is that the idea of inserting factors of the "relevant window of opportunity to act" flips the question on its head. The assessment of "imminence" should logically be on a confirmable action by a potential aggressor, yet this criterion inverts the focus onto a potential victim's own considerations and capabilities. This is backwards. Second, the enormously imprecise and incalculable phrase "the likelihood of heading off" indicates no boundaries or standards for exercising deadly force across borders.

To further confuse this question of temporal limits, the Attorney General claimed in a letter to Congress of May 2013 that "lethal force may be used only when a terrorist target poses a *continuing*, imminent threat."[80] Unfortunately, the additional term here only distorts the matter. While more information is necessary to properly analyze this change, it appears as though the Obama administration is infusing *jus ad bellum* with norms from humanitarian law. That is, the continuing threat from a (suspected) combatant is meant to be sufficiently threatening to allow the crossing of a new international border to use lethal force against this individual. This is a perilous mix.

Beyond the traditional reasoning for separating *jus ad bellum* from *jus in bello* (e.g. using the former to dismiss the applicability of the latter), there are other practical concerns. For instance, this danger can be readily discerned when we consider the ramifications had India characterized the Mumbai attack of 2008 as authorizing the use of force based on a similar reading of the law. That is, the assault sanctioned unilateral attacks by India against suspected members of the Lashkar-e-Taiba group concealing itself inside a nuclear-armed Pakistan.[81] As can be seen in this extremely disquieting example, such an opening of the door to unilateral uses of lethal force across borders based on suspicions of status and future intentions alone leaves no real limits in the self-administered global system.

V. Conclusion

We once again find ourselves in a moment of great tension due to the potency of the weapons available to those willing to commit terrorist acts. The anxiety produced is palpable, and it is justified. But should fear, even if justifiable, serve as a trigger for lethal force? As we have seen throughout this survey, the historical attempts

---

html> accessed April 2014.

80   Eric Holder, Attorney General, "Letter to Members of Congress, Chairman of the Committee to the Judiciary" (May 22, 2013) (my emphasis) <http://www.nytimes.com/interactive/2013/05/23/us/politics/23holder-drone-lettter.html?_r=0> accessed Dec 2014.

81   Craig Martin, "Going Medieval: Targeted Killing, Self-Defense and the *Jus ad Bellum* Regime" in C Finkelstein, JD Ohlin et al (eds) *Targeted Killings: Law and Morality in an Asymmetrical World* (OUP 2012) at 242.

to do so inevitably lead onto an untenable territory devoid of a rule prohibiting the free use of violence. Hart explained that the central and indisputable necessity for mutual forbearances turns on the "modest aim of survival." With this idea at the core, he constructed a foundation for the "minimum content of natural law" explaining that "[t]his simple thought has in fact very much to do with the characteristics of both law and morals."[82]

International lawyers certainly did not miss this break with legality and morality when it came to the Bush Doctrine, and by extension, what came to follow as an Obama Doctrine. In 2010, Jutta Brunnée and Stephen J. Toope delved deeply into this question on the use of force in their assessment of legitimacy and legality, concluding that "a 'rule' that has no parameters, and that is subject to entirely unilateral assessment of what amounts to a 'sufficient threat' is simply a self-constructed permission to act ... The result was to posit what may superficially have looked like a legal norm, but was actually no norm at all."[83] This assessment was echoed by Martti Koskenniemi: "a fully self-judging rule on anticipatory action, such as included in the US Security Strategy of 2002, *fails* as a legal rule altogether."[84] Koskenniemi also pointed out that this very same judgment was expressed by Hersch Lauterpacht in 1933: "An obligation whose scope is left to the free appreciation of the obligee, so that his will constitutes a legally recognized condition of the existence of the duty, does not constitute a legal bond."[85]

The intention here has been to expand on these sharp evaluations with a historical view on anticipatory attacks to advocate anew for the rejection of such an unsustainable interpretation of "imminence" that erases a reciprocal restraint on violence across borders. Article 2(4) in the UN Charter, and its exception of Article 51, were clearly not intended to codify a *state of nature*. Pivotal points in our shared history elucidate the fact that morality and lawfulness for anticipatory attacks are granted only in the most immediate and verifiable circumstances. We would be wise to heed this enduring lesson.

---

82 Hart (n 9) at 191–193.

83 Jutta Brunnée and Stephen Toope, *Legitimacy and Legality in International Law: An Interactional Account* (Cambridge Univ Press 2010) at 301 and 303.

84 Martti Koskenniemi, "The Mystery of Legal Obligation," (2011) 3 International Theory at 323 (original emphasis).

85 Lauterpacht was referring to unlimited opting-out provisions in arbitration treaties, but the relevant legal assessment was nonetheless the same: as cited in ibid.

Chapter 7

Correcting the Record: Civilians, Proportionality, and the *Jus ad Vim*

Avery Plaw and Carlos R. Colon

I. Introduction

Arguably the most intensely contested aspect of the US drone strikes away from traditional battlefields has been the number of civilians who have been killed, and whether these deaths (and other harms) have exceeded what might be legally or ethically justified. Over the last decade critics and defenders of drone strikes have clashed over a number of related issues with no decisive resolution. They have disagreed about the legal paradigm that applies to these US drone strikes (law of armed conflict vs law enforcement), the conditions under which the deliberate use of lethal force would be justified (when there is an immediate threat to life vs when consistent with military necessity and the rules of distinction and proportionality), and over the number of civilians actually being killed (thousands of people and 98 percent of all those killed vs 182 and around 4 percent of those killed). But these debates have done little to alter the status quo—the strikes remain popular in the United States and unpopular in most other places; the US government has, if anything, doubled-down on defending the legality and efficacy of its policy while critics have become even more aggressive in investigating and denouncing the campaign.

In the last couple of years however, a number of scholars, including Megan Braun and Daniel Brunstetter, have launched an intervention around the idea of *jus ad Vim*—"the just use of force short of war"[1]—which has the potential to shift the terms of discussion. In essence, the goal of *jus ad vim* is to establish a set of criteria for assessing the legitimacy of limited uses of military force short of armed conflict (including drone strikes outside conventional battlefields). We will examine *jus ad Vim* here with particular focus on an article published in 2014, entitled "Rethinking the Criterion for Assessing CIA Targeted Killings: Drones, Proportionality and *Jus ad Vim*," which makes the most sustained and compelling case to date for this approach as applied specifically to drone strikes. Braun and Brunstetter's case for this new standard involves three claims:

---

1   Megan Braun and Daniel R. Brunstetter, "Rethinking the Criterion for Assessing CIA-targeted Killings: Drones, Proportionality and Jus Ad Vim" (2014) 12(4) J of Military Ethics 304.

1. "In light of the shortcomings of both the war and preventive response arguments, a theory of *jus ad vim* would be particularly helpful" in evaluating the drone strikes away from conventional battlefields and in Pakistan in particular.[2]
2. "When viewed through the lens of *jus ad vim*, the CIA's proportionality balancing in Pakistan ... is ... highly problematic."[3] Indeed, they assert that "If we judge the CIA's use of drones in Pakistan according to the standards of *jus ad vim*, then it falters on [two] counts"[4]—that is, on both of the requirements of *jus ad vim* micro-proportionality. First, it fails in regard to "the threshold of harm that can be committed against civilians [which] should be lower than in zones of war."[5] Second, it violates the *jus ad vim* principle that "the human rights of civilians living under drones cannot be violated in order to protect the rights of US citizens."[6]
3. However, "[t]his does not mean that drones are never permitted to protect American lives from terrorists operating in these regions, but ... [i]f strikes are to be justified they will need to conform to a highly restrictive standard of proportionality and exhibit strong respect for human rights."[7]

In essence, Braun and Brunstetter's formulation of *jus ad vim* offers a middle-ground between those who insist on applying the framework of war to US counterterrorism operations and those who judge those operations according to the standards of law enforcement and human rights. As Braun and Brunstetter aptly put it, their approach "provides justification for the use of force that would be unacceptable in a law enforcement context, without invoking the permissive authorities of actual war."[8]

Nonetheless, this chapter raises three reservations about the view Braun and Brunstetter advance:

1. Their formulation of *jus ad vim* and its requirements is in several respects unrealistic and unhelpful. We argue that a careful examination of the data, and reflection on the virtues and defects of conventional just war theory (JWT) criteria suggests that a rigorous application of conventional JWT

---

2 Ibid at 304 and 317.
3 Ibid at 318.
4 Ibid at 319.
5 Ibid at 318.
6 Ibid at 319. Indeed, these findings are not surprising, for even before introducing their criteria of the *jus ad vim* they had already suggested that CIA operations in Pakistan were "disproportionate on the whole" on the less demanding *jus in bello* proportionality criterion (Ibid at 315). Or, as they elsewhere remark, "the actual 23 percent level is too high for drones to be considered, as proponents claim, proportional" (ibid at 316).
7 Ibid at 318.
8 Ibid at 317.

criteria will provide a more appropriate, flexible, and realistic framework for evaluating drone strikes outside of conventional battle zones.

2. Whatever one thinks about their account of *jus ad vim*, the strikes on which Braun and Brunstetter focus (i.e. in Pakistan), or at least most of them, are not appropriate. The areas of Pakistan targeted by drones *are* clearly in a state of armed conflict and therefore the appropriate standards are those of armed conflict (i.e. *jus in bello*).

3. Finally, even if their *jus ad vim* standards are applied (erroneously) to Pakistan, their case that these drone strikes in general fail standards of proportionality is not convincing.

The following section of this chapter (section II) offers a summary of Braun and Brunstetter's formulation of, and case for, *jus ad vim*. Here we draw on a 2013 article by them setting out their conception of *jus ad vim* which closely informs their 2014 article examining the proportionality of drone strikes in Pakistan. Section III critically examines Braun and Brunstetter's assessment of the proportionality of drone strikes in Pakistan and argues that even if one does apply *jus ad vim*, it is unclear that US drone strikes in Pakistan fail to comply with applicable standards. Section IV shows that the *jus ad vim* is at any rate not applicable to the main case that Braun and Brunstetter examine—that is, to drone strikes in Pakistan (or at least to the vast majority of them including those occurring at this time). Section V, the final substantive section of this chapter, proposes an alternative "elevated JWT" framework that we think is more plausible, attractive, and applicable to Pakistan drone strikes. The concluding section summarizes our findings.

II. Braun and Brunstetter's Case for *Jus ad Vim*

In a 2013 article entitled "From *Jus ad Bellum* to *Jus ad Vim*: Recalibrating our Understanding of the Moral Use of Force," Daniel Brunstetter and Megan Braun follow Michael Walzer in arguing that traditional JWT would benefit today from recognizing a new category of "measures short of war" as distinct from "actual warfare."[9] By "measures short of war" they mean operations that are "limited in scope," like "no fly zones, pinpoint air/missile strikes, CIA operations" which "fall short of the quantum and duration associated with traditional warfare." They point out rightly that states have relied increasingly on such limited measures in the last couple of decades, especially to respond to "the rise of non-state actors such as Al-Qaeda and its affiliates, which pose significant threats to international peace and security, but do not have international legal status and operate in the

---

9 Daniel Brunstetter and Megan Braun, "From Jus ad Bellum to Jus ad Vim: Recalibrating Our Understanding of the Moral Use of Force" (2013) 27(1) Ethics & Int'l Aff 87, at 87.

porous or disputed border regions of sovereign states."[10] Yet, as they note, the conflicts with these terrorist groups and the limited measures states employ against them do not fit neatly into the traditional legal and ethical frameworks of "armed conflict" or "law enforcement" that were primarily designed to regulate wars (either between or within states) or criminal behavior (within states) respectively. These non-state actors are far more dangerous than criminal gangs and their transnational character often puts them outside the reach of the domestic law of the states they seek to attack—indeed, against whom they are, in their minds, conducting war. At the same time, the powers and privileges of belligerency sometimes assumed by states seem excessive and indeed dangerous in relation to the scope of the threat particularly when they are seen as extending to anywhere and everywhere these non-state actors operate. But these are the only two options available under the current international legal regime and within the moral theory of JWT that informs it.[11]

The *jus ad vim* is intended to offer a third possibility, at least in cases of last resort. For example, when confronted with terrorist groups operating out of a "host country [that] does not have the will and or capacity to deal with the threat they pose," Braun and Brunstetter concur with Walzer that "international policing actions, in conjunction with actions by local authorities, should be tried first."[12] But if that proves unavailing, then what can be done, and under what rules? In such cases, they argue, we would be better served by "measures short of war" subject to appropriately restraining principles that preserve human rights as much as possible, than by a wholesale declaration of war and the invocation of all the rights and powers of belligerency. So, just as JWT establishes criteria for determining when actual warfare is justly initiated (*jus ad bellum*) and conducted (*jus in bello*), we should similarly develop a criteria for determining the justness of "measures short of war," criteria which they refer to as "*jus ad vim*."[13]

In their formulation, *jus ad vim* requirements for the resort to force should include the following:

1. Just Cause: "Within the context of *jus ad vim,* a state has just cause to use measures short of war when responding to *injuria* against its interests or citizens. This includes responding to terrorist bombings, attacks on embassies or military instillations [sic], and the kidnapping of citizens. ... Imminent threats of terrorist attacks also provide just cause."[14]
2. Resort: "Some attempt at non-violent diplomatic measures must be tried before resorting to force, even if the limited levels of violence of *jus ad vim*

10   Ibid at 88.
11   Ibid at 89–91.
12   Ibid at 89.
13   Ibid at 87.
14   Ibid at 95–96.

mean that this requirement is less exacting than in the case of war ... There must be an imminent threat and conditions that rule out policing measures."[15]

3. (Macro-)Proportionality: This means "defining what constitutes a successful outcome and determining which actions will enable this outcome," on the basis of a "probability of success criterion."[16] These actions then define "the maximally just level of force ... not what level to begin with and potentially escalate from."[17]

4. Right Intention: "Right intention must therefore be directed toward upholding the rights of the Other. In this sense, right intention for *jus ad vim* means quelling a specific threat, while causing the least amount of damage possible by protecting civilians."[18]

They also suggest an entirely new criterion, which does not appear in standard JWT:

> Probability of Escalation: "If engaging in *jus ad vim* actions has a high probability of resulting in war, then one could argue that such actions are not justifiable."[19] As they specify in their 2014 article, "limited force should not run the risk of increased violence."[20]

Finally, in Braun and Brunstetter's formulation the foregoing should *not* be seen as threshold criteria that only need to be crossed once to justify an ongoing campaign against non-state actors. These are rather ongoing requirements that need to be met separately in each resort to force even against the same target:

> Whereas in war, principles such as just cause and last resort need only be satisfied at the outset of a conflict, *jus ad vim* requires that they be continually reassessed in advance of each use of force. Additionally, by moving beyond the persistent and broad nature of a general threat that one assumes in warfare, and examining individual operations geared towards a particular threat, a theory of *jus ad vim* makes it possible to assess whether the response is calibrated to the gravity of that threat. This nuanced approach proves essential when examining the proportionality of drone strikes.[21]

In the case then of a state pursuing a series of distinct operations against a non-state actor, such as a drone campaign, compliance with *jus ad vim* on Braun and

---

15   Ibid at 97.
16   Ibid at 98.
17   Ibid at 98.
18   Ibid at 100.
19   Brunstetter and Braun, "From Jus ad Bellum to Jus ad Vim" (n 9) at 98.
20   Braun and Brunstetter, "Rethinking ... " (n 1) at 318.
21   Ibid at 317.

Brunstetter's account would require each and every strike to be justified in terms of the foregoing criteria (as well as those which follow).

Assuming the foregoing conditions on the resort to force are met, Braun and Brunstetter suggest that the actual use of force under *jus ad vim* should also be regulated by elevated standards on the use of force and human rights considerations:

> [*J*]*us ad vim* demands a stricter relationship between the use of force short of war and the *jus in bello* principles of proportionality and discrimination. In addition, we assert that human rights concerns of civilians not usually considered in the proportionality calculus need to be taken into account. Such ethical constraints severely restrict the scope of proportionality balancing—the conscious decision that anticipated military advantage outweighs collateral damage—that can be employed.[22]

How would a theory of *jus ad vim* constrict the way in which states use limited force?

On the one hand, it means that there is "less moral latitude for inflicting unintended harm on non-combatants."[23] For example, Braun and Brunstetter express approval of the standard used by the US military in Afghanistan: "prior to a strike, the military operators must believe that there is a less than 10 percent chance that the attack will kill civilians."[24] In addition, they also "assert that human rights concerns of civilians not usually considered in the proportionality calculus need to be taken into account."[25] In particular, the human rights of civilians living under drones cannot be violated in order to protect the rights of US citizens. The attenuated threat produces a much more restrictive principle of proportionality, one that needs to be concerned not only with the loss of civilian life but also the more subtle harms including "property destruction, post-traumatic stress disorder, and social disorder caused by the persistent threat of drones."[26]

In sum then, the idea of *jus ad vim* is to introduce a new category of legitimate force between armed conflict and law enforcement. In Braun and Brunstetter's formulation of this standard, states must separately justify every use of force according to an elevated set of requirements (relative to conventional JWT standards) including the new requirement that there be no probability of escalation toward armed conflict. Moreover, the type of force that can be used is "severely

---

22   Ibid at 306; "Moreover, we argue that there must be 'a strict relationship between jus ad vim and the jus in bello principles of proportionality and discrimination'" (Brunstetter and Braun, "From Jus ad Bellum to Jus ad Vim" at 100–101). "This would severely restrict the scope of proportionality balancing that could be employed when using force short of war, compared to what is permissible in war" (ibid at 319).

23   Braun and Brunstetter, "From Jus ad Bellum to Jus ad Vim" (n 9) at 101.

24   Braun and Brunstetter, "Rethinking … " (n 1) at 310.

25   Ibid at 306.

26   Ibid at 319.

restricted," including by considerations of the social disorder that may result from the persistent threat of drone strikes.

III. Braun and Brunstetter's Application of *Jus ad Vim* to Drone Strikes in Pakistan

In their 2014 "Rethinking the Criteria" article, Braun and Brunstetter apply the idea of *jus ad vim* to US drone strikes in Pakistan (2004–present) with particular emphasis on the rule of proportionality. They begin their analysis by clearing some of the underbrush created by recent debate. In particular, they criticize three ways that defenders have tried to suggest that these strikes are proportional as forms of "proportionality relativism"—that is,

> the use of impertinent comparisons to argue that drones are proportionate because they cause less collateral damage than other uses of force. Such comparisons misrepresent the true meaning of proportionality as an independent assessment of the balance between the anticipated civilian harm and military gain associated with each act of force.[27]

In their analysis, "The first and most recurrent form of proportionality relativism focuses on the weapon itself—namely, mistaking precision for proportionality."[28] The second type of proportionality relativism is "the claim that drones are more proportionate as compared to the weapons and means of war prevalent in other eras of warfare."[29] The third type of proportionality relativism is "the claim that drones are proportionate compared to other modern tactics in war."[30] In this last case they object not only to comparisons of drone strikes with the average levels of civilian casualties in armed conflict over the decade, and with Pakistani army operations in the Federally Administered Tribal Areas (FATA) where virtually all drone strikes in Pakistan take place, but even to comparisons with other tactics used by the United States in the FATA (such as manned airstrikes or commando raids) during the same period that drone strikes have occurred, all on the following grounds:

> One would need to know the specifics of each operation, and *only compare similar operations*—for example, ones that targeted leaders (which may justify higher collateral damage)—as opposed to aggregate strikes. The comparison with non-drone operations also suffers from a lack of context to weigh the foreseeable military advantage.[31]

---

27   Ibid at 305.
28   Ibid at 307.
29   Ibid at 308.
30   Ibid at 309.
31   Ibid at 309 (our emphasis).

In essence, according to Braun and Brunstetter, *any* comparison between operations using different weapons is "relativistic" unless the specific military advantages (for example, the functions and capabilities of the target) are very similar, even if both operations take place in the same place at the same time against the same enemy and are of similar scale.

With these "relativistic fallacies" cleared away, Braun and Brunstetter proceed to advance a case against the proportionality of drone strikes in Pakistan appealing to *jus ad vim* criteria of their own conception, and employing what they see as a more precise measure of proportionality:

> As a proxy for proportionality, we look at the collateral damage ratio—the strikes that produced collateral damage compared to total strikes—to determine the percentage of strikes that caused collateral damage. This focus more closely captures the individualized calculation required by the principle of *jus in bello* proportionality compared to the average used by Plaw. It allows us to gain an understanding of the scope of expected collateral damage—what we call proportionality balancing—that the CIA deems as acceptable when contemplating any particular strike.[32]

This measure, at least in the way they choose to calculate it, produces troubling results in relation to the proportionality of drone strikes in Pakistan: "[a]ccording to data collected by the New America Foundation, the USA conducted 343 drone strikes in Pakistan from 2004 to 2012. Of these, 263 strikes killed only militants, while 80 killed civilians and/or unknowns" or, as they summarize later, "23 percent of CIA strikes caused collateral damage."[33] They compare this to a similar measure in Afghanistan, where the US military has set itself a maximum collateral damage threshold of 10 percent and, they suggest, has achieved an "actual collateral damage rate of 1 percent,"[34] and conclude that,

> 23 percent should be considered excessive. While the military's operations [in Afghanistan] are arguably in compliance with the proportionality criterion, the CIA drone program, if held to the same standard, would be disproportionate on the whole.[35]

This assessment appears to be based on standard *jus in bello* proportionality, because their *jus ad vim* standards are only introduced directly thereafter. Of course, the drone strikes also violate the more elevated *jus ad vim* standard of proportionality that Braun and Brunstetter propose, and on at least two grounds— not only that 23 percent of strikes cause collateral damage which is simply too

---

32   Ibid at 311.
33   Ibid at 311 and 315.
34   Ibid at 315.
35   Ibid.

high, but also because of the rights violations and "subtle harms" (e.g. "property destruction, post-traumatic stress disorder, and social disorder") visited upon civilians:[36]

> the ubiquitous presence of drones in Pakistani airspace, coupled with *nearly one in four strikes that kill civilians*, has reportedly had serious social and political repercussions that tangentially affect the proportionality calculus.[37]

It is striking that in the italicized portion of this last remark Braun and Brunstetter use the term "civilians" to describe all of those killed in the 23 percent of US drone strikes in Pakistan—which they initially defined as killing "civilians and/ or unknowns."[38]

There are a number of features of Braun and Brunstetter's analysis that we find troubling. These concerns, which we will elaborate in more detail below, can be summarized as follows:

1. They tend to conflate the categories of "civilians and/or unknowns killed" with "non-militants killed," or "collateral damage," or even "civilians killed," which produces misleading results, especially when they compare them with counts of civilian deaths elsewhere. When the categories are properly distinguished, drone strikes in Pakistan appear to meet the standard for proportionality used in Afghanistan which Braun and Brunstetter praise.
2. The way they calculate leadership targeting tends to undercount the number of strikes aimed at leaders (by considering only those in which leaders were actually hit), and exaggerate the number aimed at low-level militants (by assuming everyone who is not a high-value target [HVT] must be low level).
3. They conflate "weapons releases" with "strikes," which again tends to exaggerate drone precision in Afghanistan in comparison with Pakistan.
4. Moreover, their preferred measure, the "collateral damage ratio," is inherently limited because it only takes account of when civilians were killed, not how many civilians were killed.

A. Conflation of Civilians and Unknowns

To get their figure of 23 percent of drone strikes in Pakistan causing collateral damage, Braun and Brunstetter combine drone strikes which resulted in civilian fatalities with those which killed "unknowns"—that is, "those who are not identified in news reports definitively as either militants or civilians" as defined by their own

---

36   Ibid at 319.
37   Ibid at 314 (our emphasis).
38   Ibid at 311.

source, the New America Foundation.[39] They then somewhat misleadingly refer to the 23 percent of strikes as killing "non-militants (civilians or those whose identity is unknown)"[40]—seemingly assuming that no unknowns could possibly be militants (contrary to their sources' definition of the term which indicated that that they might be "either militants or civilians").[41] They then repeatedly use the term "non-militant" (without the parenthetical explanation) to describe the combination of civilians and unknowns killed in those 23 percent of drone strikes in Pakistan through the remainder of the article, except in a few instances where they use even more troubling terms.

For example, they also refer to those killed in the 23 percent of drone strikes in Pakistan as "collateral damage" (again implying that all unknowns must actually have been civilians): "the data indicate that 23 percent of CIA strikes *caused collateral damage*, which is a far higher percentage than what the US military tolerates [i.e. in Afghanistan]."[42] Most strikingly of all, in one instance (as shown in the quotation at the end of the previous section) Braun and Brunstetter refer to those killed in these 23 percent of drone strikes explicitly as "civilians": "coupled with *nearly one in four strikes that kill civilians*."[43] This completes the gradual ark from these 23 percent of drone strikes in Pakistan being described as killing "civilians and unknowns" to killing "non-militants" to causing "collateral damage" to specifically "killing civilians."

Braun and Brunstetter then compare the 23 percent to the US military's 10 percent risk tolerance for civilian casualties of in Afghanistan (which effectively means that no more than 10 percent of strikes are expected to result in civilian casualties) and remark that "the observed outcome of collateral damage from CIA drone operations [in Pakistan] is more than twice as high."[44] However, in making this comparison they are in fact *comparing civilian casualties with civilian plus all unknown fatalities*. This is a far more serious case of comparing quite different things than many of those that they complain of in their section on proportionality relativism.

Unknowns are obviously not properly classified as non-militants, or collateral damage, or as civilians. The point of the designation is, as the New America Foundation's definition explicitly states, that these individuals "are not identified in news reports definitively as *either militants or civilians*," or as they put it elsewhere, these individuals "were described in a manner that made it *ambiguous*

---

39   The New America Foundation also labels unknowns as those who "were described in a manner that made it ambiguous whether they were militants or civilians." See <http://securitydata.newamerica.net/drones/pakistan/key-findings> accessed Oct 2014.

40   Braun and Brunstetter, "Rethinking … " (n 1) at 311.

41   Ibid at 319, table on 312.

42   Ibid at 315 (our emphasis).

43   Ibid at 314 (our emphasis).

44   Ibid at 315.

whether they were militants or civilians."[45] In fact, it is possible that these unknowns could all be militants, and therefore comprehensively misrepresented in Braun and Brunstetter's 23 percent statistic. Moreover, these potentially misrepresented "unknowns" account for the great majority of this 23 percent (representing 56 of the 80[46] cases, although 13 cases appear to have a mixture of civilians and unknowns). Unknowns should be kept as a distinct category and not gradually elided with civilians.[47]

What is even more striking is that if the "unknowns" are separated from the civilian casualties in the New America Foundation (NAF) data that Braun and Brunstetter themselves use (2004–2012), so that the comparison is actually between straight civilian counts on both sides, then the proportion of strikes that killed civilians in Pakistan is actually 9.038 percent. This result is below the US military's 10 percent threshold in Afghanistan, which Braun and Brunstetter laud, showing that the drone strikes in Pakistan technically met their preferred standard of proportionality. If we were to include the 2013–2014 strikes in the count (which were incomplete when Braun and Brunstetter completed their article), the civilian casualty percentage drops to 8.29 percent (see Table 7.2). Moreover, the NAF data are confirmed by other databases. For example, based on the aggregate data, from 2004 to 2014, the Center for the Study of Targeted Killing database shows that the overall proportion of drone strikes in Pakistan resulting in civilian casualties is 8.5 percent.[48]

---

45   See <http://securitydata.newamerica.net/drones/pakistan/key-findings> accessed Oct 2014 (our emphasis).

46   We give their total of 80 for the sake of consistency, but were only able to find 75 cases involving civilians or unknowns killed.

47   However, those impressed by Braun and Brunstetter's analysis, might at this point want to appeal to the rule of humanitarian law that if a party to the conflict is uncertain whether a target is a combatant or not, they must treat the target as a civilian (see Article 50(1) of the First Additional Protocol to the Geneva Conventions; see also the ICRC "Interpretive Guidance on the Notion of Direct Participation in Hostilities under International Humanitarian Law," 90(872) IRRC at 996, paragraph 8). But this move relies on an important confusion, for what is uncertain to us as observers (or the NAF in particular) need not be uncertain to planners or soldiers in the field. This is the source of the whole difficulty in assessing the proportionality of drone strikes in Pakistan—we don't know what they do and do not know. So the category of unknowns to us need not be the same as the category of unknowns for them. In fact, all of these unknowns could not only be militants but be known to be militants by those planning, approving, and conducting targeting operations. Of course, they also may not be. But the point is that we cannot assume that people's whose affiliations are unknown to us should be treated be treated as civilians by US planners because we cannot know if they are unknown to the planners. So appeal to the rule regarding uncertainty does not provide a compelling reason why those unknown to us should always be classed as civilians for purposes of estimating whether drone strikes violated proportionality.

48   Available at <targetedkilling.org/strikes> accessed Oct 2014.

Table 7.1   Drone strikes in Afghanistan

Years	A. Total strikes[1]	B. Strikes killing civilians	C. Civilian count	D. Civilians killed/ strike	E. Civilians killed/ strike that killed civilians	F. % of strikes that killed civilians
2013[2]	N/A (339[3] if prior pattern continues)	from 19 strikes	45 dead (14 injured)	N/A (0.135 if prior pattern continues)	2.4	N/A (5.6% if prior pattern continues)
2012[4]	294[5] (245 to Oct. 31)	from 5 strikes	16 dead (3 injured)	0.054	3.2	1.7%
2011	238	from 1 strike	4 dead (2 injured)[6]	0.017	4	0.4%

Notes: [1] Chris Woods and Alice K Ross, "Revealed: US and Britain launched 1,200 drone strikes in recent wars" *The Bureau of Investigative Journalism* (Dec 4, 2012) available at <http://www.thebureauinvestigates.com/2012/12/04/revealed-us-and-britain-launched-1200-drone-strikes-in-recent-wars/> accessed Sept 2014. [2] United Nations Assistance Mission in Afghanistan, "Afghanistan, Annual Report 2013: Protection of Civilians in Armed Conflict" (2014) at 46, available at <http://unama.unmissions.org/Portals/UNAMA/human%20rights/Feb_8_2014_PoC-report_2013-Full-report-ENG.pdf> accessed Sept 2014. [3] This is an approximation based on extrapolating the rate of growth from the prior two years. [4] United Nations Assistance Mission in Afghanistan, "Afghanistan, Annual Report 2012: Protection of Civilians in Armed Conflict" (2014) at 34, <http://unama.unmissions.org/LinkClick.aspx?fileticket=K0B5RL2XYcU%3D> accessed Sept 2014. [5] The figure 294 is based on extrapolating from the reported 245 to Oct 31, 2012 to cover the last two months of the year. [6] Nick Hopkins, "Afghan civilians killed by RAF drone" *The Guardian* (July 5, 2011) available at <http://www.theguardian.com/uk/2011/jul/05/afghanistan-raf-drone-civilian-deaths> accessed Sept 2014.

At this point Braun and Brunstetter could fall back on their observation that, in contrast with the Pakistani drone strikes, the Afghan strikes are in fact *far* below the 10 percent threshold of civilian casualties, indeed below 1 percent. So even if it is true that less than 10 percent of the Pakistan strikes are known to kill civilians, the Afghan strikes are still significantly lower. A difficulty arises however with this recourse, because there is some evidence that the Afghan percentage has climbed sharply in recent years while the Pakistani percentage has dropped. Indeed, a number of conspicuous facts emerge from an examination of recent data coming out of Afghanistan, including that the total number of civilians killed in Afghanistan in both 2012 and 2013 exceeded the number of civilian deaths in Pakistan (in 2013 possibly by a factor of more than 10—compare Table 7.1 Column C and Table 7.2 Column E). Moreover, the Afghan percentage of civilian casualties exceeded the

Table 7.2 Drone strikes in Pakistan based on NAF dataset

Year	A. Total strikes	B. Strikes killing civilians	C. % of strikes killing civilians	D. Total killed	E. # of civilians killed	F. % of civilians killed
2014	16	0	0	94–118	0	0
2013	27	1	3.7	132–165	3–5	2.68
2012	48	3	6.5	222–361	5–5	1.712
2011	73	7	9.59	392–604	57–65	12.248
2010	122	3	2.46	609–1,027	14 17	1.95
2009	54	7	12.96	352–723	64–74	12.82
2008	36	6	16.66	219–347	20–34	9.54
2007	4	0	0	48–77	0	0
2006	2	2	100	88–100	87–99	98.93
2005	3	2	66.6	14–15	6–6	40
2004	1	1	100	4–9	2–2	28.57
Total	386	32	8.29	2,174–3,546	258–307	10.02

1 percent threshold in 2012 at 1.7 percent, while the accuracy of drone strikes in Pakistan has increased markedly in recent years, so that in 2012, for example, it essentially equaled the Afghan rate at 1.7 percent (see Table 7.1 Column F and Table 7.2 Column F).

Unfortunately, the US Air Force stopped distinguishing data on drone strikes in Afghanistan late in 2012, so that there is no clear data on the number of strikes in 2013, but if we fill in this gap by extrapolating the rate of the growth from prior years, and use it to estimate total strikes, we get some striking percentages for 2013. Indeed, the Pakistani percentage of strikes causing civilian casualties (3.7 percent) appears to have been considerably lower than the Afghan (5.6 percent).

Here Braun and Brunstetter may want to stress that we cannot be sure of all of these numbers, most notably the number of strikes in Afghanistan in 2013 and for the last two months of 2012; also, in Table 7.1 we have integrated data from a couple of different sources, including the United Nations Assistance Mission in Afghanistan (UNAMA) reports on civilians killed by drone strikes in Afghanistan (which almost *tripled* from 2012 to 2013) and data on drone strikes reported by the US Air Force. This concern is quite legitimate, but it points to a defect of using the "collateral damage ratio" rather than a count of civilian casualties, especially in Afghanistan. We really have very little hard data on the number of strikes in Afghanistan. Braun and Brunstetter rely on a report that indicates the strikes causing civilian fatalities remained below 1 percent. Quite apart from the accuracy of that number (it does not appear to hold for 2012, for example), it is simply too vague to sustain much analysis. Moreover, UNAMA 2013 specifically warns that: "The number of civilian casualty incidents from drone strikes may be higher as UNAMA is not always able to confirm which type of platform was used during an aerial operation (fixed-wing, rotary or remotely-controlled) that resulted in civilian casualties."[49] This remark suggests that it is possible that the percentage of strikes that killed civilians in Afghanistan may have been higher than 1 percent all along.

B. Distortions of Leadership Targeting and the Number of Strikes aimed at Low-Level Militants

When discussing leadership targeting, Braun and Brunstetter continue to freely use the terms of "non-militants" and "civilians" to refer to the combination of civilians and unknowns killed. The most striking feature of the following two quotations, for example, is that the same "6 percent" of total drone strikes in Pakistan that are initially described as killing "only civilians" are later acknowledged to actually refer to "civilian/unknown" deaths:

> Of the 343 drone strikes in Pakistan from 2004 to 2012, 300 strikes or 87 percent did not kill leaders. Of these, 238 killed only suspected militants (69 percent of total strikes), 42 killed suspected low-level militants and civilians (12 percent), while 20 killed *only civilians (6 percent)*.[50]

They also add:

> Even assuming some level of civilian casualties is justified when targeting leaders, can one say that CIA drone policy is proportional when *6 percent of total drone strikes cause only civilian/unknown deaths*, while an additional 12 percent targeting low-level militants cause some civilian/unknown deaths?[51]

At the end of the passage quoted above, they also again employ the term "non-militants" to refer to the combination of civilian and unknown deaths: "2 percent of strikes killed leaders, while other militant deaths comprise 76 percent of total deaths, and *non-militant deaths make up 22 percent*."[52] While confusing, eliding the numbers this way does make for some dramatic claims:

1. non-militant deaths make up 22 percent of total deaths.
2. 6 percent of total drone strikes cause only civilian (or non-militant) deaths.
3. 12 percent of strikes targeting low-level militants (LLMs) caused civilian (or non-militant) deaths.

---

49 UN Assistance Mission in Afghanistan (n 50) at 46.
50 Braun and Brunstetter, "Rethinking ... " (n 1) at 314 (our emphasis).
51 Ibid at 314 (our emphasis).
52 Braun and Brunstetter, "Rethinking ... " (n 1) at 314 (our emphasis).

Table 7.3   Type of casualties based on NAF dataset, 2004–2014

Type of casualties	N	%
Total	2,174–3,546	
Leaders	64	2
Other militants	1,684–2,869	78
Unknowns	199–334	9
Civilians killed	258–307	10
Civilians killed alongside other militants	171–194	6.4
Civilians killed alongside unknowns	117–140	4.49
Civilians killed in strikes that killed no militants or unknowns	37–50	1.55

If all three claims hold true in regards to civilian casualties then maybe it does beg the question as to whether the CIA drone policy is proportional. However, after disaggregating all strikes that killed civilians from the unknowns and reexamining the three claims they lose some of their initial force. Once a recalculation is completed,

1. The claim of 22 percent of all fatalities being non-militant deaths from 2004 to 2012 drops to 10 percent civilian casualties (see Table 7.3).[53]
2. The claim that 6 percent total drone strikes cause only non-militant deaths drops to 2.3 percent civilians killed (representing nine strikes out of 386, which together killed an average of 44 civilians or 1.55 percent of total deaths—see Table 7.3).
3. In regards to their final point concerning strikes targeting low-level militants, which also cause some civilian deaths, it drops from 12 percent to 4.4 percent when looking at cases that only involved civilians (not unknowns). Specifically, out of 386 only 17 that hit LLMs also caused civilian fatalities resulting in 183 civilian deaths or 6.4 percent of total fatalities (see Table 7.3).

A related concern about the way that Braun and Brunstetter present their data relates to the proportion of targets they assume are LLMs. The problem is that they only consider strikes to have been directed against leadership targets when a clearly established HVT was killed. On that basis, Braun and Brunstetter suggest that strikes killing leaders "constitute just 13 percent of total strikes" and that "leaders represent only 2 percent of total deaths."[54] Braun and Brunstetter assume that all other militants not identified as a leader must be LLMs, and that these LLMs comprise 76 percent of total casualties (which is the residuum once 22 percent of

---

53 Unknown casualties make up 9 percent, and we were unable to identify the other 3 percent. This is likely to due to updates to the NAF database.

54 Braun and Brunstetter, "Rethinking ... " (n 1) at 312.

civilian/unknowns and 2 percent HVTs are removed). On this basis they "suggest that the vast majority of strikes target low-level militants."[55]

At a glance, this analysis seems plausible, but a few things should be considered when interpreting the data. The fact is that it is difficult to determine who an operational planner was targeting in an operation, especially if the target's death is not reported or cannot be confirmed. The secrecy surrounding the drone program makes it difficult to always determine who is being targeted. We are also not always privy to the intelligence establishing whether some targets are designated as HVTs and why.

Due to these challenges, there is a large disparity in the data of the overall number of HVTs killed. The NAF database provides a list of the 64 Al-Qaeda, Al-Qaeda-affiliated, and Taliban group leaders who have been killed in the CIA drone campaign in Pakistan (2 percent of total casualties in Pakistan from 2004 to 2014), meanwhile the Long War Journal provides a list with 117 names (4 percent of total casualties in Pakistan from 2004 to 2014 according to their numbers).[56] The differences in the numbers reflect that there is currently not even any general criteria on what should be considered a HVT. This problem also applies to those considered LLMs. Is the claim that 76 percent of total casualties (between 1,437 and 2,519 militants) were low-level fighters accurate? The data presented considers any militant not identified as a leader a LLM. Even if we were to accept reports in the media at face value there is an issue in separating militant casualties into two categories (HVTs and LLMs) because by design it overlooks the militants in between both criteria. The data forces us to accept that all those not identified as leaders are foot soldiers by default, which is not always the case.

For instance, the 76 percent reported by Braun and Brunstetter also includes local commanders and senior militants who were not listed as HVTs. In fact, the NAF database does not list Maulvi Liaqat, deputy commander of the banned pro-Taliban group Tehreek Nifaz-e-Shariah-Muhammadi with ties to Ayman al-Zawahiri, as an HVT, meaning he is also part of the LLM total. Although one can argue over whether or not Liaqat was a high-profile target, it is more difficult trying to make a case that he is a LLM.

The proportionality calculus also involves more than just trying to determine the status of targets hit; you must also factor in the strikes that intended to hit a particular target but missed. This is admittedly difficult because of the secrecy of the drone program. Nevertheless, based on the data publically available we are able to at least confirm that drone strikes that miss intended targets do occur. This is important to note because the proportionality calculus is based primarily on

---

55 Ibid at 313–314.

56 For NAF's list of leaders killed and "Senior al Qaeda and Taliban leaders killed in US airstrikes in Pakistan, 2004–2014" see "Drone Wars Pakistan: Leaders Killed" at <http://securitydata.newamerica.net/drones/pakistan/leaders-killed>; for Long War Journals list of leaders killed see <http://www.longwarjournal.org/pakistan-strikes-hvts.php> both accessed Oct 2014.

expectations (i.e. comparing the expectations of operational planners concerning risk to civilians and direct military objectives). In other words, the proportionality calculus is something that occurs before a drone strike takes place and not after. Therefore, in a strike that was intended to hit a target but missed, the intended target is still part of "anticipated gains" and this aspect is ignored in Braun and Brunstetter's figure of 76 percent of strikes only targeting LLMs.

To illustrate this point we will briefly mention two drone operations. On November 7, 2005, news reports emerged that days earlier (November 5) four militants, apparently Arabs, were making a bomb that exploded, blowing up a room of their mud compound, not only killing them but also a woman and three young girls.[57] It was later revealed that the incident was actually a drone attack reportedly targeting Al-Qaeda operational commander Abu Hamza Rabia, who managed to escape with an injured leg.[58] He ended up being killed in another drone strike days later.

In another case, a drone attack on January 13, 2006, killed 8 to 18 civilians (depending on the source). In this strike Ayman al-Zawahri was the intended target, but left the targeted compound with Maulvi Liaqat (who would be killed in the next drone attack) before the missiles struck. Abu Khabab al-Masri was also reportedly targeted (he was falsely reported as killed).[59] Altogether, of the five strikes that killed civilians between 2004 and 2007, three managed to kill HVTs (Nek Muhammad, Abu Hamza Rabia, and Maulvi Liaqat), while the remaining two missed their intended targets (Abu Hamza on November 5, 2005, and Zawahiri and Abu Khabab on January 13, 2006).

The number of failed attacks against a specific target can also mount up quickly when the target proves especially elusive. For instance, Jane Mayer of *The New Yorker* reports that the CIA carried out as many as 16 attempts to kill Baitullah Mehsud with drone strikes.[60] Although there were 16 attempts, only one of these strikes appears as an HVT strike in the Braun and Brunstetter metric while the remaining strikes are overlooked.

Our aim here is not to deny that a large number of casualties are low-level fighters, but rather to show that the situation is a bit more complex when it comes to proportionality and the types of casualties. Only looking at the militant causalities

---

57   "8 foreigners killed in Waziristan Blast," *Dawn* (Nov 7, 2005), available at <http://www.dawn.com/news/164477/8-foreigners-killed-in-waziristan-blast> accessed Oct 2014.

58   "Senior Al Qaeda commander killed," *Dawn* (Dec 3, 2005), available at <http://www.dawn.com/news/168309/> accessed Oct 2014.

59   "Official: Al Qaeda weapons expert likely killed in US airstrike," CNN.com (July 29, 2008) available at <http://edition.cnn.com/2008/WORLD/asiapcf/07/29/pakistan.strike/index.html> accessed Oct 2014; Carlotta Gall and Ismail Khan, "American Strike in January Missed Al Qaeda's No.2 by a Few Hours" *New York Times* (Nov 10, 2006) available at <http://www.nytimes.com/2006/11/10/world/asia/10pakistan.html?ref=abufarajallibbi&pagewanted=all&_r=0> accessed Oct 2014.

60   Jane Mayer, "The Predator War," *The New Yorker* (Oct 26, 2009) available at <http://www.newyorker.com/reporting/2009/10/26/091026fa_fact_mayer> accessed Oct 2014.

in terms of leaders and low-level fighters overlooks the militants who are between both categories. Moreover, the inability to accurately determine the intended targets and their HVT status and to identify strikes that missed their marks also affect the proportionality calculus. The key issue is that the assumption that any attack that did not kill an HVT must have been aimed only at LLMs encourages an underestimate of how much military advantage planners anticipated (and should have anticipated)—especially if the collateral damage ratio is the sole way of assessing proportionality.

C. Conflating Weapons Releases and Strikes

A third difficulty in Braun and Brunstetter's data arises over the way they enumerate strikes. Their examination of Afghan data seems to use number of missile releases as an equivalent of strikes to establish that the 1 percent threshold is not crossed.[61] They then compare this number to the number of "strikes" recorded in Pakistan. Here is a passage where they elide missile launches and strikes:

> According to a report released by the UN in February 2013, of the 294 *weapons released* by US military drones in Afghanistan in 2011, only one caused civilian casualties. In 2012, *weapon releases* rose to 506, five of which killed civilians, resulting in 16 civilian deaths (UNAMA 2013). In both years, less than 1 percent *of strikes* killed civilians.[62]

The key point to note is the shift in terminology from "weapon releases" in the first and second sentences to "strikes" in the third. But it is by no means clear that missile releases are equivalent to strikes (which often involve multiple weapons releases, often from multiple drones, and sometimes in several waves). In consequence, numbers of weapons releases in Afghanistan are not necessarily comparable with numbers of drone strikes recorded in Pakistani data. The result of trying to compare across these two very different types of numbers would likely be to vastly inflate the apparent number of strikes (really weapons releases) in Afghanistan, and correspondingly to artificially diminish the proportion of "strikes" which result in collateral damage.

D. The Inherent Limits of the "Collateral Damage Ratio"

Overall, we do not think that the "collateral damage ratio" (which focuses only on how many strikes did or did not kill civilians or unknowns), which Braun and Brunstetter rely on, is very illuminating on its own,[63] and this is not just a matter of the difficulty of accurately gathering the numbers required (as illustrated above), but also because of the crudeness of the measure itself. While we see some value

---

61 Braun and Brunstetter, "Rethinking … " (n 1) at 310–311.
62 Ibid (our emphasis).
63 Ibid.

in adding this measure to other indicators, we are skeptical of the claim that alone it captures proportionality better than alternatives. The weakness is easily illustrated with a hypothetical case. If, for example, there is only one strike in Afghanistan which killed civilians, but it killed 100 civilians, and there are 10 strikes in Pakistan which killed one civilian each, the collateral damage ratio on which Braun and Brunstetter rely will suggest that the Pakistani collateral damage is 10 times worse (and entirely miss the fact that 10 times more civilian deaths have been caused in Afghanistan).

If this imaginary scenario seems like nitpicking, consider the following facts. Observing the drone campaign in Pakistan, two strikes killed civilians in both 2005 and 2006. In terms of the number of strikes causing collateral damage, both years are equal but there is a big difference in regards to the number of civilians killed. In 2005, six civilians were killed while in 2006, although the same number of strikes caused collateral damage, the total civilian count went up to 93. Yet this difference is entirely missed in the collateral damage ratio. Additionally, at least six strikes killed civilians in 2007, three times more than 2006, but only resulted in 27 civilian causalities (three times fewer casualties than 2006). Yet the collateral damage ratio would be three times higher in 2007, other factors being equal.

It is also troubling that the numbers used in the Afghan collateral damage ratio (i.e. strikes causing civilian casualties) cannot be broken out in increments of time so that variations can be observed, for example, by year. For instance, once you break the Pakistani numbers down by year, it is clear that there has been a very sharp drop in the number of strikes causing civilian casualties (from 39.2 percent for the five years from 2005 to 2009, to 4.4 percent from 2010 to 2014 thus far—see Table 7.2 Column C). And this is obviously over a far larger number of strikes (278 over the last five years, as against 99 in the prior five). At any rate, we present data both on numbers of strikes causing civilian casualties and numbers killed broken out by year in Table 7.2 for a more comprehensive picture. In essence, while Braun and Brunstetter's measure of the number/proportion of strikes in which there is collateral damage gives a summary of cases in which it is possible that proportionality was violated, it does nothing to identify cases in which violations were more likely (where the number or proportion of civilians killed was especially large) or the way this varies over time. That is why their measure is best supplemented by others including, but not limited to, civilian casualty averages over time.

In pointing all of this out, we do not want to suggest that all of the US drone strikes in Pakistan have been proportional. We do not believe that there is sufficient evidence to draw definitive conclusions either on individual strikes or on the campaign as a whole. Indeed, our concern is that Braun and Brunstetter are too quick to draw conclusions based on scanty and at times confused or confusing data. Their approach of comparing drone strikes in Pakistan with those in Afghanistan adds value only in combination with other means of assessing Pakistan strikes (such as the overall numbers of civilian casualties, how the results compare to other weapons, and tactics used in the same location or period) and permits only the most preliminary and provisional conclusions.

IV. The Problem with Applying *Jus ad Vim* to the Pakistan Case

Independent of the problems with the measures that Braun and Brunstetter use and the way that they present the data, we have a much broader concern about the whole attempt to apply *jus ad vim* criteria to US drone strikes in Pakistan. In essence, we do not think *jus ad vim* really applies to the Pakistan case, or at least to the great majority of US drone strikes in Pakistan. The reason is because there actually is a non-international armed conflict in the FATA of Pakistan, and there has been for years, and that when the US uses force (especially with Pakistani consent), then the applicable moral standards are those of conventional JWT (as the legal standards are those of humanitarian law). This claim can be justified in two ways, either as a civil war within Pakistan or as spillover of the non-international armed conflict in Afghanistan. To our minds there was an armed conflict beginning in 2004, as the Pakistani army carried out major operations against militants in the area and the militants struck back, including intense, sustained fighting with considerable casualties (in the White Mountains, for example), the capture of hundreds of militants, negotiated ceasefires between militants and governments breaking down into renewed fighting, and other features characteristic of civil war.[64]

The "Final Report on the Meaning of Armed Conflict in International Law" (2010) of the International Law Association's Committee on the "Use of Force" identified the following general criteria for establishing the existence of an armed conflict triggering the application of international humanitarian law:

> The Committee confirmed that at least two characteristics are found with respect
> to all armed conflict:
> (1) the existence of organized armed groups
> (2) engaged in fighting of some intensity.[65]

There remains controversy over how to operationalize these conditions, but there seems little serious doubt that Al-Qaeda, and the groups that coalesced in 2007 to form the Tehrik-e-Taliban were both organized and armed. There has also clearly been "fighting of some intensity" in the region over the last decade, with over 2,200 security personnel killed, along with over 17,800 militants and over 4,400 civilians (see Table 7.4).

At this point it is difficulty to deny the fact of an armed conflict in the area. However, there is disagreement over when exactly the conflict began, which is exacerbated in turn by the absence of commonly accepted standards for what

---

64   A nice overview of the 2003 to 2006 situation in FATA is provided in the South Asian Terrorism Portal's Pakistan Assessment 2006 at <http://www.satp.org/satporgtp/countries/pakistan/assesment2006.htm> accessed Oct 2014.

65   Use of Force Committee, "Final Report on the Meaning of Armed Conflict in International Law" International Law Association, at 2.

Table 7.4   **South Asian Terrorism Portal reports on fatalities resulting from clashes between militant and security forces in FATA**

Year	Civilians	Security forces	Militants	Total
2013[1]	319	198	1,199	1,716
2012[2]	549	306	2,046	2,901
2011[3]	488	233	2,313	3,034
2010[4]	540	262	4,519	5,321
2009[5]	636	350	4,252	5,238
2008[6]	1,116	242	1,709	3,067
2007[7]	424	243	1,014	1,681
2006[8]	109	144	337	590
2005[9]	92	35	158	285
2004[10]	132	195	300	627
April 2002–April 2003	2	14	18	34

Notes: [1] See <http://www.satp.org/satporgtp/countries/pakistan/Waziristan/index.html> accessed Oct 2014. [2] <http://www.satp.org/satporgtp/countries/pakistan/Waziristan/index.html> accessed Oct 2014. [3] Based on data collected by the Institute for Conflict Management in New Delhi available at <http://www.satp.org/satporgtp/icm/index.html> and published at South Asian Terrorism Portal, "Fatalities in FATA, 2009–2011," <http://www.satp.org/satporgtp/countries/pakistan/Waziristan/index.html> accessed Oct 2014. [4] South Asia Terrorism Portal, *FATA Assessment 2010*, available at <http://www.satp.org/satporgtp/countries/pakistan/Waziristan/index.html> accessed Oct 2014. [5] Ibid. [6] South Asia Terrorism Portal, *Pakistan Assessment 2009*, available at <http://www.satp.org/satporgtp/countries/pakistan/assesment2009.htm> accessed Oct 2014. [7] South Asia Terrorism Portal, *Pakistan Assessment 2008* available at <http://www.satp.org/satporgtp/countries/pakistan/assesment2008.htm> accessed Oct 2014. [8] Ibid. [9] Ibid. [10] South Asian Terrorism Portal, FATA Timeline 2004, available at <http://www.satp.org/satporgtp/countries/pakistan/Waziristan/timeline/2004.htm> accessed Oct 2014.

qualifies as fighting of "some intensity." While acknowledging that there is room for disagreement here, we turn to a widely cited, relatively conservative standard, which is intended to be as uncontroversial as possible—that used in the Correlates of War survey. The Correlates of War criterion for the existence of non-international armed conflict is "1,000 battle-related deaths per year (twelve-month period beginning with the start date of the war) among all the qualified war participants."[66] On this view, the violence in the FATA reached the level of a civil war in 2007 and has remained in this condition since. Even one of the most

---

66   Meredith Reid Sarkees, "The COW Typology of War: Defining and Categorizing Wars" available at <http://www.correlatesofwar.org/COW2%20Data/WarData_NEW/COW%20Website%20-%20Typology%20of%20war.pdf> accessed on Oct 2014.

influential critics of the legality of drone strikes outside of conventional battle zones, Mary-Ellen O'Connell, acknowledges that FATA has been in a state of civil war since 2009:

> In the spring of 2009, Pakistan used major military force on its territory to respond to increasing challenges to central authority. In other words, the nature of the only armed conflict in Pakistan is an internal or non-international armed conflict. In such a conflict, some IHL [international humanitarian law] rules applicable in international armed conflict do not apply, but these are generally rules respecting detention. The core IHL rules respecting targeting are the same in international and non-international armed conflict. These are the rules of distinction, necessity, proportionality, and humanity.[67]

The vast majority of drone strikes take place after 2009 and therefore under the IHL rules.

So we think that there is a strong case that the vast majority of US drone strikes have taken place in a legal context of armed conflict (since 2009 and probably far before, at least as far back as 2007). But our interest here is primarily in identifying appropriate moral standards rather than splitting legal hairs, and seen in this light we think that it is even clearer that situation is one of war. This is especially obvious when one considers that the violence in FATA is intimately intertwined with the civil war in Afghanistan, which since 2003 has also claimed tens of thousands lives.[68] Moreover, both the US and the armed groups against which it is fighting believe themselves to be at war, and use military force against each other regularly as opportunity arises.

So from a moral standpoint (even more importantly than from a legal one) it seems difficult to deny that FATA has been involved in an armed conflict throughout the vast majority (if not all) of the US drone strikes, and that the appropriate moral criteria for assessing the legitimacy of operations are those embedded in the JWT principles for regulating the use of force in armed conflict. Likewise, it seems especially difficult today, in light of the sustained, intense conflict over years and with major Pakistani operations now in North Waziristan, to deny that there continues to be a state of armed conflict in the FATA (at least pending the negotiation of a durable ceasefire). In consequence it seems difficult to argue that the relevant standards for morally evaluating military operations

---

67   Mary Ellen O'Connell, "Unlawful Killing with Combat Drones: A Case Study of Pakistan, 2004–2009" in Simon Bronitt (ed) Shooting to Kill: The Law Governing Lethal Force in Context (Forthcoming; Notre Dame Legal Studies Paper No. 09–43 at 21) available at <http://ssrn.com/abstract=1501144> accessed Oct 2014.

68   See for example, "Afghanistan Civilian Casualties," *The Guardian*, available at <http://www.theguardian.com/news/datablog/2010/aug/10/afghanistan-civilian-casualties-statistics> accessed Oct 2014.

including drone strikes are anything except those prescribed by *jus in bello* under JWT.

V. Questioning the Idea of *Jus ad Vim*

Finally, while we think that there is merit in the goal of formulating rigorous ethical standards for the use of force outside conventional battlefields, we are skeptical of whether this requires the invention of a whole new category of JWT (i.e. *jus ad vim*). As Braun and Brunstetter themselves acknowledge, JWT was not formulated exclusively to regulate what we now characterize as large-scale international conflict, but rather with an eye to all uses of force:

> *Although jus ad bellum is commonly understood to include any use of force*, the distinction between limited force and war that arose in the nineteenth century when lawyers distinguished between formal acts of war and force short of war, such as reprisals and emergency acts in times of necessity, has renewed importance in current armed conflicts.[69]

Their initial claim, that JWT was (and is) often seen as extending to all uses of force finds support among contemporary just war theorists. Alex Bellamy, for example, firmly endorses this view in his important 2006 history of JWT, *Just Wars: From Cicero to Iraq*: "My view is that the [JWT] tradition applies to all forms of political violence whether or not they earn the label of war."[70] Our suggestion is that if JWT traditionally regulated all uses of force, and only ceased to do so in regards to uses of force short of war, like reprisals or extraterritorial operations against non-state actors when these were made illegal in the nineteenth century, then as these uses of force are again recognized as legitimate state actions (at least in certain cases) it makes sense that they should again be subject to traditional JWT. Of course, many JWT principles are sensitive to context, including the principles of necessity, last resort, probability of success, and proportionality, so that the requirements that will apply to a limited use of force in response to a relatively small threat will naturally be elevated. We therefore suggest that the first question that should be asked before we leap to new intermediate categories of JWT like *jus ad vim* is whether the elevated requirements implied by classical principles will be adequate to regulate uses of force short of war as they have in the past. We suggest that if the principles are interpreted rigorously that they can provide a more plausible, appropriate, and far reaching regulative framework than the introduction of a whole new and untried category of *jus ad vim*. For example, one important advantage is that these conventional standards would continue to apply to drone strikes in

---

69 Braun and Brunstetter, "Rethinking ... " (n 1) at 316–7 (our emphasis).

70 Alex Bellamy, *Just Wars: From Cicero to Iraq* (Polity Press 2006) at 5.

Pakistan despite the environment reaching the threshold of armed conflict which renders *jus ad vim* constraints moot. We imagine these principles as looking something like the following:

Jus ad Bellum Criteria:

(1) **Just Cause**: Uses of force short of war must be responding to an initial attack or attempted attack, and the explicit or manifest intention and capability to continue. The just cause may be to end an ongoing harm and/ or to disrupt an ongoing campaign of attacks, but not solely for revenge.

(2) **Last Resort**: Law enforcement and the use of military force by the state on whose territory terrorists are based must be preferred to the use of foreign military force where they may reasonably be expected to realize the just cause.

(3) **Macro-Proportionality**: The force used must be limited to what is required to accomplish the just cause, and must be designed to minimize violations of sovereignty (i.e. seeking consent, or using drones as opposed to boots on the ground, etc.).

(4) **Probability of Success**: The use of force must have some reasonable chance of disrupting or preventing future attacks (i.e. it cannot be purely symbolic or purely retaliatory).

Jus in Bello Criteria:

(1) **Distinction**: Targets must generally be critical to the military functioning of the targeted group, typically HVTs such as leaders or those with skills essential to the conduct of attacks. However, this constraint may be loosened as specific attacks loom and/or are initiated, as any and all members of the military wing of a group may be involved in, and indeed essential to, an operation. At the extreme, as an attack unfolds, the principle of distinction overlaps that which is applicable to conventional armed conflict.

(2) **Proportionality**: Military operations should proceed only with high confidence that no civilians will be killed (and harm to civilian objects will be justified by the military gains). As known attacks become more imminent, however, this requirement must be relaxed to reflect the heightened danger potentially being averted, at the extreme overlapping the conventional *jus in bello* criteria as the attack begins.

Evidently, some of these criteria overlap in whole or in part with Braun and Brunstetter's criteria, but they also differ sharply in some important ways. At the level of the just resort to force, they include something very like Braun and Brunstetter's criteria of "just cause" and "last resort." However, they reformulate right intention and macro-proportionality to say simply that the action(s) taken are clearly designed to fulfill the just cause. In other words, the state invoking the right to use force does not seek to advance sundry purposes unrelated to that cause. So if a citizen is kidnapped, for example, the actions must be designed to bring about their release. However, if the kidnapping is part of an expressly or manifestly continuing campaign of violence against the state and its citizens (say a series of bombings directed against civilians), the state may also frame its just cause as suppressing the ongoing violence either by destroying or incapacitating the source of the threats. Operations designed to obtain this goal would then be permissible provided that they qualified as "last resort" and were consistent with the *in bello* requirements.

Probably the most important divergence at the level of resort to force is that the foregoing standards exclude Braun and Brunstetter's invented requirement that there be no "probability of escalation" in order to resort to force in the first place ("that limited force should not run the risk of increased violence"[71]). But we see this as advantageous. In the first place, virtually all uses of force run a risk of triggering some escalation and hence may be ruled out by this criteria unless they overwhelm and intimidate or incapacitate the opponent—but this type of result tends to be ruled out by Braun and Brunstetter's macro-proportionality requirement (which exhorts a minimal use of force). In the second place, it is extremely difficult to imagine how the degree of risk of escalation can be measured with much confidence since these perceptions are themselves a key element in the strategic interaction between the potentially conflicting parties. In effect, this criterion turns on something that is unknowable to the state acting on *jus ad vim*—that is, how strongly the targeted group will respond. Indeed, the targeted group may not know itself and would likely work very hard to hide it if it did (since knowing its real intentions would be to the state's advantage). Therefore this criterion asks too much of the state acting on *jus ad vim* and invites too much controversy, tending to render the whole framework unwieldy. Moreover, this criterion threatens to make the right of attacked states to respond in self-defense hostage to the very groups that have attacked them, which only have to threaten escalation to block a just use of force in reply. In addition, we suggest that this rule is unnecessary, as there are already many constraints on the use of force in other *jus ad bellum* criteria (such as last resort and macro-proportionality), which will tend to restrain the cycle of escalations.

At the level of the conduct of force (*jus in bello*), the foregoing criteria overlap with Braun and Brunstetter's in focusing on the two main JWT criteria (proportionality and distinction) and in setting higher standards for limited uses of force against limited threats than would be characteristic of conventional battlefields. However, they also recognize that the level of threat arising around a limited use of force scenario may rise to levels comparable with armed conflict and that states must have commensurate flexibility to act in protection of their own citizens as circumstances change. The key difference in the foregoing criteria (from those which Braun and Brunstetter envision under *jus ad vim*) is that as the levels of danger vary, so do the principles regulating the use of force all within

---

71   Braun and Brunstetter, "Rethinking … " (n 1) at 318.

the conventionally established JWT framework (unlike their conception of a sharp break between *jus ad vim* and *jus in bello*). This flexibility has the advantage of not requiring states to shift from a *jus ad vim* to a conventional *jus in bello* approach to address urgent threats or increased violence, and allowing them also to shift back as violence and threat levels diminish. Moreover, on our approach, states would not have to justify every use of force separately on the full range of *jus ad vim* criteria. Once the *ad bellum* criteria are met, the state may use force as necessary to defend itself within the general *ad bellum* conditions and subject to specific *in bello* requirements.

The other main difference from Braun and Brunstetter's *in bello* criteria is that the foregoing criteria do not incorporate human rights criteria (like fear, social disruption, etc.) into the proportionality calculus, but rather set an elevated threshold with reference to the standard *in bello* components (i.e. harm to civilians and civilian objects). One concern that we have about the incorporation of these human rights criteria is how they could be integrated into a proportionality calculus. Obviously, the creation of fear among civilians for example is a by-product of strikes in general, and it is not obvious how it can be quantified in any particular strike, and how it could be weighed against other considerations (e.g. military goals, possible civilian deaths), at least without effectively prohibiting strikes. More importantly, we are not sure that these types of considerations should be factored into the use of force once one leaves the realm of law enforcement. In order to get to the (micro-)proportionality calculation, one must be considering a defensive action in reaction to an attack (or at least an imminent attack), an action designed to prevent further attacks, and the resort to force must be a last resort (lacking other plausible alternatives). At the point that all these conditions are met, the state must have latitude to act even if a foreseeable consequence will be social disruption in the areas in which terrorist attacks are being planned.

Insofar as a state's primary responsibility is the protection of its own citizens, it would seem to be under a moral obligation to take action that promises to disrupt further attacks unless these considerations are counterbalanced by equally compelling considerations—such as endangering civilian *lives*. This is already required in the form of the principle of proportionality. However, whether protecting foreign civilians from fear or social disruption provides an equally weighty consideration is certainly questionable. Moreover, it is vastly more questionable when the moral responsibility for attracting these unfortunate and unintended consequences onto local civilian communities lies not with the state but with terrorist groups waging war on the state and deliberately hiding among civilians.

VI. Conclusion

As the foregoing comments indicate, the JWT regulative framework we propose is on balance less restrictive than that suggested by Braun and Brunstetter under the aegis of *jus ad vim*. We think that our framework will permit states to exercise force more effectively against terrorist groups, and correspondingly to prevent terrorist attacks. It also makes it more likely that states will be willing to adopt these standards. At the same time, there remain important restrictions on how states can use force outside of conventional battlefields (as compared to on them), with a particular emphasis on the protection of civilian lives. We also think it advantageous that the balance we strike occurs within the traditional framework of JWT (i.e. *jus ad bellum* and *jus in bello*), and avoids the introduction of entirely new categories of justified force (i.e. *jus ad vim*). This not only avoids additional clutter and confusion in an area already ripe with controversy, and builds instead on principles that states have already accepted, but also offers consistent standards that apply to drone strikes whether or not a local threshold of armed conflict has been reached (as it has in the FATA).

In sum, we suggest that what we have offered here marks a more plausible and more attractive framework for evaluating the exercise of force outside conventional battlefields. Moreover, while there may well have been some cases where the United States has violated our proposed standards on the use of force, it is not at all clear that it has done so in any general or systematic way. Indeed, when compared to other tactics that have been or could be employed in the area, drone strikes appear to endanger far fewer civilians than the military alternatives, as we have shown elsewhere. While this is only one type of data among several that should be considered in forming a judgment on the ethical justification of US drone strikes (including the collateral damage ratio), it is certainly a pertinent one which does not warrant dismissal as a relativistic fallacy.

Finally, it is encouraging that these different data points, from comparison with alternative tactics and weapons, to the actual numbers and proportions of civilians killed, to the collateral damage ratio (at least once unpacked and clarified) suggest that drone strikes in Pakistan have shown considerably improved precision in recent years. Indeed, drone strikes in Pakistan appear to meet even the elevated proportionality standard (if not necessarily all of the others) of Braun and Brunstetter's *jus ad vim* framework even if it is neither wholly plausible nor technically applicable to these strikes. While plenty of ethical concerns and questions remain, this is surely a positive sign.

Chapter 8

From Just War to Clean War: The Impact of Modern Technology on Military Ethics

Delphine Hayim*

I. Introduction

For centuries, just war theory (JWT) has served as a tool to evaluate the justice of the use of armed force within the international arena. Its principles drive belligerents' behavior and affect how the rules of war are implemented during hostilities. The theory has evolved over time and has integrated changes in warfare, particularly technical innovations. The present-day advances in military technologies, and in particular the growing use of unmanned devices, are profoundly altering the attitude toward war. Around 70 states and some non-states groups are currently acquiring autonomous weapon systems.[1] Governments are investing immense resources in setting up programs aiming at creating more complex devices able to "hunt, identify, authenticate and possibly kill a target" without human intervention.[2] Worldwide spending on uninhabited aerial vehicles is expected to double over the next decade, totaling over 89 billion.[3] The path to robotized wars seems unavoidable, given that technology provides for a significant, if not decisive, military advantage to the side possessing it.

Since the 9/11 attacks, the annihilation of the terrorist plague has become the new just cause for Western democracies for taking up arms. However, states have also committed grave abuses in the name of counterterrorism, which has negatively impacted their reputation. The twentieth century sanctioned human dignity as the supreme value; action that hurts individual rights is thus widely perceived as immoral and illegal. In this context, war needs renewed rationales to be put before national constituencies and public opinion. Technology and robotics offer a unique opportunity to leaders to portray war as an activity that is no longer

* My deepest thanks to Professor A Bianchi for his sound advice and his indispensable support; and to Melanie Wahl for her help, her expertise, and her encouragement.

1 UK Ministry of Defense's Development, Concepts and Doctrine Centre, *Global Strategic Trends out to 2040*, (2013); Office of the US Secretary of Defense, *Unmanned Systems Integrated Roadmap 2013–2038*, (2013).

2 Stew Magnuson, "Robo Soldiers" (2007) The Free Library <http://www.thefreelibrary.com> accessed April 2014.

3 Teal Group Corporation <http://tealgroup.com> accessed May 2014.

barbarous. Novel communication strategies try to demonstrate that armed conflicts can be fought with minimal violence. States have always sought to develop more effective and precise means of warfare capable of reducing casualties and damages, but contemporary systems present remarkable characteristics in terms of sophistication, which has led to a "revolution in military affairs."[4] These trends produce consequences on JWT, which must, once again, face challenges to its relevancy. One cannot deny that the "clean war" illusion is seductive. Cleanliness is indeed "connected with the creation of a discourse of normality and legitimacy, by contrast to dirt which is essentially linked to disorder and chaos."[5] In a changing military environment, understanding to what extent technological advancements shape theoretical thinking, war rhetoric, and law's observance is worthy of scrutiny.

This chapter examines the implications of the concept of clean war for the legal and moral justifications for the use of force. It does so mainly in the light of US policies, considering that America makes an extensive use of unmanned systems and displays "a boundless confidence in the power of science and technology to promote progress."[6] Is clean war supplanting the classical just war tradition, thus imposing new military ethics? Or, does it represent an additional prerequisite for a war to be just? *A priori*, unmanned and distance weaponry is not intrinsically unethical. As the following discussion attests, basic principles of JWT, as well as the laws of armed conflict (LOAC) are useful in assessing the legality and the morality of its use. It is first and foremost the manner in which individuals choose to employ these systems that will determine their conformity with generally recognized standards.

The discussion about the potential shift from a just war paradigm to a clean war pattern takes place at the crossroads between legality and morality. Any attempt to normatively appraise the legitimacy of a war entails both a moral and a legal assessment. The two dimensions significantly influence each other. International law rules, which now comprehensively regulate the use of armed force, are the results of the consolidation of collective aspirations and expectations that are inherent to any judgment about human activities, such as war. Simultaneously, international law orients the mental framework with which we tend to appreciate belligerent decisions and behaviors. The constant dialog between legal standards and the various values underlying philosophical reflections concerning armed violence underpins

---

4   Thomas G Mahnken, "Weapons: The Growth and Spread of the Precision Strike Regime" (2011) 140 Daedalus 45, at 46: "In each case [military revolution], new combat methods arose that displaced previously dominant forms of warfare by shifting the balance between offense and defense, space and time, and fire and maneuver." The expression "revolution in military affairs" was first coined by Charles Dunlap, *Technology and 21st Century Battlefield: Recomplicating Moral Life for the Statesman and Soldier* (Diane Publishing 1999).

5   Elspeth Van Veeren, "Clean War, Invisible War, Liberal War: The Clean and Dirty Politics of Guantanamo" in Hillary Fotitt and Andrew Knapp (eds), *Democracies at War* (Bloomsbury Academic 2013) at 93.

6   Jack M Beard, "Law and War in the Virtual Era" (2009) 103 Am J Intl L 409, at 411. US arsenal comprises 12,000 unmanned military systems and 24 billion USD have been allocated to unmanned vehicles in 2013 according to US Secretary of Defense (n 1) at 3.

the choice to move beyond disciplinary boundaries in this chapter. If the idea of cleanliness is really about to replace the justice criterion, morality and law will have to accommodate this change through a rebalancing of their basic requirements.

The objective is not to examine all pros and cons of new warfare devices, but identifying some of their advantages and flaws will be valuable for our endeavor. The first part provides an overview of JWT and its interrelations with military innovations. It explores how technology and the zero-casualty mentality have brought the concept of clean war into the debate related to the normative evaluation of organized violence. It then demonstrates how political and military circles manipulate the potentialities of unmanned systems to convince world opinion of their positive aspects. Scientific evidence, expertise, and medical terminology have bolstered this persuasion campaign. The second section isolates the effects of the clean war promise on JWT tenets. The ultimate purpose of the chapter is to apprehend if the framework has to be reinterpreted in accordance with this set of new requirement or, if clean war renders the old paradigm obsolete and unsuited to modern conflicts.

II. From Just to Clean War

A. Preliminary remarks

Just war theory was conceived as a set of principles regulating when and how sovereign entities could resort to armed force. The theory developed out of "articulated norms, customs, professional codes, legal precepts, religious and philosophical arrangements that form our judgments on military conduct."[7] Catholic theologians captured its main precepts and systematized them into an coherent doctrine explaining how to wage wars in a sufficiently "Christian manner."[8] The secularization of these principles from the sixteenth century onwards led to their transcription in international law and political philosophy.[9] With the codification of the LOAC and the prohibition of aggressive and punitive military operations, just war has been progressively equated with "just according to the law." This contribution will not, however, take an extra-legalistic stance, since law alone cannot solve every ethical question linked to violence.

The just war tradition shows features of flexibility, which have allowed it to adjust its criteria to fluctuating conditions and diverging interests.[10] For instance,

---

7 Michael Walzer, *Just and Unjust Wars* (3rd edn, Basic Books 2000) at 44.

8 St Augustine first introduced defined criteria for just wars in *The City of God and Christian Doctrine* (WB Eerdmanns 1997) Book XIX. For St Thomas Aquinas, JWT was an answer to whether it is "always a sin to make war": *Summa Theologiae* (Marietti 1962–1963) IIa-IIae, question 10, article 1.

9 Geoffrey Best, *War and Law since* 1945 (Clarendon Press 1949) at 289.

10 Mark Evans, "Moral Theory and the Idea of Just War" in Mark Evans (ed), *Just War Theory a Reappraisal* (Edinburgh University Press 2005) at 1.

the notion of just cause has meant something different depending on the period. Humanitarian interventions and national liberation struggles have only recently emerged as justifications for wars. The notion of competent authority has been also shaped by the rise of non-state actors being capable of acting as formal combatants. Given its adaptive dimension, JWT should be able to accommodate current technological developments.

Despite interminable disputes about the consistency and the operability of JWT, post-Cold War militarism prompted a "renewed and sustained engagement with the tradition."[11] The Operation Desert Storm was presented as the "war to re-establish war," which would free America from the Vietnam legacy.[12] Regardless of the shifts in theories and practices, belligerents persistently attempt to validate their activism with reasons of law and ethics in "response to a deep-seated spiritual need of the human nature to base political actions on just and equitable grounds."[13] The campaigns in Iraq, Kosovo, Libya, which involved an important deployment of unmanned weapon systems, were defended in terms of just war rhetoric doubled with the language of surgical strikes. The fundamental canons of JWT are still considered as guidelines informing decisions as whether it is appropriate to use force in a certain case (*jus ad bellum*), whether a certain act committed during hostilities is acceptable (*jus in bello*), and whether the war is conducted in a way allowing peace once the hostilities are over (*jus post bellum*).[14] Besides identifying good reasons to fight, the theory's main objective is to oppose violence, without promoting an idealistic pacifism. War may sometimes constitute the only solution, but must remain a last resort response to the gravest injustice.

As for explaining the technical side of this work, an unmanned weapon system can be defined as "any remotely-controlled powered [aerial, naval, or ground] vehicle or robot that does not carry a human operator ... and can carry a lethal and non-lethal payload or otherwise can aid the military in some functional way."[15]

---

11   Ibid at 6.

12   Tom Engelhardt, "The Gulf War as Total Television" in Susan Jeffords, Lauran Rabinovitz (eds), *Seeing Through the Media* (Rutgers University Press 1994) at 92.

13   Joachim von Elbe, "The Evolution of the Concept of Just War in International Law" (1939) 33 Am J Int L 665, at 685.

14   Christian Mellon, "Just War: The Catholic Church Updates its Legacy" in Gilles Andréani and Pierre Hassner (eds), *Justifying War? From Humanitarian Intervention to Counterterrorism* (Palgrave MacMillan 2008) at 54.

15   Jai C Galliott, *Unmanned Weapons. The Ethical and Social Implications of Twenty-First Century Military Robotics* (Lambert Academic Publishing 2013) at 8. A weapon system is "any integrated system, usually computerized, required for the control and operation of weapons of a particular kind." *Encyclopedia Britannica* <http://www.britannica.com> accessed May 2014. A weapon is "any device, munition, substance, object or piece of equipment capable of directing an offensive force toward a combatant or a military objective." William Boothby, *Weapons and the Law of Armed Conflict* (OUP 2009) at 34. Three levels of autonomy are distinguished: remotely-piloted systems (human-in-the-loop), which perform some tasks autonomously, but cannot select targets and deliver force

Technical innovation has constantly fed reflection about the ethics of war, even though just war founding fathers could not have foreseen the rise of military robots. Technology and science radically alter the experience and the imagery of war by creating optimistic expectations regarding its morality and legality. Modern devices presumably amount to features of accuracy and efficiency, intended to decrease the risk of casualties and damages. Furthermore, the high-tech possessing party may potentially win the war faster, with a reduced level of force. These capabilities weigh on political calculations concerning military strategies and tactics. War will never be a decent business, but technology could possibly help humanize it, through the improvement of the overall performance on the battlefields. Notwithstanding theses new elements instilling the paradigm, just war rationales have not disappeared from the scene of war's ethics and rhetoric.

B. The Robotics Answer

The Vietnam debacle represented a break in the way American society looked at war. It corresponded to a "loss of trust and unity and the partial dissolution of shared myths and common values."[16] The trauma provoked by pictures of massacres and the awareness of the absurdity of this campaign compelled the US to shift to more restrained forms of violence and to moderate their bellicosity. Since, American leadership has tried to shape its foreign policy according to public demands as well as to develop doctrines and technologies that would deliver less lethality, while maintaining maximum efficacy (see section III of this volume). Major military innovations have come about because of operational problems or policy needs requiring tailored responses. Technology appeared as the answer to the twentieth-century culture of aversion to the imagery of dead bodies. The zero-casualty strategy has become the democratic standard that politicians and militaries must comply with to gain popular support. The fight against terrorism endorses this latter observation: As the justice of the cause has been questioned and the methods widely criticized, wars to come must, at least, spare humanity as much as possible.

War has always been motivated by different objectives: Conquest of territories, control over natural resources, or defense against aggressors. Today, belligerents also fight for marginal political or ideological purposes and minor identity claims that only require partial mobilization. Favoring uninhabited vehicles over human soldiers reflects a promptness to fight, but in the pursuit of causes that are not

---

without human command; automated systems (human-on-the-loop) that can select targets and deliver force following a preprogrammed integrated software, but under the oversight of a human agent that keeps the power to override the system's actions; autonomous systems (out-of-the-loop), which can perform any task, including the use of lethal force, without human input. International Human Rights Clinic, *Losing Humanity: The Case against Killer Robots*, Human Rights Watch and Harvard Law School (Nov 2012) at 2–8.

16   Arnold R Issacs, *Vietnam Shadows: The War, Its Ghosts and Its Legacy* (John Hopkins Univ Press 1997) at 6.

worth dying for. Unmanned devices are perfectly suited for conducting small- or medium-scale hostilities that do not involve the whole nation or the whole military apparatus. The Obama Doctrine expresses this stance by favoring "air power and surgical strikes rather than boots on the ground."[17] Furthermore, dropping bombs from a drone or sending robots to remote areas to track terrorists entails another significant advantage: The protection of the secrecy shrouding operations in states that are not officially at war with the United States. Covert operations, permanent war against "terror," and attacks on safe havens would not have been achievable without the advent of remote weaponry.[18]

Technology and military practices are intimately intertwined. Conceiving means and methods that cause harm and destruction, while minimizing casualty rates by increasing the distance between the parties, has boosted innovation and shaped the conduct of hostilities. Sun Tzu already stated 2,500 years ago that "to subdue the enemy without fighting is the supreme excellence."[19] Spears, arrows, crossbows, canons, aviation, mines, missiles are all devices that moved military staff away from the battlefield. Germany began to test automated systems during World War I, but the first uninhabited aerial vehicles were used in Vietnam as reconnaissance tools. The Operation Desert Storm saw the utilization of a few armed remote-piloted planes and tanks.[20] The push toward autonomy combined with the spread of precision-strike weapons and the improvements of information technologies have produced the ongoing military revolution. Nowadays, unmanned systems represent a common characteristic of warfare, even though there is still a human taking the final decision to shoot. Governments remain cautious when mentioning the future deployment of deadly force by robots and prefer to confine them to defensive functions until technology attains an acceptable level of predictability. Yet, engagement without human intervention is duly considered.[21] Robotized wars may seem to belong to science-fiction scripts, but it could be only a matter of years before it becomes a reality. Independent systems, such as the ship-based Phalanx, the aerial Aegis, the land-based C-RAM, the SSMS (surface-to-surface missile system) ground vehicle, the K-Max helicopter, the Sentry-tech station, and the SRG-1 robot are already operational, but are mostly used in foreseeable contexts

---

17   Ross Douthat, "The Obama Synthesis" *New York Times* (Jan 12, 2013).

18   Asaro, "Military Robotics and Just War Theory" in Gerhard Dabringer (ed), *Ethica Themen: Ethical and Legal Aspects of Unmanned Systems, Interviews* (Institut für Religion and Frieden 2010): Drones provide a means "for … US military presence in Pakistan which would not be possible otherwise, without either the overt consent of the Pakistani government [or] an act of war by the US against Pakistan's sovereignty" at 111.

19   Sun Tzu, *The Art of War* (Wordsworth 1990) at 105.

20   William Boothby, "How Far Will the Law Allow Unmanned Targeting to Go?" in Dan Saxon (ed), *International Humanitarian Law and the Changing Technology of War* (Martinus Nijhoff Publishers 2013) at 46–47.

21   US Secretary of Defense (n 1) at 66 § 4.6 and UK Ministry of Defense, *Joint Doctrine Note 2/11: the UK Approach to Unmanned Aircraft Systems* (2011).

or outside densely populated areas and limited to protective functions or offensive ones against clearly identifiable military objects.

The issue of artificial intelligence lies beyond the scope of this contribution, but intelligent weapon systems are expected to be soon capable of learning, adapting to changing circumstances, and observing the LOAC.[22] In about 10 years it is envisaged that robots will be able to freely decide when and how to apply lethal force.[23] While the world should get ready to cohabit with cyborgs, soldiers should prepare for wars against faceless enemies.[24] The next section will consider some of the rhetorical techniques used by decision-makers for persuading their audience that unmanned systems will not dehumanize wars. Quite to the contrary, they are transformed into virtuous agents.

C. The Bloodless War Illusion

With advanced technology, war tends to look more like a surgical scalpel than a "bloodstained sword."[25] Middle Ages thinkers already presented war as a poisonous remedy against diseases, whose salutary power was solely attributed to just interventions.[26] Medical jargon helps depict the choice of armed force as a reasonable and necessary course of action. Present-day bombing campaigns are called surgical strikes, giving the impression that they are acts of healing and not massive enterprises of destruction.[27] War is the appropriate treatment against the evil enemy. These enemies, Hussein, Milošević, Gadhafi, Bin Laden, Abu Bakr al-Baghdadi, are subject to demonization and portrayed as inferior beings not deserving to live.[28] The dirty language used to qualify the opponents contrasts with

---

22 Chantal Grut, "The Challenge of Autonomous Lethal Robotics to International Humanitarian Law" (2013) 18 JCSL 5, at 6. About intelligent robots, John McGinnis, "Accelerating AI" (2010) 104 Nw U L Rev 1253.

23 Noel Sharkey, "Robot wars are a reality" *The Guardian* (Aug 18, 2007); US Air Force, *Global Horizons. Final Report, United States Air Force Global Science and Technology Vision*, Air Force Chief Scientists Report AF/ST TR 13–01 (2013).

24 Peter Asaro, "How Just Could a Robot War Be?" in Adam Biggle, Katinka Waelbers, Philip AE Brey (eds), *Current Issues in Computing and Philosophy* (IOS Press 2008) at 50; Ronald C Arkin, *Governing Lethal Behavior in Autonomous Robots* (CRC Press 2009) at 173; Peter S Singer, *Wired for War: The Robotics Revolution and 21st Century Conflict* (Penguin Books 2009).

25 Michael Ignatieff, *Virtual War: Kosovo and Beyond* (Vintage 2000) at 215.

26 Johannes de Legnagno, *Tractatus de Bello, de Represaliis et de Duello*, Erskine Holland (ed), (first published 1360, Classicus of International Law 1917) at 218–224.

27 Donald Rumsfeld, Secretary of Defense with Joint Chiefs of Staff General Myers, *Briefing on Enduring Freedom* (October 7, 2001) and "President Obama Lays Out ISIS Strategy," Washington, NBC.com (Sept 10, 2014) <http://www.nbcnews.com> accessed September 2014; both compare terrorism to a cancerous tumor.

28 Albert Bandura, "Moral Disengagement in the Perpetration of Inhumanities" (1999) 3:3 Person & Soc Psycho R 193, at 200.

the antiseptic dimension of contemporary conflicts. The rhetoric of cleanliness leads to the emergence of a set of "beliefs, knowledge and associated practices" supporting the standardization of violence, which allows leaders to keep their hands clean, by denying the muddy aspects of the wars they are waging.[29] Any use of force requires the construction of a consensus about its commanding necessity. Sanitizing terminology participates in building agreement, by "modeling perceptions, affecting emotions and structuring cognition."[30] The acceptability of any military project will be reviewed in the light of these reshaped expectations and collective standards of judgment: "The moral reality of war is not fixed by the actual atrocities committed by soldiers but by the opinions of mankind."[31]

Through technical expertise and discursive strategies, bloodless wars become possible. According to specialists, basing their claims on statistics and compelling evidence, unmanned systems permit states to triumph more decisively than ever before, without excessive sacrifices. The speediness and endurance of automated devices, as well as their ability to access remote areas, strengthen the case that they perform better than humans. Robots can accomplish dull, dirty, dangerous, and deep jobs, such as prolonged surveillance, reconnaissance missions, demining, borders' control, operations in complex urban, contaminated, or deep-water zones.[32] The potential capture of human staff, which usually deteriorates the morale of the troops and erodes domestic support, is not a pressing issue anymore.

The cost-saving dimension of the new generation of weapons is also frequently put forward. On the one hand, they decrease the overall rate of casualties and on the other hand, they reduce the need for human input in daily operations and the expenses associated with it, like training, salary, health care, and pensions. This latter argument is welcome in a time of economic downturn, but tends to underestimate the procurement and maintenance costs of these devices.[33] The strongest claim in favor of clean war is definitely the unprecedented level of targeting accuracy of remote-piloted systems. Endowed with improved sensors and capacities, they can be very precise. Therefore, less lethal force is needed to achieve the war's objectives. Greater risks can also be taken to verify the military nature of the target, since machines can be easily self-sacrificed. Ultimately, the computerization of the assessment process of an attack should provide for better estimation of expected results and chances of success. In sum, fewer human losses and material damages, less money spent, increased prospects of fast victory. Viewed this way, the use of unmanned systems

---

29   Van Veeren (n 5) at 90; Michael Walzer, "Political Action: The Problem of Dirty Hands" (1973) 2:2 Philo and Public Affairs 160, at 173.

30   Richard Jackson, *Writing the War on Terrorism. Language, Politics and Counterterrorism* (Manchester Univ Press 2005) at 21–22.

31   Walzer, *Just and Unjust Wars* (n 7) at 15.

32   Gary Marchant et al, "International Governance of Autonomous Military Robots" (2011) XII Col Sc and Tech R 272, at 275.

33   Stockholm International Peace Research Institute, *Trends in the World Military Expenditures–2013 Fact Sheet* (SIPRI, 2014) <http://www.sipri.org/> accessed May 2014.

is almost morally compulsory.[34] Seduced by this technological hopeware, public opinion is willing to accept the humanizing dimension of these devices and forget their primary function, which is killing people.[35]

Technology is supposedly able to mitigate the probability of human failures and unpredictable behaviors that could hurt innocent individuals. This explains why the terms "target" and "collateral damage" replace stained words such as death, injury, and devastation. This belongs to the plan aiming to sterilize the horrors of war with neutral rhetoric.[36] Mistakes causing civilian casualties and destructions are either perceived as deliberate acts or presented as accidents due to some seldom-technical defaults.[37] Precision-guided weapons have raised an unrealistic optimism about their level of performance—every attack must hit the target. Outraged voices are heard when it is not the case.[38] It should be recalled that incidental damages are conditionally permissible during hostilities and that no weapon is completely flawless.[39] Nonetheless, military planners have no other choice than to wage wars as humanely as possible. It remains that harmless people will continue to suffer, despite the emphasis put on the extreme care taken to minimize their pains. By framing innocent deaths as technical errors, it is easier to elude questions as to whether the attack was proportionate and directed at a lawful objective and whether the precautionary measures were sufficient.[40] Any honest viewpoint would recognize that uninhabited systems are first "effective tools that protect the homeland by enabling targeted killings of terrorists with minimal downsides of collateral impact."[41]

This campaign of persuasion has been successful in rendering wars less visible in the Western world. This representation has also imbued minds through the muzzling of the media urged to construct images of combats looking like video games.[42] While society does not face the horrors of hostilities as it did before, and

---

34   Bradley Jay Strawser, "The Moral Landscape of Unmanned Weapons" in Bradley Jay Strawser (ed), *Killing by Remote Control. The Ethics of Unmanned Military* (OUP 2013) at 17.

35   The term "hopeware" is used by Noel E Sharkey, "The Evitability of Autonomous Robot Warfare" (2012) 94 Intl R of Red Cross 787, at 796.

36   Jackson (n 30) at 132.

37   Patricia Owens, "Accidents Don't Just Happen: The Liberal Politics of High-Technology Humanitarian War" (2003) 32 J Intl Studies 595, at 604–606.

38   Editorial, "How Precise is Our Bombing" *New York Times* (March 31, 2003).

39   Walzer, *Just and Unjust Wars* (n 7): "the killings of soldiers and nearby civilians are to be defended only insofar as they are the product of a single intention, directed at the first and not the second" at 153–156.

40   Nicholas J Wheeler, "Dying for Enduring Freedom: Accepting Responsibility for Civilian Casualties in the War against Terrorism" (2002) 16 Intl Rel 205, at 212.

41   Peter S Singer, "Do Drones Undermine Democracy" *New York Times* (Jan 21, 2012)   <http://www.nytimes.com/2003/03/31/opinion/how-precise-is-our-bombing.html> accessed Nov 2014.

42   Eric Louw, "The War Against Terrorism: A Public Relations Challenge for the Pentagon" (2003) 65 Gazette 211, at 220–226.

appears more disinvested from war's affairs, public scrutiny, which traditionally acts as a limitation on military activism and a strong check on executive's action, appears to be vanishing. When the bloody effects of violence are not immediately available, citizens tend to loosen their tolerance threshold and forget that people are currently dying on battlefields.

The clean rhetoric played a key function in defending the 1991 Gulf War, presented as the most discriminate war ever fought, even though unmanned systems represented only 8 percent of the total ordinance.[43] Unfortunately, casualty rates were much higher than anticipated. Kosovo, Iraq-2003, Afghanistan, and Libya are archetypes of the bloodless warfare phenomenon, in which healing terminology produced a "specific understanding of war."[44] It helped picturing the use of force as a coherent decision implying minimal risks. Technology has the power to "play to American fantasies," which disregard the brutality of war. Armed conflicts become entertainment to be watched like sport games.[45] When wars resemble action movies more than calamities, will public outrage continue to restrain them? If some states possess the technology to go on war with impunity, will they still be reluctant to fight? The discursive construction of new types of war, waged against new enemies posing new threats requiring new means and methods to be defeated, tries to ground the justice of taking up arms by erasing the dirty consequences of hostilities. Cleanliness generates significant impacts on the legal and ethical boundaries of just wars. The next section tries to grasp how the promise of clean conflicts redefines the comprehension of wartime justice.

III. Technology and Just War

The justice of a war is appraised "first, with reference to the reasons states have for fighting; secondly, with reference to the means they adopt."[46] JWT includes a third tenet, albeit a bit underestimated until recently: The extent to which post-conflict conditions guarantee the fulfillment of the main war's objective, namely the restoration of peace. To determine if new warfare devices have remodeled the burden of justification for deploying armed force, the following questions must find answers: Is the decision to use unmanned systems morally and legally defendable? And, could cleanliness excuse some departures from JWT standards and the LOAC? This part explores the ethics of contemporary wars and the issue of law's observance, with respect to *jus ad bellum*, *jus in bello*, and *jus post bellum*, even if the most important challenges of robotics systems seem to concern *jus in bello*.

---

43   US Department of Defense, *Conduct of the Persian Gulf Conflict. Interim Report to the Congress* (1991) at 12–2/3.

44   Ignatieff (n 25) at 161.

45   Colin McInnes, *Spectator-Sport War: The West and Contemporary Conflict* (Lynne Rienner 2002) at 27.

46   Walzer, *Just and Unjust Wars* (n 7) at 21.

Although they are not an exact copy of JWT, the laws regulating the use of force are infused with its philosophical underpinnings. Just war provides "a model of interpretation in the application of international law."[47] Hugo Grotius first organized the doctrine into a legal code by translating its principles into binding rules.[48] With the quasi-complete codification of its laws and customs, war became a comprehensive legal institution. Because of the intermeshing between JWT and international law, clean wars must be evaluated with regards to both corpuses.

A. Jus ad Bellum: *Reconsidering the Threshold of War*

Human beings are paradoxical. They value peace, but constantly think about ways to make wars more palatable. Several commentators assert that the advent of remotely piloted systems exacerbate aggressive tendencies by lowering the barriers to entry into war. The decision to launch a military campaign might be easier if a limited number of soldiers are sent to the fields and if public scrutiny is weaker.[49] The concrete consequences of this greater warlike propensity remain unclear. It can be argued that more unjust wars or operations based on a dubious cause will break out. This can be evidenced by the distortion of the notion of immediacy in counterterrorism activities.[50] On the contrary, presumably moral wars, such as those aiming at protecting civilians against genocide or political repression, could receive the support of foreign states, even in the absence of a major threat to their interests.[51] The Kosovo and Libya campaigns would not have been undertaken without the deployment of unmanned vehicles. Whatever the disagreements, high-tech weaponry bears on leaders' willingness to engage in war. Since it is not necessary to commit the nation to colossal economic investments and human sacrifices, war can be presented as a quasi-banal policy measure.

The traditional global involvement of all the sectors of a society obviously requires a compelling just cause to go to war. The current military revolution opens up the possibility to fight with impunity and, thus, fosters misperceptions about what is realistically achievable on the ground. The lures of riskless warfare and the chimera of distance may feed decision-makers with overconfidence about the true stakes of an operation.[52] The human revulsion toward killing and getting killed and

---

47   Michael N Schmitt, "The Confluence of Law and Morality: Thoughts on Just War" (1992) 31 USAF Mil L & L War Rev 91, at 103.

48   Hugo Grotius, *De Jure Belli ac Pacis* (Louis R Loomis tr, Walter J Black 1949).

49   Ignatieff (n 25) at 179–180; Robert Sparrow, "Killer Robots" (2007) 24 J Applied Phil 62, at 64–65; Singer, *Wired for War* (n 24) at 316.

50   See Chapter 6 by Steven J Barela in this volume.

51   Zack Beauchamp, Julian Savulescu, "Robot Guardians: Tele-operated Combat Vehicles in Humanitarian Military Intervention" in Bradley Jay Strawser (ed), *Killing by Remote Control: The Ethics of Unmanned Military* (OUP 2013) at 106.

52   Mary L Cumming, "Automation and Accountability in Decision Support System Interface Design" (2006) 32:1 J of Techno Studies 23, at 26.

the fear of losing loved ones have always curbed violence, but, with the physical and psychological detachment of citizens and soldiers from hostilities, these concerns become less central. The concept of bloodless war is powerful enough to relax the just cause criterion. A state eager to fight is responsible for proving that the justice of its cause surpasses the downsides that hostilities would bring; the expected low rate of damages puts a significant recalibration on this appreciation.

An imprudent or ill-founded campaign can be turned into a sensible and acceptable one, if there is little danger for the initiating state. An unjust cause can appear worth the risk if no dead bodies invade the public space, if the economy does not overly suffer, and if the battle is promised to be a fast success. It is not to say that the requirement for a just cause becomes irrelevant, but clean war stretches out its content. As mentioned above, the definition of a just cause has changed over history. Theologians considered punitive campaigns and holy wars as valid reasons for fighting. With the United Nations Charter, self-defense and actions under the authorization of the Security Council seems to be the only uncontested ones. Notwithstanding this variability, any recourse to armed force looks for defenses. The advent of autonomous systems does not replace, but rather remodels this prerequisite.

Another pillar of JWT is the one of right intention: Proper and commendable motives must guide just combatants, who should exclusively fight in the pursuit of their cause. Ultimately, the use of force is only defendable when it aims "at advancing the good [peace] and avoiding the devil."[53] Remote-piloted devices make covert operations easier to carry out. This lack of transparency may hide the true intentions of the belligerent, especially when security and intelligence agencies act in parallel with the military. Besides the difficulty of decrypting inner intents and motivations, the secrecy surrounding drones attacks, the selective media coverage, and the public disengagement—think about the US operations in Yemen and other safe havens—help in concealing the war's real objectives.

JWT then provides that the deployed violence has to be proportionate to the threat it is meant to overcome and the values it seeks to preserve. The good projected to be achieved should be greater that the harm expected to be done. According to the principle of proportionality, it is unjust to wage a massive war to remedy a small wrong-doing. The introduction of accurate unmanned systems can skew the calculation, because one side could supposedly win a fast and cheap war without causing many damages. This nice picture does not however correspond to the facts as they have been reported: Irrespective of the controversies concerning the exact figures, drones attacks have already killed innocents and led to significant destructions.[54]

---

53   St Thomas quoted by von Elbe (n 13) at 669.

54   The Bureau of Investigative Journalism, *Covert War on Terror the Data* (BIJ, May 8, 2012); International Human Rights and Conflict Resolution Clinic and Global Justice Clinic, *Living under Drones: Death, Injury and Trauma to Civilians from US Drone Practices in Pakistan*, Stanford Law School and NYU School of Law (Sept 2012); *The*

Just war thinking starts with this idea of proportionality as a balancing exercise between the ends and the means.[55] Yet, the ends can be easily exaggerated or falsified. The magnification of the threat and the enemy's demonization may prompt overly harsh responses.[56] Another argument evidences the interdependence between technology and proportionality. When the discrepancy in terms of military capabilities is big, the "evil" that the high-tech side hopes to exterminate may be insignificant compared to the capacities used against it. As long as national troops are safe, any method can be resorted to.[57]

JWT asks the belligerent, before entering into war, to evaluate the probability of attaining its objectives. The reasonable chance of success standard implies that launching a war with the knowledge that violence is likely to be futile is unjust. Some campaigns—operations in densely populated areas or those presenting a chemical or biological risk—would not be endorsed without unmanned systems, since they would represent an unacceptable danger to soldiers or entail very high costs in comparison with their potential benefits. The deployment of robots could be the best and unique way to defeat the adversary in hazardous environments. Consequently, the prospect of success is enhanced. The operations in Kosovo and Libya demonstrate that goals can be fully achieved through the use of drones; but, the rationales behind these interventions were mostly of humanitarian nature, and not inspired by key national interests. Could we affirm the same about Afghanistan, Iraq, and the global fight against terrorism? Faceless technology can make lasting peace more difficult to establish: By creating inequality, it nourishes resentments among targeted populations. The US strategy of surgical strikes aimed at cutting off Al-Qaeda leadership can look like a victory given the absence of attacks on its soil. Yet, it is probably too early to announce the complete annihilation of the menace, while the rise of ISIS (Islamic State of Iraq and Syria) can be said to reveal the relative inefficiency of such counterterror measures.

Technology is also relevant for the concept of right authority. The just decision to launch a war falls upon a legitimate or lawful authority. Given its varying content, the expression has to be contextualized. It can be generally defined as an authority that is exercised in a fashion that fosters justice.[58] Today, actors other than sovereign states have the ability to act as formal belligerents. The concept of authority also differs from one country to another, according to the existing

---

Long War Journal <http://www.longwarjournal.org>, the Drones Wars UK <*dronewars. net*> and the New America Foundation's data <http://natsec.newamerica.net> all accessed May 2014; see also Chapter 7 by Avery Plaw.

55   Walzer, *Just and Unjust Wars* (n 7) at 129.

56   Kateri Carmola, "The Concept of Proportionality: Old Questions and New Ambiguities" in Mark Evans (ed), *Just War Theory a Reappraisal* (Edinburgh University Press 2005) at 99.

57   Jai C Galliott, "Uninhabited Aerial Vehicles and the Asymmetry Objection: A Response to Strawser" (2012) 11:1 J of Milit Ethics 58, at 63.

58   Schmitt (n 47) at 97.

political system. In a democracy, it includes, besides formal and legal powers, a minimum threshold of public deliberation. As explained, clean war rhetoric has been effective in disengaging societies from military affairs. The limitation of public participation in the debate endows governments with more leeway in relation to the use of force. To eradicate terrorism, the US Congress granted the government with quasi-discretionary war powers, even though it is not the first time in American history that the executive, with or without the Congress' approval, possesses the right to choose the country's destiny.[59] Barack Obama argued that the legislature did not have to authorize the operation in Libya, because it was not involving "sustained fighting or active exchanges of fire with hostile force" or "the presence of US ground troops, US casualties or a serious threat thereof."[60] It follows that the democratic understanding of right authority is partly devoid of its substance, letting the President and its close guard free to determine when and how military force can be deployed and if the level of violence is such that it requires the nation's approbation.

Decision-makers duly take into account *jus ad bellum* standards of JWT. Autonomous systems generate a supplementary requirement for a war to be just: It must be clean, in terms of material costs, human lives, and representation. Nonetheless, the reinterpretation of the tenets may lead to more cavalier approaches to *jus ad bellum*, as exemplified by the doctrine of preventive self-defense. With clean effects, war is much cheaper to sell, and this apparent easiness to go to war challenges the post-World War II presumption of *jus contra bellum*.

B. Jus in Bello: *Reinterpreting the Laws of War*

Just war theorists consider belligerency as a rule-governed and virtue-driven activity.[61] *Jus in bello* constitutes an equilibrium between the principles of humanity and military necessity.[62] It offers protections to individuals affected by the effects of war and regulates combatants' behaviors with regards to the means and the methods they employ. Modern technology conditions the respect for moral principles and the enforcement of legal rules during hostilities, though the problem of distance weaponry is not a novel one. For instance, Pope Innocent II prohibited crossbows in 1139, because of their inherent inhumanity. Still, remoteness is at times also conceived as a tool to limit war's passions and reckless actions. The distance separating the operator from the battlefield eliminates some of the

---

59   For instance, the 1941 War Powers Act, the 1973 War Powers Resolution and the 2001 Authorization for the Use of Military Force.

60   White House, "United States Activities in Libya" *Washington Post* (June 17, 2011).

61   Walzer, *Just and Unjust Wars* (n 7) at 36–41.

62   *Instructions for the Government of Armies of the United States in the Field (Lieber Code)*, General Order N° 100 (April 24, 1863): "Art. 14: Military necessity (...) consists in the necessity of those measures which are indispensable for securing the ends of the war, and which are lawful according to the modern law and usages of war."

psychological and physical factors—fear, fatigue, senses' disorder, stress, hatred, contact with the enemy—that can bear on the propensity of soldiers to transgress their obligations.[63] Accordingly, automated devices could potentially reinforce the compliance with the laws of war.

The impact of the "impersonalization of the battle" is a heated debate.[64] Human emotions and reactions could restrain aggressive attitudes, due to sentiments of empathy, pity, and mercy felt by combatants. The buffer zone may simplify the decision to shoot, because of the lack of awareness concerning the nature of the target.[65] Inversely, the safety of the operator might contribute to their ability to make sound decisions.[66] Whatever the environment, our natural inhibition not to kill each other has never prevented us from committing atrocities. These controversies explain why new weapon systems must be evaluated by comparison with average human performance and not in the light of standards of perfectly ethical fighting, although our faith in science generates reluctance to forgive machines that commit mistakes.[67] This section attempts to establish if technological superiority favors or impedes the enforcement of the LOAC.

No one knows for sure what will be feasible in the future, considering the rapid evolution of military robotics and the secrecy of related research. Yet, the main ambition of technology is to improve human capabilities. Although it is claimed that robots will never be able to apply the LOAC, new devices offer indisputable advantages in this regard.[68] Apart from reducing the need for military staff and limiting incidental damages, high-tech systems can also gather detailed statistics and data that are crucial for military planners. Computerized programs and complex software are now able to collect, analyze, and disseminate information with very high efficiency and speediness. Precise figures might enhance the transparency of operations. States can use information to rebut accusations of misconduct. The constant surveillance over soldiers and commanders' behaviors can in turn help in identifying illegal acts and orders and thus facilitate prosecutions and sanctions. Finally, the available information permits greater care from military commanders with respect to their obligation "to take all feasible measures within their power to prevent, suppress or repress breaches of the Conventions and the Protocol"

---

63 Arkin (n 24) at 29–30.

64 John Keegan, *The Face of Battle* (Penguin Books 1977) at 320–322.

65 Hannah Arendt, "Introduction" in Glenn J Gray, *The Warriors, Reflections on Men in Battle* (Harper and Row 1967).

66 Office of the Surgeon General, US Army, *Final Report: Mental Health Advisory Team IV* (Nov 17, 2006) showing the negative impact of human psychological factors on LOAC's observance during Operation Iraqi Freedom.

67 Markus Schulzke, "Robots as Weapons of Just War" (2011) 24 Philo Technol 293, at 296.

68 Armin Krishnan, *Killer Robots* (Ashgate 2009) at 99; Robert Sparrow, "Robotic Weapons and The Future of War" in Paolo Tripodi, Jessica Wolfendale (eds), *New Wars and New Soldiers: Military Ethics in the Contemporary World* (Ashgate 2011) at 121.

committed by the armed forces.[69] All these valuable aspects may suggest that the fog of war will be soon a relic of the past. Nonetheless, the subsequent evaluation of the impact of autonomous devices on *jus in bello* core principles could contradict this assumption.

Article 35 of Additional Protocol I bans inherently inhumane means and methods and those that cause unnecessary or superfluous sufferings and severe damages to the environment. The principle of humanity, expressed in the Martens Clause, is a basic pillar of the LOAC and "has proved to be an effective means of addressing the rapid evolution of military technology."[70] However, uninhabited systems are not inhumane as such. Therefore, their legality has to be considered with regards to other LOAC standards. Article 36 of API sets out the obligation for the parties to determine whether the employment of a new mean or method is in conformity with the Protocol's prescriptions. The commentary goes on by saying: "[t]he use of long distance, remote control weapons, or weapons connected to sensors positioned in the field, leads to the automation of the battlefield in which the soldier plays an increasingly less important role. The counter-measures developed as a result of this evolution ... exacerbate the indiscriminate character of combat ... if man does not master technology but allows it to master him, he will be destroyed by technology."[71] This control check over warfare technologies is thus an imperative and a constant duty.

The principle of distinction states that there are two kinds of people involved in wars—combatants and non-combatants—and two sorts of things—military objectives and civilian objects. Individuals not directly taking part in hostilities and objects not contributing to the war effort are immune from attacks.[72] As international jurist Emmerich de Vattel expounded in the eighteenth century, "for whatever causes a country is ravaged, we ought to spare those edifices which do honour to human society, and do not contribute to increase the enemy's strength ... What advantage is obtained by destroying them? It is declaring one's self an enemy to mankind."[73] The same applies for non-combatants. The principle of discrimination stems from this rule. It entails the prohibition of deliberately targeting protected categories and the obligation to take reasonable steps to avoid civilian casualties and destructions.[74] The LOAC logically outlaw means and

---

69   Article 86 (failure to act) and Article 87 (duty of commanders) of Additional Protocol I to the 1949 Geneva Conventions Relating to the Protection of Victims of International Armed Conflicts (signed June 8, 1977, entered into force Dec 7, 1978) 1125 UNTS 3 (hereinafter AP I).

70   International Court of Justice, Advisory Opinion, *Legality of the threat or use of nuclear weapons* (July 8, 1996) at para 78.

71   ICRC Commentary of Article 36 of AP I (1987) <http://www.icrc.org> accessed May 2014.

72   Article 48 of AP I.

73   Emer de Vattel, *The Law of Nations or the Principles of Natural Law in Four Books* (Joseph Chitty tr, T. & J. W. Johnson 1758), Book III, Chapter 9, §168, at 361.

74   Articles 51 to 56 of AP I.

methods that are intrinsically indiscriminate or used in an indiscriminate way.[75] It is never admissible to intentionally kill innocent civilians, but this does not imply that it is always acceptable to kill enemy combatants. Only actual and immediate threats can be purposely targeted, which excludes wounded, surrendering, and disarmed soldiers and imposes the duty to capture the enemy if possible. Determining the role of an individual and the function of an object corresponds to a difficult exercise, especially in guerilla-type conflicts. New systems allow more precision in verifying the nature and the status of a target by getting closer to it or spending hours observing its life patterns. The amount of information provided by multiple sources, coupled with advanced surveillance systems, can indeed substantially improve the faculty to distinguish and to discriminate.

Both the LOAC and JWT set out the proportionality imperative, which proscribes any violence that is unnecessary to achieve the war's goals.[76] The incidental harm done to civilian individuals and objects must not be excessive in relation to the anticipated and direct military advantage that an attack would bring. This process of comparing human life with military benefits is riddled with subjective and context-dependent judgments. Proportionality should nevertheless guide all stages of the attack, from its planning to its launching, and regulate troops' behaviors. Technology is capable of maximizing this proportionality ratio: When a situation presents a real danger, the temptation to use more force to save lives is greater. With unmanned systems, fewer soldiers are fighting and, as a result, a lower degree of force is needed to protect them and achieve the same goals. Two other advantages flow from new devices in relation to proportionality: Their hitting-accuracy and the complex programs able to evaluate expected collateral damages through multiple-variables testing.

According to the LOAC, civilian deaths and damages are thus conditionally permissible. They are justified by the projected military value of an action. This assessment has to be made before the attack, on the basis of the principle of precaution. The principle first stipulates the obligation to verify, during the preparation phase, if the target is a military objective. Then, those who decide upon the attack shall estimate its proportionality and ensure that the means and methods chosen minimize the risks to civilians. Military commanders are under the requirement to refrain from launching any operation that would cause excessive collateral damages and to cancel or suspend it when appearing to be in breach of proportionality.[77] They must take all "feasible" measures to respect these prerequisites. The meaning of feasibility largely depends on available technology. Situations have to be gauged in reference to the information that is reasonably accessible in given circumstances.[78] If advanced capacities provide data that could

---

75   Article 51§4 of AP I.

76   Articles 51§5(b) and 57§2(a) iii and iv of AP I.

77   Articles 57 and 58 of AP I.

78   Feasible is understood as what "is practicable or practically possible, taking into account all circumstances ruling at the time, including humanitarian and military considerations." *UK Reservation to API* <http://www.icrc.org> accessed May 2014. Or,

contribute to reducing mischief, this might suggest that the high-tech side has no other option but to use them.[79] The obligation of sparing innocents puts one party under a stricter constraint of precaution.[80] Yet, automated systems are neither perfect—errors, bugs, communication failures, and cyber-attacks can all occur—nor can they entirely dissipate the chaotic unpredictability of battlegrounds. The issue of distance calls into question the real level of distinction and discrimination allowed by remote-guided devices. Criticism has also been voiced toward the accuracy of the intelligence on which the targeting is premised, given the documented civilian casualties caused by wrongly directed unmanned attacks.[81] Expectations cannot be too high: The strong party endures some pressure, but cannot be bound to fulfill unachievable obligations.

New military tools participate in redefining the obligations related to the targeting process as well as the reasoning behind war's decisions. Proportionality calculation usually integrates every life involved, soldiers and civilians. Belligerents employ unmanned systems primarily to protect their own troops. A new logic is thus emerging from technological possibilities and "zero-death" mentality, as the balance irremediably goes toward one's own soldiers. The risk of harm dramatically shifts from national combatants to enemy soldiers and civilians.[82] This "risk-transfer militarism" generates a form of inequality in the way life is valued.[83] Ultimately, it would lead to more damages than with conventional weapons. The language of surgical strikes hides the probable disproportionate effects and the uncertainty inherent to the usual battlefields' trade-offs. The tendency to systemically prioritize the security of one's fighters over the principles of distinction and proportionality is morally and legally questionable.

Jus in bello rests on some idea of equilibrium that should exist between belligerents. Although the LOAC and JWT do not explicitly ask for perfect symmetrical relations, a minimum level of fairness and equality, in terms of capabilities and ethics, seems necessary to ensure respect for the rules. Collateral damages are hardly acceptable if they do not correspond to a situation of mutual exposure to danger. Military necessity grounds the right to kill other combatants, but it is not an absolute prerogative. The wartime right to self-defense depends on the reciprocal acceptance of potential lethal risks equally threating belligerents.

---

as "the necessary identification measures [to be taken] in good time in order to spare the population as far as possible." ICRC Commentary of Article 57 of AP I <http://www.icrc. org> accessed May 2014.

79   Bradley Jay Strawser, "Moral Predators: The Duty to Employ Uninhabited Aerial Vehicles" (2010) 9:4 J of Military Ethics 342.

80   Gabriella Blum, "On a Differential Law of War" (2011) 52 Harv Intl L J 163.

81   Human Rights Watch, *Precisely Wrong: Gaza civilians killed by Israeli drone-launched missiles* (2009); see also (n 54).

82   Walzer, *Just and Unjust Wars* (n 7): "And if saving civilian lives means risking soldiers' lives, this risk must be accepted" at 156.

83   Martin Shaw, "Risk Transfer Militarism. Small Massacres and the Historical Legitimacy of War" (2002) 16:3 Intl Rel 343, at 352.

The capacity for one side to kill without being killed reverses the premise. While there are fewer soldiers fighting, sometimes even none, the enemy combatant has no one to strike back at. This means that they cannot be harmed, since they do not fit within the definition of a military objective. In this case, no belligerent can exercise its right to kill and war loses its *raison d'être*.[84] For the time being, this is only a fictional scenario. Even with the growing deployment of automated machines, war still seems to be a preferred means of dispute settlement.

The moral equality between combatants is a basic precept of JWT in its modern expression. Soldiers have the same rights and obligations, regardless of their nationality or allegiance. They are not responsible for the decisions of their leaders. This status of "moral innocence" is not contingent upon the justice of the war they fight for. Detaching the political ends from the ethics of an operation justifies the dissociation between *jus ad bellum, jus in bello*, and *jus post bellum*.[85] Due to the unprecedented asymmetry resulting from complex systems, this independence is put under severe jeopardy. New technology gives one side a "godlike power to call down destruction from the skies to its enemies."[86] David fights against an emotionless and mechanized Goliath. With robots, warfare is no longer a "shock between two hostile bodies in collision [but an] action of a living power against an inanimate mass."[87] What are the options left to the low-tech side in the absence of equal peers to fight against? It can surrender, but the recourse to illegal practices— human shields, misuse of protective emblems, hiding in civilian areas, and suicide bombings—is at times the preferred path. Huge discrepancies can create calls for brutal resistance, terrorism, and measures of reprisals against enemy civilians, on foreign and home soil.[88] As there are no combatants to shoot, the weak belligerent hits anyone who is perceived as hostile. On the basis of reciprocity, the strong party will harshly respond, taking for granted that it is justified to transgress the LOAC to crush the criminal enemy. More violence and a higher number of civilian casualties would follow.[89] The blurring of the combatants' identity alters the principle of distinction: The core differentiation between immune civilians and soldiers becomes pointless. As long as dead bodies are not too numerous on TV screens, public opinion will not firmly react to these unjust behaviors. At least, the outrage will not be such as to be able to interrupt an ongoing campaign. Thanks to this illusionary cleanliness, the ends come to justify the means.

---

84   Paul Khan, "The Paradox of Riskless Warfare" (2002) 22:3 Philo and Publ Pol Q 2, at 2.

85   See Grotius, *De Jure Belli ac Pacis* (n 48).

86   Robert Sparrow, "Building a Better Warbot: Ethical Issues in the Design of Unmanned Systems for Military Operations" (2009) 15 Sc and Engineering Ethics 169, at 179.

87   Carl von Clausewitz, *On War* (first published 1832, Penguin 1982) at 104.

88   Khan (n 83) at 6, as confirmed by the recent decapitations of US, UK, and French citizens in Iraq and Algeria.

89   Suzy Killmister, "Remote Weaponry: The Ethical Implications" (2008) 25:2 J of App Philo 121, at 131.

The extensive use of uninhabited systems produces some effects on the very essence of war. Classical ethics presupposes the existence of interpersonal relationships on the battlefield. Yet, when mechanization commands, individuality is removed. This trend runs contrary to the anthropomorphic dimension of conflicts.[90] New devices concurrently create a novel category of soldiers. They are contractors, scientists, technicians, acting more as game players than heroic soldiers. Because they do not face the reality of the battle, they do not have to exhibit courage, loyalty, camaraderie, or sense of sacrifice. Doubts can be raised as to the ability to instill these virtues into civilian operators. Secular military ethos, which promotes *temperamenta belli*, is originally drawn from concepts of honor and chivalry rooted in the representation of war as a man-to-man confrontation. If human death disappears from war, it will not be seen as an exceptionally brutal business, but as a normal and easily manageable dispute settlement mechanism. The readiness to die for the nation used to be a symbol of bravery; today, it stands for irresponsibility, even insanity.[91] Post-heroic warfare will turn to be a valueless activity, as traditional armies are replaced by virtual troops fighting a "push-button war."[92]

Technology has the capacity to make wars less bloody and restrain the amount of violence needed for defeating the enemy. Unfortunately, unmanned vehicles have not so far improved the situation of people affected by hostilities. As of today, there are still men are on the ground, but for how long? It might be also conceivable that the proliferation of military robots and their accessibility to "rogue" states, terrorists groups, or malevolent individuals will force belligerents to go back to conventional ways of fighting. History proves that no military advantage lasts forever. Technology innovations benefits and challenges *jus in bello* enforcement. The level of compliance with the laws of war ultimately rests on the eagerness to use these devices in conformity with accepted JWT standards and to maintain a little humanity in the midst of barbarity.

C. Jus post Bellum: *Reevaluating the Prospects for Peace*

Justice after war consists in achieving the cause(s) that initially triggered the commencement of the conflict. The intended end of any use of force is a conclusive victory. *Jus post bellum* stipulates long-term objectives, namely the establishment of a sustainable peace. Notwithstanding their theoretical separation, *jus ad bellum* and *jus in bello* have some implications in this regard: A war that is fought without

---

90   Robert Sparrow, "War without Virtue" in Bradley Jay Strawser (ed), *Killing by Remote Control. The Ethics of Unmanned Military* (OUP 2013) 84.

91   Michael Walzer, "Kosovo" in William J Buckley (ed), *Kosovo: Contending Voices on Balkan Interventions* (Eerdmans 2000) at 344.

92   Hans Morgenthau, *Politics Among Nations* (first published 1948, 7th ed, McGraw-Hill Humanities 2005) at 250. On post-heroism, Christian Enemark, *Armed Drones and the Ethics of War. Military Virtue in a Post-Heroic Age* (Routledge 2014).

a just cause or with a deliberate policy of violations of the LOAC can hardly satisfy *jus post bellum* criteria. Inversely, a just war, though it can produce unjust outcomes, will more likely lead to a just peace. JWT has not always explicitly discussed *just post bellum*, but it embodies the contemporary conception of the victor's responsibilities.[93] As already stressed, a campaign, in which one side possesses technical superiority, can be faster and more decisive. This is not to say that "blitz" and uncontestable victories guarantee an undisputed peace. Mighty militaries can of course successfully destroy enemy lines, while ultimately failing to attain their objectives due to their inability to win the "hearts and minds of those who live there."[94] Gaining local support is even more critical in counterinsurgency operations in which rebel fighters are often seen as heroes.

Drone attacks have dramatic effects on the daily life of individuals, at a physical, material, and psychological level. Intensive bombings can cause the radicalization of some segments of targeted groups. General Petraeus asserted that the US drones policy is nurturing terrorism. In his opinion, people who are not originally militants will possibly become extremists when living in constant fear.[95] Why should people accept the triumph of a party that did not think it was necessary to send humans on the ground, but was very prone to safely shoot at its enemies? Besides being treated as subhumans only deserving to face machines, the defeated ones may perceive the winner either as arrogant or as a cowardly bully.[96] "Drone" even became a word in Urdu to talk about Americans not fighting with honor.[97] In this unfair fight, mistakes resulting in civilian casualties are hardly tolerated by the weak ones. Anger, revenge, and mistrust make it hard, sometimes impossible, to build up peaceful relationships with local actors and can eventually reignite the conflict into a deadlier one. The peace imperative stresses the importance of conducting the war in a way as to "avoid the danger of provoking reprisals and causing bitterness that will long outlast the fighting."[98]

Robots could certainly be useful in unsafe post-conflict contexts by bringing resources and medical assistance to people living in remote locations. Furthermore, automated peacekeepers will not be "raping, pillaging, degrading, taunting or stealing food from the local population, as might occur with humans fueled by

---

93   James Turner Johnson, "Moral Responsibility After Conflict" in Eric Patterson (ed), *Ethics Beyond War's End* (Georgetown University Press 2012) 21.

94   Joseph Cummins, *Why Some Wars Never End: The Stories of the Longest Conflicts in History* (Fair Winds Press 2010).

95   Center for Civilians in Conflict, Columbia Law School Human Rights Clinic, *The Civilian Impact of Drones: Unexamined Costs, Unanswered Questions* (2012) at 23.

96   Aaron M Johnson, Sydney Axinn, "The Morality of Autonomous Robots" (2013) 12:2 J Mil Ethics 129, at 136.

97   Peter W Singer, "The Ethics of Killer Applications: Why it is so hard to talk about morality when it comes to new military technology?"(2010) 9:4 J of Military Ethics 299, at 306.

98   Henry Sidgwick, *Elements of Politics* (first published 1891, Cosimo 2005) at 264.

adrenaline, emotions and hatred."[99] The United Nations has recently announced that the use of unmanned vehicles in peacekeeping operations will constitute one of its future goals, given their efficiency in monitoring ceasefires and protecting humanitarian convoys.[100] Aside these positive aspects, machines are nonetheless considered as more threatening than humans. The presence of soldiers, once hostilities are terminated, might be welcome, since they can serve as food and supplies agents. Well-trained military professionals are also able to create some "benevolent appearance" by exchanging with people in their own language.[101]

The post-conflict situations in Iraq and Afghanistan illustrate the difficulty of achieving just peace following an upstream just war. The combination of dubious causes, falsified intentions, contested authority, and unlawful practices are destructive for any process of national reconstruction, by irremediably shaping perceptions of the environment. The use of uninhabited vehicles during an occupation or a transitional period can surely spare a few lives on the winner's side, but are unlikely to definitely calm the resentments of local populations.[102]

IV. Conclusion

Is the demise of human referents from battlefields ethically acceptable? Or, does the end of human monopoly over violence represent a step back in the light of the sustained efforts made by philosophers, decision-makers, and practitioners to attenuate the dirty sides of conflicts? By removing soldiers from fighting areas and, thanks to their accuracy, unmanned military systems should result in fewer casualties for all warring parties. Because of this superficial cleanliness, technology is claimed to support the moralization of armed conflicts.[103] Some disagree by saying "because war is a clash of opposing wills, the human dimension is central. War may involve technology but it is waged by people."[104] However, as in other societal realms, the trend to military robotization appears inescapable. Therefore, shall we continue to develop means that could potentially diminish sufferings

---

99   Patrick Lin, George Bekey, Keith Abney, *Autonomous Military Robotics: Risk, Ethics, and Design* (US Department of Navy, Office Research 2008) at 54 <http://www.isn. ethz.ch> accessed April 2014. One may think about the recent allegations against French peacekeepers accused of having sexually abused children in Central Africa.

100   Better World Campaign, The UN's use of Unmanned Aerial Vehicles in the DRC: US Support and Foreign Policy Advantages (May 2013) <http://www.betterworldcampaign. org> accessed May 2014.

101   Schulzke (n 67) at 304.

102   Brian Orend, "Jus Post Bellum" in Stahn Carsten, Jann K. Kleffner, *Just Post Bellum: Towards a Law of Transition From Conflict To Peace* (TMC Asser Press 2008) 31.

103   Ronald C. Arkin, "The Case for Ethical Autonomy in Unmanned Systems" (2010) 9:4 J Mil Ethics 332, at 333.

104   Admiral Mike G Mullen, US DoD, *Capstone Concept for Joint Operations: Version 3.0* (Jan 15, 2009) available at <http://www.thefreelibrary.com> accessed May 2014.

or shall we make wars so pitiless that we would not engage in it? Considering the violent dimension of history, we may be tempted to select the first option. A firm willingness to respect the standards regulating wartime situations should, however, imperatively underpin this choice.

It remains complex to speak about ethics when it comes to killer robots. Still, all military devices have raised moral and legal issues. Nothing is then really new "under the sun of warfare," but the contemporary setting brings about challenges that have to be the focus of renewed thinking.[105] War may seem to be a sanitized activity, but "few, if any, military analysts have ever used the term clean war. Even fewer ... would seriously claim that there is likely to be anything approaching an antiseptic war, where there are no casualties, no harm and no destruction of consequence."[106]

In the twenty-first century, a good war should not only be just, but must also be clean in terms of effects and outcomes. This chapter has tried to demonstrate that the criterion of cleanliness explains some deviations from the classical just war paradigm and entails a redefinition of its principles. Actions, which do not comply with the old scheme, are presented as unfortunate accidents or inevitable results of contemporary conflicts. Transgressions are tolerated as long as the hostilities are (quasi-)bloodless, at least from one's own perspective. We should not expect every actor to deal with war in the same fashion or to have similar interpretations of the principles governing the use of force. Yet, some kind of agreement exists on violence's justifications. JWT corresponds to "a methodology ... within a commonly accepted set of boundaries," aiming at the normative evaluation of military conduct.[107] Regardless of the technology, just war is a relevant ethics, which has showed enough flexibility to adapt to changes and enough stability to attenuate the consequences of wars.[108]

The conquest of battlefields by robots is not about to stop, since states will not give up the military and economic benefits flowing from military innovation. It falls upon lawyers, scientists, militaries, and politicians to set up a regulatory frame imposing a reasonable use of new devices in accordance with the LOAC and JWT. To this end, they must reevaluate what is, and will be, feasible and ethical to do in war, by anticipating unexpected effects of technology. Improved capabilities, particularly unmanned systems, create puzzling dilemmas with regards to decisions, strategies, tactics, doctrines, and military institutional organization. Clean war

---

105 Emmanuel Goffi, "The Use of Warbots: A Matter of Legal Responsibility" (2013) 8 Dynamiques Internationales <http://dynamiques-internationales.com> accessed May 2014. See also on this issue, the second CCW Meeting of Experts on Lethal Autonomous Weapons Systems, which took place in April 2015 at the United Nations in Geneva. See <http://www.unog.ch/80256EE600585943/%28httpPages%29/6CE049BE22EC75A2C1257C8D00513E26?OpenDocument> accessed May 2015.

106 MLR Smith, Sophie Roberts, "War in the Gray: Exploring the Concept of Dirty War" (2008) 31 Stud in Conflicts & Terrorism 377, at 381.

107 Ibid at 382.

108 See (n 10).

rhetoric has almost succeeded in convincing public opinion of the necessity of autonomous vehicles, while generating high expectations about the intended death rate. Although the current transformations within military affairs present some novel characteristics, technology has always impacted our perception of wars and related ethical standards. As it has been rightly underlined, "no technical advances by itself made a revolution; it is how people responded to technology that produced seismic shifts in warfare".[109]

The concept of clean, civilized, safe, pure, sterilized, or curative war gives the impression that belligerency is almost a virtuous activity. Unfortunately, using force is still about "blowing up things and people".[110] Those who deploy force with quasi-impunity rely on discursive stratagems and a strict media control to protect their citizens from pictures of mutilated bodies. Is the world moving toward a "factory of death and clean-killing"?[111] Technology's potentialities and zero-casualty strategies do not guarantee a safer world. Because riskless and bloodless wars are misleading illusions, JWT should resist the temptation of those idealistic promises. Sooner or later, we will consider reassessing our faith in the salutary power of scientific progress. Syria, Ukraine, Gaza, Iraq: Boots are on the ground and humans are still bloodthirsty. Consequently, "[when] things go wrong with humanity, it is not always appropriate to just reach for technology to fix the problems".[112] Technology might help us preserving our humanity, but will never be a substitute for it.

---

109   Max Boot, War Made New (Gotham Books 2006) at 10.
110   "Interview with Peter W. Singer" (2012) 94: 886 Intl Rev Red Cross 467, at 472.
111   Sharkey, "The Evitability of Autonomous Robot Warfare" (n 35) at 788.
112   Ibid at 796.

SECTION III
Through the Lens of Efficacy—
Empirical Validity

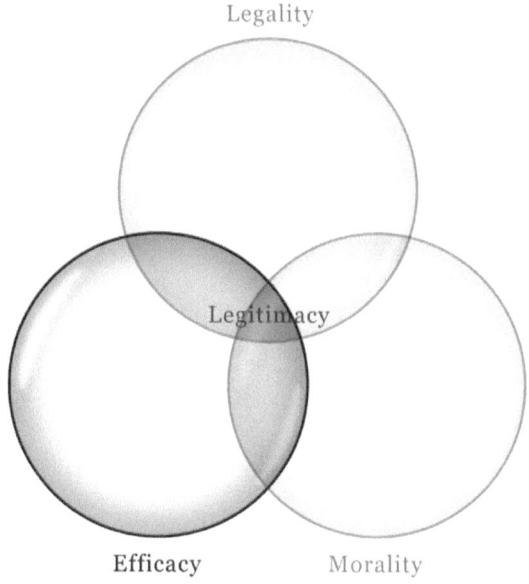

Legality

Legitimacy

Efficacy   Morality

Chapter 9

Data on Leadership Targeting and Potential Impacts for Communal Support

Jenna Jordan

I. Introduction

Leadership decapitation, which includes the killing or capturing of terrorist leaders, has become a priority for United States counterterrorism policies. Both academics and policy-makers have argued that this leadership decapitation is an effective strategy for combating terrorism. They saw the killing of Osama bin Laden by special operations forces on May 2, 2011, in Abbottabad, Pakistan, as a victory for the war on terrorism. However, according to the data analyzed here, despite the success of such attacks on militant leaders, continued decapitation efforts are not likely to result in a long-term organizational demise and may in fact have counterproductive consequences, emboldening or strengthening the organization. Through empirical research and analysis this chapter will focus on two primary questions: (1) Does leadership decapitation lead to the demise of terrorist organizations? and (2) What factors account for organizational resilience to leadership targeting?

Since the terrorist attacks of September 11, 2001, the US has successfully carried out many leadership strikes against high-level Al-Qaeda operatives in an effort to bring about the organization's demise. Drone strikes have been employed with increasing frequency in targeting Al-Qaeda leaders and other militants in Pakistan, Afghanistan, and Yemen. According to the data collected on drone strikes in Yemen and Pakistan by Peter Bergen and Katherine Teidemann at the New America Foundation, the US carried out 1 strike in 2004, 101 in 2010, 85 in 2011, 104 in 2012, and 51 in 2013, and 39 in 2014.[1] In June 2012, Abu Yahya al Libi, Al-Qaeda's deputy leader, was killed in a CIA drone strike in Pakistan. He was highly experienced and served an important operational function within the

---

1 For Peter Bergen and Katherine Tiedemann's data at the New American Foundation see <http://securitydata.newamerica.net>. For data from the Bureau of Investigative Journalism see <http://www.thebureauinvestigates.com/category/projects/drone-data/>. This data has been updated as of June 9, 2015. However, it should be remembered that the New America website updates constantly and is likely to have done so by the time of this publication. Consequently, this is study is analyzing the best available data as of that moment.

organization. His death was seen a significant blow to an already weakened Al-Qaeda.[2] Nine months prior, Anwar al-Awlaki, a Yemeni-American cleric linked to a number of terrorist plots in the West, was killed in Yemen on September 30, 2011, by a Hellfire missile fired from an American drone.[3] Other key leaders have been consistently targeted. On August 22, 2011, Atiyah Abd al-Rahman, believed to be the organization's second highest leader and played a key operational role, was reportedly killed in a drone strike in Pakistan.[4] He served as a link between bin Laden and members within lower ranks of the organization. A drone strike also killed Ilyas Kashmiri, a senior member of Al-Qaeda and the operational commander for Harakat-ul-Jihad al-Islami (HuJI, an Islamist militant organization largely active in Pakistan, Bangladesh, and India) in South Waziristan on June 3, 2011. This is by no means a complete list; rather, I wish to illustrate the frequency with which the US has targeted Al-Qaeda leaders and operatives over the past few years, specifically through the use of drone strikes.[5]

The Obama administration has placed leadership targeting at the center of its strategic framework on counterterrorism. In a May 2013 address to the National Defense University, President Obama announced that rather than military action, targeted operations to dismantle terrorist networks would be a critical aspect of the administration's counterterrorism strategy. The administration has stated a preference for capturing terrorist leaders, however in certain cases it is either not possible or poses considerable risks. In these circumstances, the administration argues that targeted killings should be employed. While it is clear that a large number of these attacks have been successful in killing their intended targets, there is considerable debate regarding the effectiveness of decapitation, and more specifically the use of unmanned aerial vehicles, in destabilizing militant

---

2   Declan Walsh and Eric Schmitt, "Drone Strike Killed No. 2 in Al Qaeda, U.S. Officials Say" *The New York Times* (June 5, 2012) <http://www.nytimes.com/2012/06/06/world/asia/qaeda-deputy-killed-in-drone-strike-in-pakistan.html?pagewanted=all> accessed Dec 2014. In the aftermath of al Libi's death, Peter Bergen argued that Zawahiri is likely the only remaining influential leader in Al-Qaeda. See <http://articles.cnn.com/2012–06–05/opinion/opinion_bergen-al Qaeda-whos-left_1_abu-yahya-aqap-drone-strikes?_s=PM:OPINION> accessed Dec 2014.

3   Mark Mazzetti, Eric Schmitt and Robert F. Worth, "Two-Year Manhunt Led to Killing of Awlaki in Yemen" *The New York Times* (Sept 30, 2011) available at <http://www.nytimes.com/2011/10/01/world/middleeast/anwar-al-awlaki-is-killed-in-yemen.html?pagewanted=all> accessed Dec 2014.

4   Mark Mazzetti, "C.I.A. Drone Is Said to Kill Al Qaeda's No. 2" *The New York Times* (Aug 27, 2011) <http://www.nytimes.com/2011/08/28/world/asia/28qaeda.html> accessed Nov 2014.

5   Drone strikes are not the only tactic for targeting terrorists. Raids and special operations forces, for example, have also been used in many high profile instances of targeting. Moreover, drone strikes do not only target militant leaders; mid and lower level operatives have also been the target of strikes through UCAVs. This study however, focuses exclusively on the use of drone strikes against militant leaders.

groups. Despite the lack of consensus regarding whether targeting leaders will actually weaken terrorist organizations, it is likely to remain a key component of counterterrorism policies.

Leadership decapitation, however, is not always successful. This chapter will explore the efficacy of leadership targeting from both a quantitative and qualitative perspective. I argue that leadership targeting is not an effective strategy and further that drone strikes are likely to have adverse consequences. The chapter will proceed as follows. In the second section, I will look at the extant literature on leadership targeting. Next, using data on such targeting, I identify the conditions under which decapitation results in organizational decline and argue that decapitation is not an effective counterterrorism strategy. In the fourth section, I present a theory of organizational resilience to explain why certain organizations are more resilient than others to attacks on their leadership. In the fifth section, I utilize this theory to assess both the strategic effectiveness of drone strikes and their potential for counterproductive consequences, with a focus on Al-Qaeda. Finally, I conclude with a discussion of policy implications.

II. Existing Studies on Decapitation

There is considerable debate over leadership targeting, and specifically the use of unmanned aerial vehicles strikes in killing militant leaders.[6] Many scholars have argued that decapitation successfully weakens terrorist organizations. Unsurprisingly, many in both the academic and policy communities saw the killing of Osama bin Laden as a tactical and strategic victory in the struggle against Al-Qaeda. Even those who have concerns about the potential for backlash have argued that decapitation should continue to be a key component of counterterrorism policy.[7] Much of the optimism regarding leadership targeting comes from theories of charismatic leadership, which posit that the susceptibility of terrorist organizations to leadership targeting is a function of individual qualities inherent to the leader.[8] These qualities contribute to the belief that leaders are irreplaceable because they have special qualities from which their legitimacy arises and that their authority is critical to a group's operational success.[9] These theories, however, overpredict the

---

6   For a fuller articulation of extant research on leadership targeting see Jenna Jordan, "Attacking the Leader, Missing the Mark: Why Terrorist Groups Survive Decapitation Strikes" (2014) 38 Int'l Sec, at 7.

7   Steven R. David, "Fatal Choices: Israel's Policy of Targeted Killing" (2002) 51 ME Sec and Pol Studies.

8   See Max Weber, *The Theory of Economic and Social Organization* (Free Press 1964); Max Weber, *The Theory of Economic and Social Organization* (Free Press 1964) at 358.

9   It is possible that charisma can be routinized or institutionalized which would strengthen a group's resilience to leadership targeting. See Michael Freeman, "A Theory of

success of leadership targeting and overlook variables relating the organization to the community in which it functions.

First, quantitative studies on leadership targeting have reached different conclusions regarding its ability to weaken a terrorist group or result in its decline. While I argue that decapitation is not an effective counterterrorism strategy, other studies have found that it can reduce the duration and intensity of armed conflict. Similarly, Patrick Johnston concludes that leadership decapitation is effective. Focusing on insurgent organizations, he shows that it increases both the chance of war termination and the probability of government victory, and decreases the intensity of militant violence, and the frequency of insurgent attacks.[10] While Johnston's statistical findings are robust he does not provide an explanation for why decapitation is effective. Finally, Aaron Mannes looks at 60 cases of decapitation against 71 terrorist organizations and compares terrorist activity in the period before and after a terrorist attack.[11] While most of the results were not statistically significant, he finds a decline in the number of terrorist attacks and an increase in the fatality of attacks, particularly in the case of religious organizations. However, he only examines targeted groups, not those who have not undergone targeting, undermining his conclusions regarding the effectiveness of decapitation.

Second, other studies have examined how leaders affect organizational cohesion and resilience to leadership attacks. Bryan Price finds that decapitation increases the mortality rate of terrorist organizations and argues that leadership succession is difficult due to the clandestine and values-based nature of terrorist organizations.[12] He posits that clandestine organizations are usually headed by charismatic leaders, who are less likely to institutionalize their operations for both strategic and personal reasons. While this study offers an important theoretical analysis of terrorist organizations by treating all terrorist organizations as values-based, the theory does not account for variation in the resilience of terrorist organizations to leadership targeting. Michael Freeman takes a different approach to understanding

---

Terrorist Leadership (and its Consequences for Leadership Targeting)" (2014) 26 Terrorism and Pol Violence. For a discussion of Max Weber on the institutionalization of charisma see Jenna Jordan, "When Heads Roll: Assessing the Effectiveness of Leadership Decapitation" (2009) 18 Sec Studies, at 719.

10   Patrick Johnston, "Does Decapitation Work? Assessing the Effectiveness of Leadership Targeting in Counterinsurgency Campaigns" (2012) 34 Intl Sec, at 50. However, his study focuses only on insurgent campaigns and includes groups that would not be classified as having engaged in terrorism.

11   Aaron Mannes, "Testing the Snake Head Strategy: Does Killing or Capturing its Leaders Reduce a Terrorist Group's Activity?" (2008) 9 J of Intl Policy Solutions. He leaves out the smallest organizations, those with under 100 members, in order to exclude groups that committed fewer than 10 attacks. However, he includes small groups that committed highly significant operations, such as Aum Shinrikyo.

12   Bryan Price, "Targeting Top Terrorists: How Leadership Decapitation Contributes to Counterterrorism" (2012) 34 Intl Sec, at 14.

the role of a terrorist leader. He argues that the ability of decapitation to weaken a terrorist organization is a function of whether the targeted leaders have an operational or inspirational role; he concludes that such organizations are the most likely to fall apart after decapitation.[13]

Third, critical information can be obtained during the capture of terrorist leaders. Targeting operations can result in the discovery of documents revealing information about group activity. For example, during the arrest of Abimael Guzman, leader of Sendero Luminoso, authorities found documents that resulted in the arrest of other Sendero members.[14] The arrest of Basque Homeland and Freedoms' (ETA) leader Francisco Mugica Garmenia revealed that the organization was more sophisticated than authorities had previously thought.[15] Captured leaders can provide information about location, capabilities, personnel, and operations.[16]

Fourth, other studies, focusing on organizational structure have found that decentralized organizations are much harder to destabilize than hierarchical organizations.[17] In a study of the global Salafi jihad, Marc Sageman argues that the Salafi jihadists' decentralized and local nature makes them difficult to target. Given the structure of such networks, it is necessary to identify key hubs within the organization and individuals who may be the top leaders.[18] Audrey Cronin has argued that Al-Qaeda's organizational structure, which is both hierarchical and decentralized, is particularly stable.[19]

Finally, those who have argued that decapitation is not likely to be effective have focused on the potential for counterproductive consequences, such as the creation of a martyrdom effect, a surge in recruitment, the occurrence of retaliatory attacks, an increase in group resolve and strength, and a surge in the frequency and intensity of attacks.

---

13   Freeman (n 9).

14   Cynthia McClintock, *Revolutionary Movements in Latin America: El Salvador's FMLN and Peru's Shining Path* (United States Institute of Peace Press 1998) at 92.

15   William Drozdiak, "Avowed 'Decapitation' of Basque Group in Doubt as Olympics Near" *The Washington Post* (May 3, 1992) First and A16.

16   Audrey Kurth Cronin, *How Terrorism Ends: Understanding the Decline and Demise of Terrorist Campaigns* (Princeton University Press 2009) 17.

17   See Kathleen Carley, "A Theory of Group Stability" (1991) 56 American Sociological Review 331; Kathleen Carley, Ju-Sung Lee and David Krackhardt, "Destabilizing Networks" (2002) 24 Connections 79; Daniel Byman, "Do Targeted Killings Work?" (2006) 85 For Aff; Carley, Lee and Krackhardt James D. Thompson, *Organizations in Action: Social Sciences Bases of Administrative Theory* (Transaction Publishers 2007).

18   See Marc Sageman, *Understanding Terror Networks* (University of Pennsylvania Press 2004).

19   Cronin, *How Terrorism Ends ...* (n 16) at 177.

III. Does Decapitation Work?

Drawing upon findings of an earlier study on leadership targeting, this section examines data on leadership attacks and whether leadership decapitation is an effective counterterrorism strategy. The study also identifies conditions under which decapitation is more or less likely to result in organizational decline.[20] Looking at 298 instances of leadership targeting from 1945 to 2004, I used cross tabulations and logit analysis[21] to determine whether decapitation is an effective strategy and whether a group's susceptibility to decapitation is correlated with its age, size, or type; the leaders level of leadership; and whether the leader was arrested or killed.

There are a number of different ways to measure the efficacy of leadership targeting. Leadership targeting can result in organizational decline, cause short-term destabilization, degrade the ability of a group to carry out frequent or highly lethal operations, or increase the organization's rate of decline. As Steven J. Barela argues in his chapter on strategic efficacy, there are fundamental challenges to evaluating efficacy, and even more problematically in identifying both the threat and the specific enemies in some cases. Despite these challenges, it is important to try to develop standards by which to evaluate whether current counterterrorism strategies are reducing terrorist activity or creating an enabling environment for further recruitment. In this study, the dependent variable is measured according to whether the group was inactive for two years following an instance of leadership removal.[22] If the group experienced a decline in activity but it resumed attacks within a two-year period, then it was coded as a failure.[23] While this is a fairly restrictive criterion for success, it is important to consider both short- and long-term consequences of leadership targeting.

The data suggest that overall leadership targeting is not an effective counterterrorism strategy and may in fact be counterproductive. In order to establish

---

20   These results were initially published in a 2009 article. See Jordan, "When Heads Roll" (n 9).

21   The bivariate cross tabulations analyze the relationship between the independent variables and organizational decline. This study also used a logistic regression analysis, which measures the relationship between a categorical dependent variable and multiple independent variables.

22   I based this two-year criteria on the United States Department of State designation of Foreign Terrorist Organizations (FTO). "Designations are valid for two years, after which they must be redesignated or they automatically expire. Redesignation after two years is a positive act and represents a determination by the Secretary of State that the organization has continued to engage in terrorist activity and still meets the criteria specified in law." See <http://www.state.gov/r/pa/prs/ps/2001/5265.htm> accessed Dec 2014.

23   For example, following the arrests and suicides of its two founders, the Baader-Meinhof gang became known as the Red Army Faction. While Baader-Meinhof no longer existed as a group, the Red Army Faction had the same goals and was formed from the remnants of the Baader-Meinhof.

a baseline rate of decline for terrorist organizations, I looked at organizations that had and had not undergone attacks on their leadership. Decapitation was successful in 17 percent of the 298 instances of leadership removal. However, in order to assess whether decapitation is an effective strategy, it is necessary to look at the rate of decline for groups that did not have their leaders targeted. The data show that organizations that have not undergone decapitation have a higher rate of decline. While 53 percent of decapitated terrorist groups fell apart, 70 percent of groups that had not experienced decapitation were no longer active, a 20 percent lower rate of decline.

I also identified the conditions under which decapitation results in organizational decline. The data suggest that organizational age, type, and size are important predictors of organizational susceptibility to decapitation. Younger and smaller organizations are more likely to cease activity after the removal of a leader. Religious organizations, and to a lesser degree separatist organizations, are more resilient to decapitation, while ideological organizations are more susceptible to collapse. See Table 9.1 for logit coefficients.

Table 9.1   Logit model of terrorist group fate after decapitation

Variable	Coefficient	Standard error
Age	- .529***	.180
Size	- .188*	.111
Religious	-1.420**	.632
Ideological	1.186*	.667
Separatist	0.423	.576
Arrest or death	- .282	.454
Top leader or upper echelon	-1.183	.875
Cons	.722	.966

Number of observations   294
LR Chi-squared (6)   60.29
Prob>Chi-Sq.   0.000
Note: Entries are logit coefficients
* = p<0.1, ** = p<0.05, *** = p<0.01.

First, I looked at whether the decapitation was an arrest or a death. I predicted that the death of a leader would result in organizational decline more than an arrest. Incarcerated leaders may remain active, and operatives can plan activity in order to free imprisoned militants.[24] For example, after the arrest of Baader-Meinhof leaders, the remaining members of the organization continued to carry out terrorist activity in order to free their leaders from prison.[25] This variable was not statistically significant in the multivariate (multiple variables) analysis.

Second, the removal of a top leader should result in organizational destabilization more than a member of the upper echelon. While the members of the upper echelon, as opposed to the organization's top leader, are usually more responsible for the group's operational activity, it is often easier to replace lower level leaders. As a result, removing the group's top leaders should be positively correlated with organizational decline. Additionally, removing both the top leader and members of the upper echelon should significantly destabilize an organization. However, this variable, which coded whether a leader or member of the upper echelon was targeted, was not statistically significant

Third, I classified organizations by type. Organizations were coded as religious, separatist, or ideological. Ideological organizations aim to politically transform society and include left-wing Marxist, Leninist, social revolutionary organizations, fascist organizations, and white-supremacist organizations. Organizations can also be coded as more than one type. Hezbollah and Hamas, for example, can be classified as both separatist and religious, while the Kurdistan Workers' Party (PKK) pursues both separatist/nationalist and Marxist goals. I predicted that ideological organizations would be more susceptible to destabilization as their doctrine often seems more dependent upon a specific ideological interpretation of the leader. As Audrey Cronin notes, these groups "were notorious for their inability to articulate a clear vision of their goals that could be handed down to successors after the first generation of radical leaders departed or were eliminated."[26]

Religious organizations, on the other hand, should be most resilient to destabilization after the removal of a leader.[27] Recent studies have found that religious organizations tend to be more decentralized in structure. Studies in network analysis have found that decentralized organizations are the hardest to weaken through targeting the leadership. Furthermore, religion is a highly motivating factor. Operatives in religious organizations often feel a sense of purpose that comes from being a member of the organization, eliciting a more

---

24   See Audrey Kurth Cronin, "'How al-Qaida Ends: The Decline and Demise of Terrorist Groups" (2006) 31 Int'l Sec 7.

25   Sean K Anderson and Stephen Sloan, *Historical Dictionary of Terrorism* (The Scarecrow Press, Inc 2002) at 415.

26   See Cronin "How al-Qaida Ends ... " (n 24) at 23.

27   An important caveat needs to be added here. I assume that religious groups tend to be decentralized in structure. However, there are a number of religiously motivated terrorist organizations that are centralized. For example, Hamas' leadership is more centralized, while other aspects of the organization are more decentralized. Lashkar-e-Taiba in Pakistan/Kashmir is highly centralized. For the purposes of theory building, I assume that religious groups tend towards decentralization.

passionate sense of dedication.[28] Most importantly, religious and separatist groups tend to have more support from local communities than ideological organizations. The data show that religious organizations are the least likely to fall apart after decapitation, while ideological organizations are the easiest to destabilize. Less than 5 percent of religious groups ceased activity after decapitation compared to 32 percent of all ideological organizations.

Fourth, I looked at a group's age, which was coded according to the when an organization began conducting terrorist attacks. I predicted that younger organizations are easier to destabilize than older groups. Older organizations tend to be more complex and have developed bureaucratic features, bolstering their organizational resilience, an idea that I will develop in the next section of this chapter. In a pivotal study, Arthur Stinchcombe argued that new organizations have a higher rate of failure than old organizations.[29] New organizations depend upon new roles that lack a normative basis and have not developed stable linkages.[30] Age was highly significant; the oldest organizations, those that had been active for over 40 years, were always resistant to decapitation, while decapitation was successful against 29 percent of groups that had been active for less than 10 years.

Finally, I looked at an organization's size, measured by the number of active members. Size is a difficult variable to measure. For example, there has been much debate regarding Al-Qaeda's size and its structure. Similar to age, larger groups often have to develop more complex bureaucratic hierarchies in order to efficiently manage their large number of operatives. Larger organizations often delegate operational activity to localized networks, which increases a group's resilience to targeting efforts. The data indicate that decapitation is less effective against larger groups. The smallest organizations were highly susceptible to collapse. Groups with fewer than 25 members fell apart 54 percent of the time; those with between 26 and100 members fell apart 41 percent of the time; groups with between 5,000 and 10,000 members fell apart in 36 percent of cases; and decapitation was successful against only 9 percent of groups with over 10,000 members. In the next section, I develop a theory to explain why some groups are able to resist attacks on their leadership, while others are weakened.

---

28 See David C Rapoport, "Fear and Trembling: Terrorism in Three Religious Traditions" (1984) 78 American Pol Sci Rev 658.

29 Arthur Stinchcombe, "Organizations and Social Structure" in JG March (ed) *Handbook of Organizations* (Rand McNally 1965).

30 Josef Bruderl and Rudolf Schussler, "Organizational Mortality: The Liabilities of Newness and Adolescence" (1990) 35 Admin Science Q 530.

IV. A Theory of Organizational Resilience

Decapitation can be successful, unsuccessful, and even counterproductive, and in this section I develop a theory to account for this variability. The ability of an organization to withstand attacks on its leadership and continue carrying out terrorist activity is a function of two variables—communal support and bureaucratization.[31] Decapitation is unlikely to result in the decline of groups that are highly bureaucratized or that have high levels of popular support. Organizations that are bureaucratic and have a considerable amount of popular support are the most difficult to destabilize following leadership targeting. Targeting also has the potential to result in counterproductive consequences, particularly for organizations with high levels of communal support. In the fifth section of this chapter, I argue that understanding communal support is essential to understanding the impact of drone strikes targeting militant leaders.

A. Bureaucracy and Organizational Resilience

The extent to which an organization is bureaucratized can account for whether decapitation is likely to destabilize a terrorist organization. Bureaucratized terrorists have an organized administrative staff, a division of administrative responsibilities, a division of labor, a hierarchy of authority, and a stable structure of rules, policies, and procedures. It is these features that promote stability in bureaucratic organizations and contribute to their ability to withstand the sudden removal of a leader or multiple leaders. Larger and older groups are more likely to have developed a bureaucratic authority, which can account for the finding that they are more resilient to leadership attacks.

Bureaucratic organizations have specific features that increase organizational stability and efficiency. This form of hierarchy can make groups more resilient to an attack on their leadership and can facilitate a clear process of leadership succession, reducing the potential for instability. First, bureaucratic organizations are diversified, with a clear delineation of duties and power. As groups become larger, more complex, and more specialized, they are more likely to have diversified functions, increasing their stability.[32] Second, bureaucracies develop stable rules and routines, which can enhance stability and efficiency. Organizations will struggle until they are perceived as reliable and accountable, which according

---

31   The theoretical analysis presented in this section appears in *International Security.* See Jordan, "Attacking the Leader, Missing the Mark: Why Terrorist Groups Survive Decapitation Strikes."

32   Grinyer and Ardenki find a positive correlation between size and diversification. As an organization grows in size, it is likely be more diversified and structurally complex. See Peter H Grinyer and Masoud Yasai-Ardekani, "Strategy, Structure, Size, and Bureaucracy" (1981) 24 The Acad of Management J 471, at 475. See also Jodi Vittori, "All Struggles Must End: The Longevity of Terrorist Groups" (2009) 30 Contemp Sec Pol 444, at 445.

to Hager, happens once "the organization has established routines."[33] Routines can help groups to survive turnover in leadership and increase the capacity for learning.[34] It follows that groups that have experienced an external shock, such as leadership decapitation, should be able to adapt in ways that ensure organizational stability and efficiency.

There are difficulties in identifying the extent to which a terrorist organization is bureaucratized. Not only do groups have an incentive to keep information secret regarding their structure, terrorist groups also face a trade-off between efficiency and secrecy.[35] However, many groups organize themselves bureaucratically. Terrorist organizations, such as Al-Qaeda in Iraq, may keep documentation on financial activities, membership, and other group activities. Groups can also have separate political, military, and social wings in order to carry out separate and distinct functions. Hamas and the Liberation Tigers of Tamil Eelam (LTTE) are two examples of organizations that have separate wings in order to carry out different organizational functions. As explained above, this is a central feature of bureaucracy. In addition to functionally separate branches, bureaucratic organizations also tend to have a hierarchy of authority, organized administrative staff, and standard operating procedures.

Certain types of organizations are more likely to be bureaucratized than others. Older and larger groups are more likely to have developed bureaucratic features, increasing their stability, effectiveness, staying power, and ability to survive an attack on their leadership.[36] Stinchcombe's "liability of newness" thesis supports the notion that older and larger groups are more stable and resilient. He argues that because older and larger groups should be more stable, a higher number of new organizations fail than old organizations.[37] There is a body of literature in the business management

---

33   Mark A. Hager, Joseph Galaskiewicz and Jeff A. Larson, "Structural Embededdness and the Liability of Newness Among Nonprofit Organizations" (2004) 6 Pub Management Rev 159, at 162.

34   Barbara Levitt and James G March, "Organizational Learning" (1988) 14 Amer Rev of Socio 319, at 320.

35   See Jacob N Shapiro, *The Terrorist's Dilemma: Managing Violent Covert Organizations* (Princeton University Press 2013).

36   See John Freeman, Glenn R. Carroll and Michael T. Hannan, "The Liability of Newness: Age Dependence in Organizational Death Rates" (1983) 48 Amer Socio Rev 69; Mark A Hager and others, "Tales from the Grave: Organizations' Accounts of Their Own Demise" (1996) 39 Amer Behav Scientist 975; Hager, Galaskiewicz and Larso Michael T Hannan and John Freeman, "The Population Ecology of Organizations" (1977) 82 Amer J of Socio 92; Robert O Keohane, "The demand for international regimes" (1982) 36 Intl Org 32; Jitendra V Singh, David J Tucker and Robert J House, "Organizational Legitimacy and the Liability of Newness" (1986) 31 Admin Science Q 17; James Ranger-Moore, "Bigger May be Better, But Older is Wiser? Organizational Age and Size in the New York Life Insurance Industry" (1992) 62 Amer Socio Rev 903.

37   See Stinchcombe (n 29). Numerous studies have found support for the empirical accuracy of the liability of newness. See Ranger-Moore Freeman, Carroll and Hannan

field that finds that organizations are more likely to develop bureaucratic traits as they grow in age and size.[38] In fact, regardless of the initial organizational blueprint of the organization, these authors argue that age and size are positively correlated with bureaucracy.[39] As an organization grows in size and complexity, bureaucratization becomes necessary to effectively manage the organization.

Managers of larger organizations are increasingly forced to delegate decision-making responsibilities. As a result, a group may be hierarchical at the upper administrative levels and largely decentralized at lower and more operational levels, allowing for greater organizational flexibility. Decentralization can also help a group in its ability to withstand counterterrorism measures. This characterizes the quasi-bureaucratic structure of many terrorist organizations. This may sound like a contradiction; bureaucracy and decentralization are often portrayed as opposites, particularly in the literature on terrorism. While Weber argues that a hierarchical system of authority is a critical feature of a bureaucracy, Mansfield notes that, "[a]t no point did [Weber] suggest, however, that centralization of decision-making in such a hierarchy was a characteristic of bureaucracy nor did he even make explicit the relationship between bureaucracy and centralization."[40] Mansfield points to "the absence of any positive relationship between centralization and bureaucratization. In general there is some evidence of negative association between these variables."[41]

Furthermore, in discussing this quasi-bureaucratic organizational structure, Michael Kenney claims that the decentralized nature of networks "does not preclude the existence of vertical decision-making hierarchies within nodes that carry out the network's most dangerous activities."[42] For example, Kenney classifies Hezbollah as a "bureaucratic network hybrid" that coordinates different groups of Shi'ite extremists. According to Kenney, Hamas is even more decentralized than Hezbollah. While Hamas is made up of a network of mosques, schools, health care centers, and paramilitary cells, it is run by a hierarchical administrative body,

---

Hager, Galaskiewicz and Larso Singh, Tucker and House.

38   See Jack A Nickerson and Todd R Zenger, "Comment: Dynamically Engineering Bureaucracy" (1999) 15 The J of Law, Econ, and Org; James N. Baron, M. Diane Burton and Michael T Hannan, "Engineering Bureaucracy: The Genesis of Formal Policies, Positions, and Structures in High-Technology Firms" (1999) 15 The J of Law, Econ, and Org. It is important to note that terrorist organizations do not necessarily increase in size as they age. However, independent of the relationship between size and age, organizations that age or increase in size are more likely to develop characteristics of bureaucracies in order to carry out an effective terrorist campaign.

39   See ibid Baron, Burton and Hannan, at 31; and Grinyer and Yasai-Ardekani (n 32) at 484.

40   Roger Mansfield, "Bureaucracy and Centralization: An Examination of Organizational Structure" (1973) 18 Admin Sci Q 477, at 478.

41   Ibid.

42   Michael Kenney, *From Pablo to Osama: Trafficking and Terrorist Networks, Government Bureaucracies, and Competitive Adaptation* (The Pennsylvania State University Press 2007) at 149.

which funds and guides clandestine cells that are given a substantial amount of autonomy in carrying out activities.

Generally bureaucratic organizations tend toward stability, yet the Weberian model overlooks other organizational transformations, such as factional splits.[43] Terrorist organizations are often susceptible to factionalization and splintering.[44] For example, the Irish Republican Army (IRA) splintered into multiple groups— the Real IRA, the Irish National Liberation Army, and the Continuity Irish Republican Army. The arrest and subsequent deaths of Baader-Meinhof leaders is often seen as the main factor contributing to the demise of the group, yet the Red Army Faction emerged from the remnants of Baader-Meinhof. Radicalization can also increase within splinter groups. Ethan Bueno de Mesquita finds that splinter groups, such as the Real IRA, can become more militant.[45] Cronin argues that this can occur when the splinter groups is "responding to the imperative to demonstrate their existence and signal their dissent."[46] While large and older organization are at risk of splintering, I argue that this process becomes less likely the more bureaucratized the organizations becomes.

B. Support and Organizational Resilience

Popular support is critical to a group's ability to withstand attacks on its leadership. It has been widely argued that effective insurgencies require vast amounts of popular support.[47] Groups with more support are likely to be seen as legitimate, further increasing their strength and effectiveness. Drawing upon this literature, I argue that terrorist organizations with high levels of communal support are better able to withstand attacks on their leadership.[48] Religious and separatist groups

---

43   Mayer N Zald and Roberta Ash, "Social Movement Organizations: Growth, Decay, and Change" (1966) 44 Social Forces, at 327–328.

44   I thank Mia Bloom for reminding me that the factionalization and splintering of terrorist organizations is an important process in organizational development and a potential consequence of counterterrorism policies such as leadership targeting.

45   Ethan Bueno de Mesquita, "Conciliation, Counterterrorism, and Patterns of Terrorist Violence" (2005) 59 Int'l Org 145, at 172.

46   Cronin "How al-Qaida Ends … " (n 24) at 26.

47   Paul Collier and Anke Hoeffler, "Greed and Grievance in Civil War" (2004) 56 Ox Econ Paper, at 563.

48   There is an important distinction between terrorist movements and insurgencies. Many of the cases that I evaluate could be classified as both terrorist organizations and insurgent organizations. While I do not theorize the difference between insurgencies and terrorism, the difference is strategic. Insurgent organizations can choose between carrying out a strategy of terrorism, guerilla warfare, or more conventional war fighting. This explains the distinction the partial overlap between studies of insurgencies and terrorist campaigns. Matthew Kocher has offered a helpful distinction between terrorist organizations and insurgencies. He suggests that the difference between the two groups is grounded in strategic choice. Terrorist organizations employ a strategy of punishment,

generally have higher levels of communal support than ideological organizations, making those organizations harder to weaken through targeting efforts.

There are a number of different mechanisms through which popular support contributes to group strength and capacity. Local support is necessary for recruitment, finances, and the acquisition of resources necessary to operate covertly. First, terrorist organizations fundamentally depend upon replenishing their membership, and popular support is essential for attracting new recruits.[49] Scott Atran claims that "without community support, terrorist organizations that depend on dense networks of ethnic and religious ties for information, recruitment, and survival cannot thrive."[50]

Second, popular support allows a group to function covertly. Community support is important for a group's ability to avoid "detection, surveillance, and elimination by the security forces of the target society."[51] Clandestine organizations need places to hide, information, and basic necessities, and groups with more support will have more access to these resources. Moreover, as decapitation can generate a sense of outrage within the community, particularly in the case of drone strikes (see below), local populations may be even more willing to provide safe places for leaders and operatives of heavily targeted organizations.

Third, communal support has an effect upon a group's willingness to carry out violent campaigns. Mia Bloom argues that when societies support killing civilians, terrorist groups have an incentive to adopt such measures.[52] Terrorist organizations care about social approval and their status relative to other organizations. Risa Brooks argues that "the militant's home constituency forces militants to adhere to societal norms about how violence is used."[53] A society's tolerance for violence and support for the militants' cause is dependent in part on the heavy-handedness of a state's counterterrorism strategies. Bloom argues "The fashion in which a state responds to the threat will also impact international public opinion and international support for the terrorist organizations and the targeted state."[54] Similarly, Scott Atran finds that traditional top-heavy strategies, such as "strategic

---

while insurgencies use a strategy of denial. Of course, this definition allows for potential overlap. For example the PKK uses denial in rural areas and punishment in urban areas. This distinction is important, but is beyond the scope of this chapter.

49   See Roger D Petersen, *Resistance and Rebellion* (Cambridge University Press 2001); Robert Pape, *Dying to Win: The Strategic Logic of Suicide Terrorism* (Random House 2005) at 81.

50   Scott Atran, "Mishandling Suicide Terrorism" (2004) 27 The Wash Q at 82.

51   Pape (n 49) at 81.

52   Mia Bloom, *Dying to Kill: The Allure of Suicide Terror* (Columbia University Press 2005).

53   See Risa Brooks, *Societies and Terrorist Violence: How Social Support Affects Militant Campaigns* (Marquette University 2011).

54   Bloom (n 52) at 94.

bombardment, invasion, occupation, and other massive forms of coercion" cannot suppress popular support for jihadist movements.[55]

Fourth, many groups depend upon the appeal and popularity of their political and ideological agenda. Local communities can lose interest in the ideology or aims of the group, undermining a group's cause.[56] This loss of ideological resonance explains the decline of many Marxist groups.[57] However, certain types of organizations should generate more popular support than others. I argue that separatist and religious groups generally have more support than ideological groups.[58] It is easier for religious and separatist groups to reproduce the doctrine upon which the organization is based, as their ideology is prevalent within the communities in which they operate. In comparison, the doctrine upon which ideological organizations are based is usually dependent upon a particular set of beliefs that is exclusive to the group, the leader, or a particular historical time period. As a result, religious and separatist groups are more likely to be seen as legitimate in the communities within which they operate, increasing their support and their capacity to carry out a violent campaign.[59]

Leadership targeting has the potential for significant counterproductive or adverse outcomes, most of which increase or strengthen support for organizations. These adverse consequences are particularly salient within the context of drone strikes on militant leaders. The removal of a leader from an organization with high levels of communal support may cause public outcry, resulting in even more sympathy, support, or publicity for the cause. This is particularly apparent in the case of leaders that serve an inspirational or spiritual role.[60] Killing leaders can also increase local and international sympathy for the organization and its cause, providing motivation for further recruitment. For example, the assassinations of Sheikh Ahmed Yassin, founder and spiritual leader of Hamas, and Abdel Aziz al-Rantisi, Hamas' co-founder, triggered massive local and international outrage. Yassin's death was condemned by the international community and triggered a huge amount of sympathy throughout Palestinian society. Civilian deaths, which have occurred during drone strikes on terrorists, can heighten sympathy for militant groups and animosity toward the state carrying out counterterrorism operations. Finally, future leaders can be more radical than prior leaders. In a very recent case,

---

55   Atran at 84.

56   Cronin, *How Terrorism Ends* ... (n 16) at 104.

57   Cronin references the decline of the Weather Underground, the Red Army Faction, and November 17 as examples of groups whose ideology became irrelevant. She also includes groups that were supported by the Soviet Union and thus became historically irrelevant, in this category. See Cronin, *How Terrorism Ends* ... (n 16) at 105.

58   The category "ideological organizations" includes both left and right-wing groups.

59   There are cases in which the opposite is true. However, I argue that this trend should be predominant.

60   Michael Freeman examines the difference between the effectiveness of targeting leaders with inspirational or operational functions. See Freeman (n 9).

the leader of the Islamic State of Iraq and Syria (ISIS), Abu Bakr al-Baghdadi has carried out a much more brutal and violent campaign than his predecessors. More radical leaders also emerged in the case of the Chechen militants and the IRA.

In the next section, I discuss how communal support is critical to understanding the strategic efficacy and the consequences of drone strikes against militants globally.

V. Drone Strikes and Al-Qaeda

The use of drone strikes in targeting terrorist leaders poses a unique challenge from a strategic perspective. While US strikes on terrorist organizations have successfully targeted many militants and their leaders, there is considerable debate regarding the use of drone strikes to bring about a group's demise. I argue that targeting leaders through unmanned strikes is not only unlikely to be effective, but it can bolster popular support for the organization, ultimately strengthening its resilience to leadership attacks. This section proceeds in two parts. First I look at data on drone strikes, and second I discuss how drone strikes affect communal support for militant organizations, specifically within the context of civilian casualties.

Many analysts and scholars argue that drone strikes are the best option available for targeting terrorist organizations.[61] They argue that drone strikes are the most discriminate means by which to target terrorists. Unmanned vehicles do not put soldiers or pilots at risk, a particularly salient point given the number of US soldiers killed in Iraq and Afghanistan.[62] Conventional military action can result in attacks directed at security forces and an increased sympathy for the militants cause. Others argue that the targeted nature of unmanned strikes limits the number of civilian casualties that can arise during conventional air strikes. However, studies on the efficacy of drone strikes to reduce militant activity and capacity have reached varying conclusions.[63]

---

61   There is a considerable debate regarding the effectiveness of drone strikes, the degree to which they result in civilian death, and adversely impact public opinion. Christine Fair argues that may be the best option available or the tribal areas of Pakistan. Christine Fair, "For Now, Drones are the Best Option" *New York Times* (Jan 29, 2013). Other studies argue that drones strikes result in considerable and unnecessary civilian casualties. See David Kilcullen and Andrew McDonald Exum, "Death From Above, Outrage Down Below" *The New York Times* (May 16, 2009); International Human Rights and Conflict Resolution Clinic (Stanford Law School) and Global Justice Clinic (NYU School of Law), *Living Under Drones: Death, Injury and Trauma to Civilians from US Drone Practices in Pakistan* (2012) (*Living Under Drones*, hereinafter); International Crisis Group, *Drones: Myth and Reality in Pakistan*, Asia Report No 247 (2013).

62   Brian Glyn Williams, *Predators: The CIA's Drone War on al Qaeda* (Potomac Books 2013) at 183.

63   For studies discussing the effectiveness of drone strikes see Patrick B Johnston and Anoop Sarbahi, *The Impact of U.S. Drone Strikes on Terrorism in Pakistan* (Berlfer

A. Organizational Degradation

Many of the arguments in favor of drone strikes do not rigorously evaluate their ability to actually weaken and destabilize militant organizations. There is disagreement regarding their ability to reduce militant activity and weaken terrorist organizations. In a recent study on drone strikes in Pakistan's Federally Administered Tribal Areas (FATA) from March 2004 to June 2010, Patrick Johnston and Anoop Sarbahi argue that drones strikes resulted in a modest decline in the frequency and lethality of terrorist attacks and a decline in the number of improvised explosive devices (IEDs) and suicide attacks.[64] James Walsh and Megan Smith offer a different measure for evaluating the effectiveness of drone strikes against Al-Qaeda. They argue that the goal of drone strikes is to degrade a terrorist group's capacity to carry out political and violent action, and the authors measure degradation as a decrease in a group's ability to generate and disseminate propaganda. If drones strikes degrade Al-Qaeda's capacity, then there should be a reduction in propaganda output. Unlike Walsh and Sarbahi's study, Walsh and Smith find that little evidence that drone strikes have undermined Al-Qaeda's operational capacity.

The extant literature on the efficacy of drone strikes focuses on targeting both militant leaders and lower level operatives; however, this chapter looks exclusively at strikes against militant leaders. While the data presented above in the chapter indicate that targeting leaders is not an effective way to undermine terrorist organizations, drone strikes pose a unique challenge in that there is even more potential for adverse outcomes. The remainder of this chapter will focus primarily on Al-Qaeda, as they have been the primary target of US counterterrorism policies. While there have been hundreds of drone strikes against terrorists, there have been 138 reported drone strikes against the leaders of terrorist organizations since 2001, with the majority occurring in Pakistan and most of them against Al-Qaeda or Al-Qaeda-affiliated organizations.[65] Figure 9.1 shows the number of drone strikes against militant leaders broken down by country. Figure 9.2 displays the number of drone strikes overall and the number of strikes against militant leaders, the focus of this study, from 2001 to 2012.[66]

Center for Science and International Affairs, JFK School of Government, Harvard University 2011); Megan Smith and James Igoe Walsh, "Do Drone Strikes Degrade al Qaeda? Evidence from Propaganda Output" (2013) 25 Terrorism and Pol Violence 311; Walsh and Schmitt (n 2).

64   Johnston and Sarbahi (n 63).

65   This data was collected from my own dataset on leadership targeting, and was supplemented and verified with data on drone strikes from databases held at the New America Foundation, the Long War Journal, and the University of Massachusetts-Dartmouth.

66   These figures aggregate strikes against Al-Qaeda to include affiliated organizations as well.

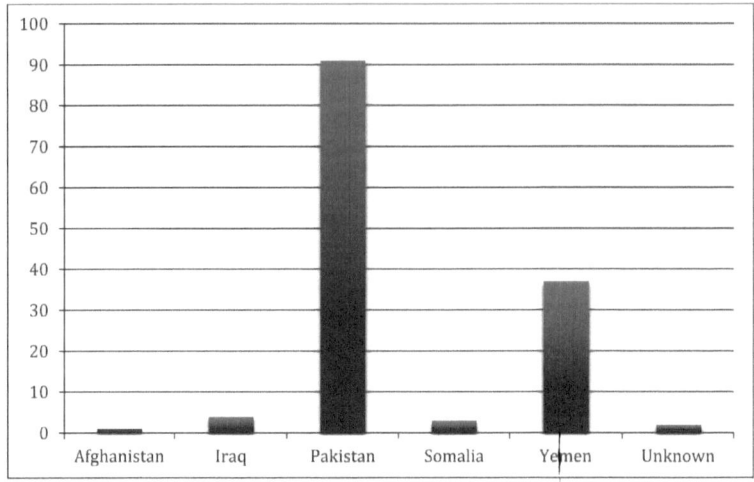

Figure 9.1   Drones strikes against terrorist leaders by country, 2001–2012

Source: Data from the New America Foundation and original data by Jenna Jordan.

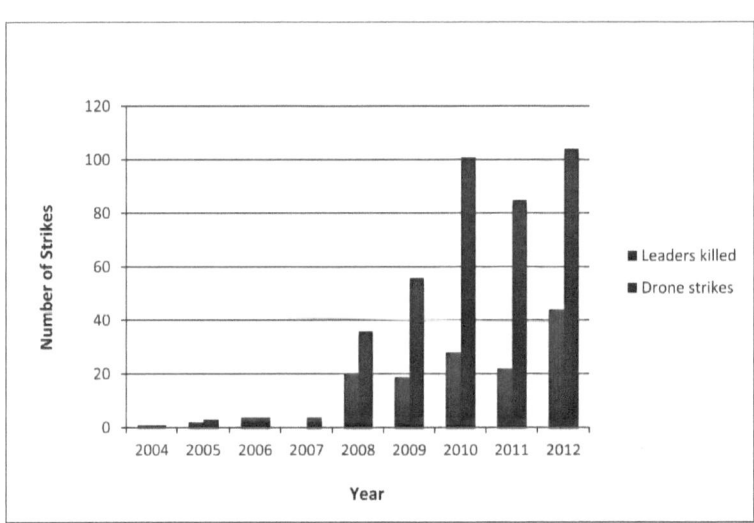

Figure 9.2   Number of drone strikes against militants and group leaders, 2001–2012

Source: Data from the New America Foundation

The statistical results presented earlier indicate that targeting groups like Al-Qaeda or the Taliban is not likely to result in a group's decline, and could actually increase support for the organizations, which increases its ability to survive targeted attacks.[67] While bin Laden's death may destabilize the group in the short term, decapitation alone is not likely to result in its demise.

First, Al-Qaeda, formed in 1988, is over 20 years of age, which should significantly increase its ability to withstand bin Laden's death. Decapitation is almost 20 percent less effective than doing nothing against groups between 21 and 30 years of age.[68] Second, Al-Qaeda is clearly a religious organization as its goals include the establishment of a pan-Islamic caliphate, overthrowing non-Islamic regimes, and expelling infidels from Muslim countries—which also tends to make terrorist groups more resistant to attacks on its leadership.[69] Finally, the issue of whether Al-Qaeda's size will work in its favor is less clear-cut, because experts disagree over its exact size.[70] Yet it is reasonable to believe that the group has over 500 militants—which would put it over the threshold at which terrorist organizations become better able to withstand decapitation. Even if 500 overestimates the number of active members, the rate of success for groups with between 100 and 500 members is still very small. In fact, the rate of collapse for decapitated groups of this size is about 45 percent less than non-decapitated groups of the same size.[71]

Looking at changes in Al-Qaeda's organizational activity provides some evidence for the claim that targeting militant leaders has not been effective in significantly reducing the frequency and lethality of Al-Qaeda attacks. In this chapter, Al-Qaeda refers to the larger umbrella organization, which includes Al-Qaeda's core and affiliated organizations. As a result, in evaluating the group's activity, I include attacks carried out by Al-Qaeda Central, Al-Qaeda in Iraq, Al-Qaeda in the Islamic Maghreb, and Al-Qaeda in the Arabian Peninsula. Figure 9.3 looks at the number of attacks carried by Al-Qaeda from 1998 to 2010, and Figure 9.4 looks at the lethality of Al-Qaeda attacks from 1998 to 2010.

---

67 See Jordan, "When Heads Roll" (n 9).

68 Ibid at 746

69 Cronin, *How Terrorism Ends ...* (n 16) at 182.

70 See Peter Bergen and Katherine Tiedemann, "The Almanac of Al Qaeda" Foreign Policy; and Peter Bergen and Bruce Hoffman, *Assessing the Terrorist Threat* (2010).

71 Jordan, "When Heads Roll" (n 9).

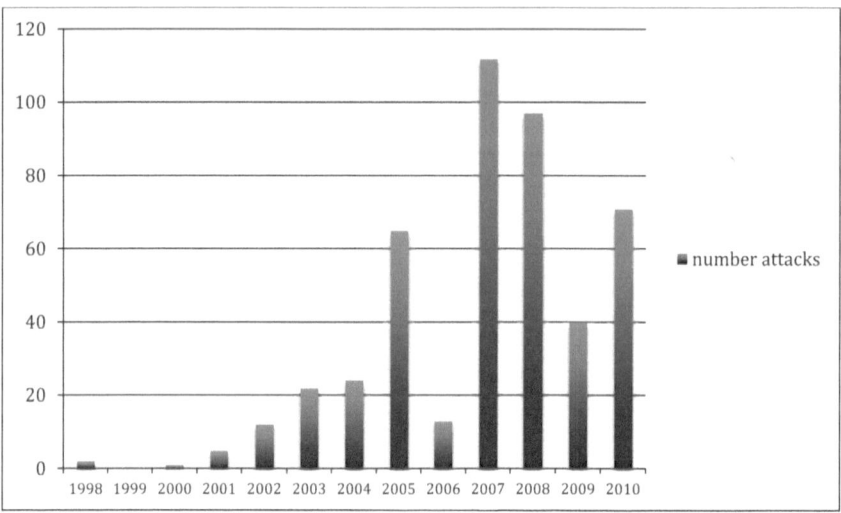

Figure 9.3   Number of Al-Qaeda attacks, 1998–2010
Source: Data from the National Consortium for the Study of Terrorism and Responses to Terrorism (2013). Global Terrorism Database.

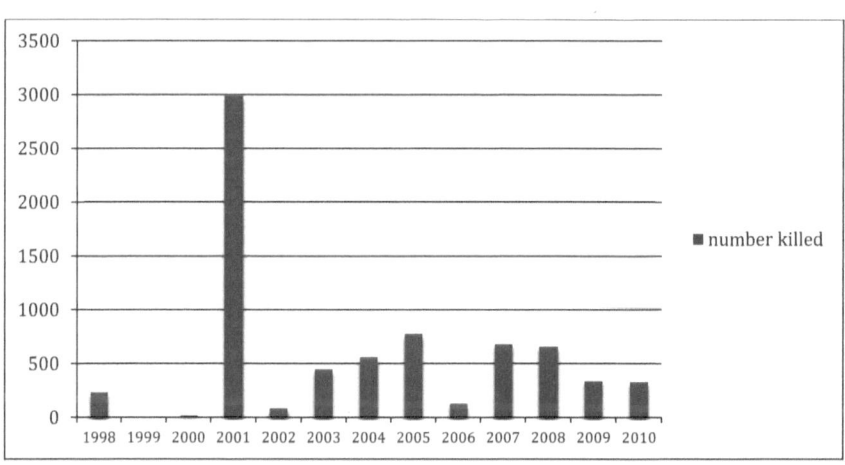

Figure 9.4   Lethality of Al-Qaeda attacks, 1998–2010
Source: Data from the National Consortium for the Study of Terrorism and Responses to Terrorism (2013). Global Terrorism Database

These figures indicate that while there have been periods of decline in the frequency and lethality of Al-Qaeda attacks, they continue to be operationally functional and lethal, despite attacks on its leadership beginning in 2001. In the remainder of this chapter, I will look at how Al-Qaeda's bureaucratic structure and popular support can explain its resilience to drone strikes.

B. Organizational Resilience

The theory of organizational resilience provides an explanation for why targeting and, specifically, drone strikes are unlikely to be effective against Al-Qaeda. Al-Qaeda is a quasi-bureaucracy, with a hierarchy in place at the upper levels of the organization. Al-Qaeda was originally a highly bureaucratized organization, yet as the organization became more decentralized, it retained elements of a well-organized group with a central command.[72] This structure facilitates an easy process of leadership succession.

Documents captured over the course of US counterterrorism operations, held in the Department of Defense's Harmony database, demonstrate Al-Qaeda's bureaucratic features.[73] Al-Qaeda's bylaws lay out the group's goals, principles, voting laws, administrative processes by which grievances can be filed, procedures for filing reports, details of the organizational structures, leadership and organizational duties, membership requirements, duties, financial policies, budgetary requirements, and policies for different committees (military, political, and security).[74] Al-Qaeda had employment contracts, which detailed duties, holidays, salaries, travel, rewards punishment, and a required an oath.[75] There are also hundreds of captured Al-Qaeda & Associated Movements (AQAM) documents that record intelligence and security information, operational activity, training, personnel, recruiting, strategy, and politics. Rosters are another important mechanism, critical for bureaucratic functioning. These documents illustrate Al-Qaeda's organizational structure, with an administrative staff, a hierarchy of authority, and clear rules, policies, and procedures.

Al-Qaeda core and many of its affiliates have taken on features of traditional bureaucracies. Al-Qaeda has taken advantage of political upheaval in places like Libya and Syria to reassert its relevance. This was particularly evident in the very public disputes between the Al-Qaeda-affiliated al Nusra Front and ISIS.[76] Zawahiri claimed that al Nusra was the only sanctioned Al-Qaeda affiliate in Syria, and that what is now the Islamic State should withdraw from the Syrian conflict and return to Iraq. The group has continued to expand to new areas through affiliated organizations emerging in context of failing states, and through developing ties with local jihadist groups. Overall, Al-Qaeda core's hierarchy,

---

72   Bruce Hoffman, *Combating Al Qaeda and the MIlitant Islamic Threat* (2006) at 3.

73   Harmony documents, original and translations, can be found through the Combating Terrorism Center available at <http://www.ctc.usma.edu/programs-resources/harmony-program> accessed Dec 2014.

74   Ibid at #AFGP-2002–600048.

75   Ibid at #AFGP-2002–600045.

76   See Liz Sly, "Al-Qaeda disavows any ties with radical Islamist ISIS group in Syria, Iraq" *Washington Post* (Feb 3, 2014) available at <http://www.washingtonpost.com/world/middle_east/al-qaeda-disavows-any-ties-with-radical-islamist-isis-group-in-syria-iraq/2014/02/03/2c9afc3a-8cef-11e3–98ab-fe5228217bd1_story.html> accessed Oct 2014.

while weakened, has adapted and withstood repeated leadership attacks. These bureaucratic features reduce the likelihood that drone strikes targeting Al-Qaeda's leadership will significantly weaken Al-Qaeda and its operational capacity.

Communal support is critical not only to a terrorist group's ability to function as a clandestine organization, but also its ability to withstand attacks on its leadership. In addition to a lack of consensus regarding their ability to weaken extremists groups like Al-Qaeda, it is important to consider whether the use of drone strikes are actually creating and radicalizing more terrorists. According to Mark Mazzetti, a new report of the Task Force on US Drone Policy conducted by a bipartisan panel found that the US "has yet to carry out a thorough analysis of whether the costs of routine secret killing operations outweigh the benefits. The report urges the administration to conduct such an analysis and to give a public accounting of both militants and civilians killed in drone strikes."[77] This section will address concerns regarding the costs of drone strikes. Looking at popular opinion regarding drones strikes, the remainder of this chapter will evaluate the potential for an increase in support and sympathy for militant groups and their cause.

One of the most widely debated aspects of drone strikes is the number of civilian casualties.[78] The New America Foundation estimated that out of 351 strikes, between 261 and 305 civilians were killed out of a total of 1,965 to 3,295 deaths, while the Bureau of Investigative Journalism estimates that out of 366 strikes, between 411 and 884 civilians were killed out of a total of 2,537 to 3,581 deaths.[79] The Pakistani government has estimated the number of civilian deaths to be between 400 and 600. These discrepancies are due in part to the clandestine nature of the organizations targeted by drone strikes, the secrecy of the program, and the difficulty of reporting in Yemen and FATA. Most datasets, including the Long War journal and the database held at the University of Massachusetts Dartmouth, tend to estimate the number civilian casualties at about 5 percent of the overall mortality rate.[80] In Chapter 11 of this volume, Steven J. Barela discusses some of the problems regarding analysis of strategic efficacy due to the secrecy of the US drone program. The United States, for example, has not released data on the number of drone strikes and civilian casualties. The discrepancy in data,

---

77   Mark Mazzetti, "Use of Drones for Killings Risks a War Without End, Panel Concludes in Report" *New York Times* (June 26, 2014).

78   There is considerable disagreement over the number of civilians or non-combatants that have been killed during drone strikes. For example the International Crisis Groups' report on drone strikes claim that, "[b]oth the international and Pakistani media often rely on figures provided by unnamed sources in the US government and/or Pakistani military, each with a vested interest in under-or overreporting civilian causalities." See International Crisis Group at 25; Christine Fair has argued that the number of civilian casualties in drone strikes has been overreported. Christine Fair, "Drone Wars" Foreign Policy; see also Christine Fair, "Drones, Spies, Terrorist, and Second-Class Citizens in Pakistan" (2014) 25 Small Wars and Insurg 205; Fair, "For Now, Drones are the Best Option" (n 61).

79   See International Crisis Group (n 78) at 8.

80   See Williams (n 62).

lack of transparency, and as Barela notes the difficulties in even identifying the "enemy," makes it hard to evaluate the efficacy of counterterrorism policies. (For and in-depth discussion of civilian casualties see Chapter 7 in this volume by A. Plaw and C. Colon.)

Regardless of the accuracy of drone strikes and the actual number of civilian casualties, media reporting on the accuracy of drone strikes has an effect upon public opinion. The Pakistani media has overstated the number of civilians killed, which according to Brian Glyn Williams inflames public opinion against drones.[81] He references an article in the *Dawn*, Pakistan's most prominent English-language newspaper, which reported that only 5 out of 44 drone strikes in a 12-month period were able to hit their intended target and that 700 innocent civilians were killed. While these claims were not backed by data, many Pakistanis believed that drones were highly inaccurate and resulted in many civilian deaths.[82] Other studies have also found that the occurrence of civilian casualties has an impact on support for drone strikes. A study by New York University (NYU) and Stanford University argues that because drone strikes have resulted in considerable civilian death, they have little local support.[83] The study concludes that of those familiar with the drone campaign, 94 percent of Pakistanis believe the attacks kill too many innocent people and 74 percent say they are not "necessary to defend Pakistan from extremist organizations."[84] However, analysts in both Pakistan and the United States have concluded that the number of civilian casualties is overreported.[85]

Recent public opinion polls indicate that there is considerable opposition to drone strikes, both in areas where strikes are conducted and more broadly. A 2013 PEW survey found that 7 percent of respondents from Turkey and Tunisia, 5 percent of respondents in Pakistan, Egypt, and Jordan, and 3 percent of respondents in the Palestinian territories approve of "the United States conducting missile strikes from pilotless aircraft called drones to target extremists in countries such

---

81   Ibid.

82   Ibid.

83   *Living Under Drones* (n 61) at 131–137.

84   Ibid.

85   For additional details on the debate over civilian casualties, beyond Chapter 7 by Avery Plaw and Carlos R. Colon in this volume, see Williams (n 62). There is also disagreement over what constitutes a civilian or combatant. The Pakistani military, for example, considers civilians who harbor Taliban militants to be militants themselves. It also considers civilians who have not fled a conflict zone ahead of a military operation to be militant sympathizers and thus targets. The US government argues that military-aged men, killed in a strike zone, are militants, unless proven otherwise. While there are considerable differences between existing databases on drone strikes, they all show a decline in the number of civilian deaths. Two more recent studies look at specific drone strikes to address some of the legal issues regarding civilian casualties. See Amnesty International, *"Will I Be Next?" US Drone Strikes in Pakistan* (2013); Human Rights Watch, *"Between a drone and al Qaeda" The Civilian Cost of US Targeted Killings in Yemen* (2013).

as Pakistan, Yemen and Somalia."[86] Another PEW survey found that "in nearly all countries, there is considerable opposition to a major component of the Obama administration's anti-terrorism policy: drone strikes. In 17 of 20 countries, more than half disapprove of U.S. drone attacks targeting extremist leaders and groups in nations such as Pakistan, Yemen and Somalia."[87] However, it is hard to get an entirely accurate view of public sentiment regarding drone strikes. Pakistani journalists can also be regularly coerced or threated by the local government's Inter-Services Intelligence (ISI) and by militants. Moreover, some opponents to drone strikes have an incentive to hide their opposition in order to avoid being seen as pro-Taliban, as a result the level of public opposition to drones could be much higher than reported. However, overall, the PEW reports indicate that there is considerable opposition to the use of drone strikes.

It is also important to consider whether drone strikes undermine counterterror operations, bolster support for the militants cause, contribute to radicalization, or increase the likelihood of further terrorist activity. A recent report of the Task Force on US Drone Policy released by the Stimson Center addressed the issue of support for drone strikes and their potential for adverse outcomes. The report claims that there is a large amount of resentment created by the use of unmanned strikes. The report references remarks by retired Army General Stanley McChystal, who claims that part of the backlash to drone strikes is due to the perception that such strikes result in a high number of civilian casualties and concerns about sovereignty, transparency, accountability, and other human rights and legal issues. The Task Force concludes that "civilian casualties, even if relatively few, can anger whole communities, increase anti-US sentiment and become a potent recruiting tool for terrorist organizations. Even strikes that kill only terrorist operatives can cause great resentment, particularly in contexts in which terrorist recruiting efforts rely on tribal loyalties or on an economically desperate population."[88]

Referring specifically the use of drone strikes in Yemen and the potential for radicalization, Robert Grenier, head of the CIA's Counter-Terrorism Center 2004–2006 argued:

> We have gone a long way down the road of creating a situation where we are
> creating more enemies than we are removing from the battlefield. That brings
> you to a place where young men [in Yemen], who are typically armed, are in
> the same area and may hold these militants in a certain form of high regard. If
> you strike them indiscriminately you are running the risk of creating a terrific

---

86   Pew Research Center, "Pew Research Global Attitudes Project," Washington DC (2014) available at <http://www.pewglobal.org/database/indicator/52/> accessed Oct 2014.

87   Pew Research Center, "Pew Research Global Attitudes Project," Washington DC (2010) available at <http://www.pewglobal.org/2010/07/29/concern-about-extremist-threat-slips-in-pakistan/> accessed Oct 2014.

88   Gen John P Abizaid and Rosa Brooks, *Recommendations and Report of the Task Force on US Drone Policy* (2014).

amount of popular anger. They have tribes and clans and large families. Now all of a sudden you have a big problem ... I am very concerned about the creation of a larger terrorist haven in Yemen.[89]

Even inflated reports of civilian casualties can be used to recruit volunteers to terrorist organizations. A report conducted by NYU and Stanford concludes that "it is clear that US strikes in Pakistan foster anti-American sentiment and undermine US credibility not only in Pakistan but throughout the region. There is strong evidence to suggest that US drone strikes have facilitated recruitment to violent non-state armed groups, and motivate attacks against both US military and civilian targets."[90]

Some scholars have argued that opposition is overstated.[91] Christine Fair argues in reference to the individuals she spoke with from South Waziristan and other FATA agencies, "*These* FATA residents are strong proponents of the drones. They report that the drones are so precise that the local non-militants do not fear them when they hear the drones above as they are confident that they will hit their target."[92] A report by the International Crisis Group highlights the significant consequences of drone strikes, such as civilian casualties, but concludes that there is minimal impact in terms of recruitment. Brian Glyn Williams argues that drones do not indiscriminately target civilians while evading their actual targets.[93]

In spite of the debate regarding civilian casualties and recruitment, it is the perception of these strikes that matters. Communal support for militant groups targeted by drones could be strengthened if there is a belief that they are indiscriminately killing civilians, violating sovereignty, and not adhering to the rule of law. Increased support for terrorist groups can strengthen the organization in a number of ways. As argued earlier in the chapter, high degrees of communal support is necessary for groups to function clandestinely and can allow an organization to withstand leadership attacks. Groups with substantial support from the communities in with they operate have an easier time finding new recruits.

89　Paul Harris, "Drone Attacks Create terrorist Safe Havens, Warns former CIA official" *The Guardian* (June 5, 2012).

90　*Living Under Drones* (n 61) at 124.

91　Christine Fair, Karl C Kaltenthaler, and William J Miller challenge reports that conclude that most Pakistanis oppose drone strikes in FATA. They argue that most surveys overlook those individuals in FATA, who are directly impacted by these strikes. They disaggregate Pakistani attitudes toward strikes in order to look at who oppose or support the use of drones in Pakistan. See Christine Fair, Karl C Kaltenhaler and William J Miller, "You Say Pakistanis All Hate the Drone War? Prove It" The Atlantic (Jan 23, 2013).

92　Christine Fair, "Drones Over Pakistan—Menace or Best Viable Option?" Huffington Post (Aug 2, 2010) <http://www.huffingtonpost.com/c-christine-fair/drones-over-pakistan----m_b_666721.html> accessed Oct 2014.

93　See Williams (n 62).

From a policy perspective, it is critical to consider whether unmanned strikes aimed at militant leaders are an effective means of reducing organizational activity and whether they bolster a group's support and foster an anti-US sentiment. The theory of organizational resilience and the data presented in this chapter indicate that drone trikes against Al-Qaeda are not likely to be an effective counterterrorism strategy and can have counterproductive outcomes by bolstering sympathy and support for the targeted organizations.

VI. Conclusion

Leadership targeting and the use of unmanned combat air vehicles have become a central feature of current counterterrorism policy, yet they are not always effective and in some cases have counterproductive outcomes. My research has found that targeting terrorist leaders is unlikely to result in the decline of religious, older, and larger organizations. These groups are likely to have considerable levels of communal support and bureaucratic structures, both of which increase a group's ability to withstand attacks on its leadership. Moreover, targeting groups that are well supported within their local communities, or that have a broad base of support, is likely to have adverse consequences. In these cases, the removal of high-level and visible leaders has the potential to result in retaliatory attacks, to further recruitment, and to increase radicalization in local communities.

The use of unmanned vehicles for targeting both leaders and lower level operatives is likely to remain a core aspect of US counterterrorism policy for some time. However, decapitation alone is unlikely to be effective against the organizations that are currently targeted by US drone strikes. Al-Qaeda and affiliated organizations show features of both support and bureaucracy that should increase their resilience to these counterterrorism measures. Applying the empirical findings regarding the efficacy of leadership targeting discussed in this chapter to the use of unmanned combat air vehicles poses some unique challenges. As Steven J. Barela argues in his Chapter 11 of this book, there are difficulties in finding and evaluating both data on drone strikes and public opinion toward these strikes. Such unmanned assaults have the potential to undermine counterterror operations, increase support for the militants cause, contribute to radicalization, or even increase the likelihood of retaliatory terrorist activity. Furthermore, the occurrence civilian deaths highlight these concerns. Opposition toward strikes from a legal, strategic, and moral perspective sentiment further reinforce these reservations regarding the potential for counterproductive outcomes. While targeting is likely to remain a prominent tactic in the fight against terrorism, these studies highlight the importance of continuing to investigate and evaluate the impact of these strategies on militant groups and on the local communities from which they emerge.

Chapter 10

Tactical Efficacy: "Notorious" UCAVs and Lawfare

Marek Madej

I. Introduction

It is a truism that technology has an impact on the way military conflicts are conducted and that technological advantages often lead to military superiority. However, in the current technology-rich environment, with—as sci-fi novelist William Gibson once (allegedly) said—the future is already present, but just not yet evenly distributed. Technology is commonly perceived as one of the most powerful guarantor of maintaining an advantage over potential enemies, but it also means that in this technological arms race states and societies have to move fast if they do not want to be left behind. Among technologies most intensively developed currently, and assessed as the most promising in securing a privileged position on the battlefields and in the global or regional balance of power, are unmanned robotized devices, including aerial vehicles or UAVs (unmanned aerial vehicles). However, although the employment of such machines in military operations has a relatively short history, they have already become as controversial in the eyes of the public opinion (all over the world) as they are promising for military strategists and defense industry engineers.

To some degree these controversies are the result of the mystery caused by the technological novelty of a new, highly "unconventional" category of weapons. Nevertheless, even more important is the way these vehicles are actually employed in military operations and so the issue of the legitimacy of the use of unmanned vehicles to exercise lethal force.

Therefore, it is hardly surprising that unmanned military equipment soon became one of the most hotly debated topics among academics, professionals, politicians, and the wider public alike. It is also rather understandable that in such circumstances these debates are focusing now on legal aspects of the problem, particularly the limits of legally justifiable use of unmanned vehicles in military operations. However, what is less debated, although equally important as a reason for growth in use by armed forces of such means of warfare, is the operational efficacy of unmanned vehicles as weapons. Specifically, a "tactical efficacy" is understood to mean the success of a "maneuver or plan of action designed as an expedient toward gaining a desired end or temporary advantage."[1] In other words, this chapter will primarily address the use of armed UAVs as a means. Nonetheless, as seen in the above definition, the "desired end" will also come into play for such an assessment.

Thus this chapter is meant to shed light on this specific aspect of the topic, and assess the (initial) usefulness of such technologies as a weapon in counterterrorist operations. To do so section II will provide a short history of the technological development and present some of the prevalent UCAV (unmanned combat aerial vehicle) equipment available today. Next, in section III, I will discuss the use of UCAVs for counterterrorism campaigns specifically and present the available data on strikes. Then in section IV, I will broach some current interpretations of existing international law to show how this operational aspect quickly bleeds over into a broader understanding of efficacy due to the growth in public attention to international legal norms. I will then discuss in section V how such increased awareness of legality has given rise to "lawfare"—resulting in a shift of short-term gains to be expected by killing suspected terrorists using strained interpretations of law to create more legal targets. To conclude, section VI will present what all of this portends for the future use of UCAVs for cross-border counterterrorism.

II. UAVs and UCAVs—A Very Brief History of Origins and Development

The term "unmanned aerial vehicle" is used to describe "a powered, aerial vehicle that does not carry a human operator, uses aerodynamic forces to provide vehicle lift, can fly autonomously or be piloted remotely and can be expendable or recoverable."[2] Currently, in many sources the term UAV is replaced by another acronym; UAS (unmanned aerial systems). It is used to show that in practice a UAV is always just an element of more complex system, consisting of a vehicle, a control element, a human element (operator), and communication and logistics architecture.[3]

Works on the military use of such machines, commonly called drones (as a form of robotized military equipment), were initiated as early as in the final years of World War II (Third Reich) or just after its end (United States). At this time, however, their tasks were largely limited to the role of being a remotely controlled target for shooting drills (rocket missiles were then perceived as much more promising). Additionally, unmanned technologies were developing very slowly and therefore, due to their low operational utility, as well as high failure rate (especially in comparison to manned aircrafts), most of military research programs in this area, conducted mainly in the

---

1   Entry for "tactical" at Dictionary.com, available at <http://dictionary.reference.com/browse/tactical?s=t> accessed Dec 2014.

2   US Department of Defense, *Unmanned Aircraft Systems Roadmap 2005–2030* (Office of the Secretary of Defense, DoD 2005) at 1. The possibility of multiple uses of UAVs differs from ballistic or cruise missiles or artillery projectiles.

3   Chris Jenks, "Law from Above: Unmanned Aerial Systems, Use of Force and the Law of Armed Conflict" [2009] N Dak Law Rev 650, at 652.

US, were gradually suspended or slashed. Some revival of unmanned technologies had been experienced in the early 60s, partially because of the shooting down of a US spying manned U2 aircraft over Soviet territory in 1960. This resulted in the introduction into the US army a small reconnaissance drone called the *Fire Fly*, which was utilized quite successfully during the Vietnam War.[4]

However, the real "new beginning" for unmanned military technologies had come with the advent of the revolution in military affairs (RMA), which took place in the late 70s and 80s. RMA, generally referring to the exploitation of the results of the information revolution and rapid development of digital technologies for military purposes, has drastically altered the picture of military relations and opened new opportunities for development. The most intensive research programs—both in a scientific and financial sense—were initiated in the US and Israel, with the focus mainly (but by no means exclusively) on reconnaissance, surveillance, and intelligence tasks. In 1985, as a result of a joint US-Israeli research program, a small tactical reconnaissance drone called the *Pioneer* (with some 100 nautical miles range) was introduced into the armed services of both countries.[5] Success of that project proved its high value during the Israeli intervention in the civil war in Lebanon and even today Pioneers are still used by the US and some other militaries spurring further research and development initiatives. Particularly in the US, these programs started to be aimed less on creation of more efficient tactical vehicles, but more on constructing UAVs with longer range and endurance (loitering time), bigger payload, and higher altitude, able to perform intelligence tasks over long distances.

In the early 90s, Americans introduced into service a reconnaissance drone called RQ-1 Predator, which has substantially longer range than its predecessor (up to 500 nautical miles), and was successfully tested on the battlefield during NATO operations in Bosnia & Herzegovina and Kosovo.

In 2001, the upgraded version of Predator was introduced: the Predator MQ-1 model. The main modification of MQ-1 was the ability to carry and fire two air-to-surface Hellfire rockets (initially designed for attack helicopters). That makes this model the first operational armed UAV, otherwise known as the UCAV.[6] Because a number of this first armed model were lost in action due to foul weather, later versions were fitted with de-icing systems, an uprated turbocharged engine, and improved avionics, thus giving birth to the more predominant MQ-1B Predator. And because this book is concerned with the legal, moral, and efficacy implications of UAVs as a weapons platform for counterterrorism operations across international borders, the UCAV sits at the center of our analysis.

---

4   Peter W Singer, *Wired for War: The Robotics Revolution and Conflict in the 21st Century* (Penguin 2009) at 54–55.

5   Richard Schwing, *Unmanned Aerial Vehicles: Revolutionary Tools in War and Peace* (US Army War College 2007) at 7.

6   UAS Roadmap (n 2) at 4.

Figure 10.1   MQ-1B Predator

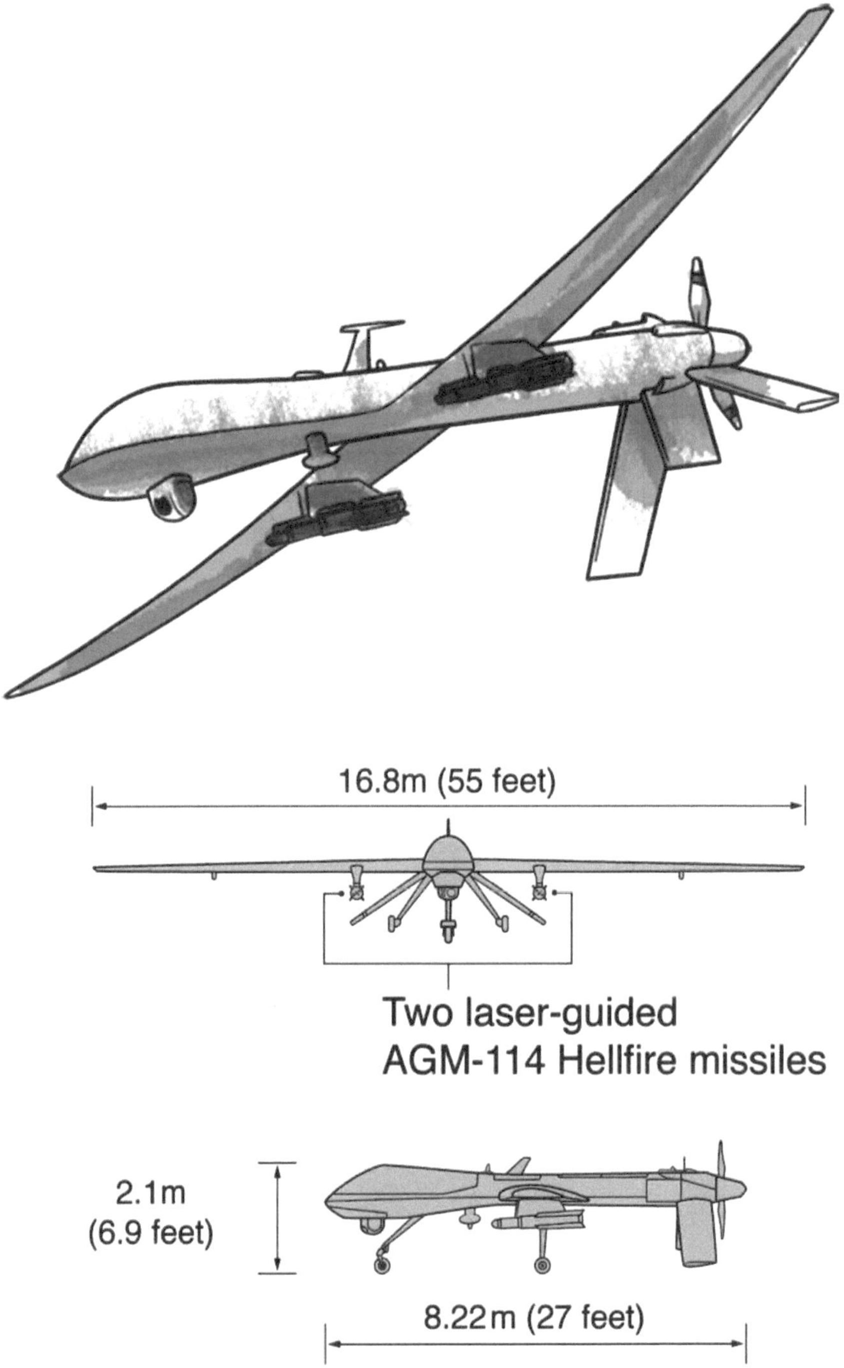

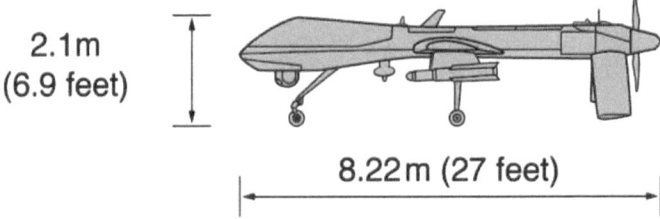

16.8m (55 feet)

Two laser-guided
AGM-114 Hellfire missiles

2.1m
(6.9 feet)

8.22m (27 feet)

General Characteristics of the MQ-1B Predator[7]

Primary Function: Armed reconnaissance, airborne surveillance, and target acquisition
Contractor: General Atomics Aeronautical Systems Inc.
Power Plant: Rotax 914F four-cylinder engine
Thrust: 115 horsepower
Wingspan: 55 feet (16.8 meters)
Length: 27 feet (8.22 meters)
Height: 6.9 feet (2.1 meters)
Weight: 1,130 pounds (512 kilograms) empty
Maximum takeoff weight: 2,250 pounds (1,020 kilograms)
Fuel Capacity: 665 pounds (100 gallons)
Payload: 450 pounds (204 kilograms)
Speed: Cruise speed around 84 mph (70 knots), up to 135 mph
Range: Up to 770 miles (675 nautical miles)
Ceiling: Up to 25,000 feet (7,620 meters)
Armament: Two laser-guided AGM-114 Hellfire missiles
Crew (remote): Two (pilot and sensor operator)
Unit Cost: US$20 million (includes four aircraft with sensors, ground control station and Predator Primary satellite link) (fiscal 2009 dollars)
Initial operational capability: March 2005

The success—defined by finding and eliminating targets—of the MQ-1B in military and anti-terrorist actions in the first years of the "war on terror" led to further development of UCAV technologies. This resulted in an upgraded, more capable (greater payload, longer endurance, longer range, etc.) successor to the MQ-1B Predator: the MQ-9 Reaper.[8] Most significantly, the increased payload allows for a substantial expansion of armaments. That is, while the Predator carried only two air-to-surface Hellfire rockets, the Reaper drone can carry up to four Hellfire II anti-armor missiles and two laser-guided bombs (GBU-12 or EGBU-12) and 500lb GBU-38 JDAM (joint direct attack munition) making it the most heavily armed UCAV to date.

---

7   United States Air Force, MQ-1B Predator, General Characteristics, Published July 20, 2010, available at <http://www.af.mil/AboutUs/FactSheets/Display/tabid/224/Article/104469/mq-1b-predator.aspx> accessed Dec 2014.

8   Other R&D projects were advancing simultaneously on Intelligence, Surveillance and Reconnaissance (ISR) drones. For example, since 2002 the US military has been using Boeing's *Global Hawk* surveillance UAVs with a range of some 5,400 miles and altitude of up to 20,000 feet. At the same time, the *Global Hawk* is one of the most expensive UAVs—the price of one vehicle in basic version is some USD141 million: Jeremiah Gertler, "US Unmanned Aerial Systems" (2012), CRS Report for Congress R42136, at 36–39, available at <https://www.fas.org/sgp/crs/natsec/R42136.pdf> accessed July 2014.

Figure 10.2   MQ-9 Reaper

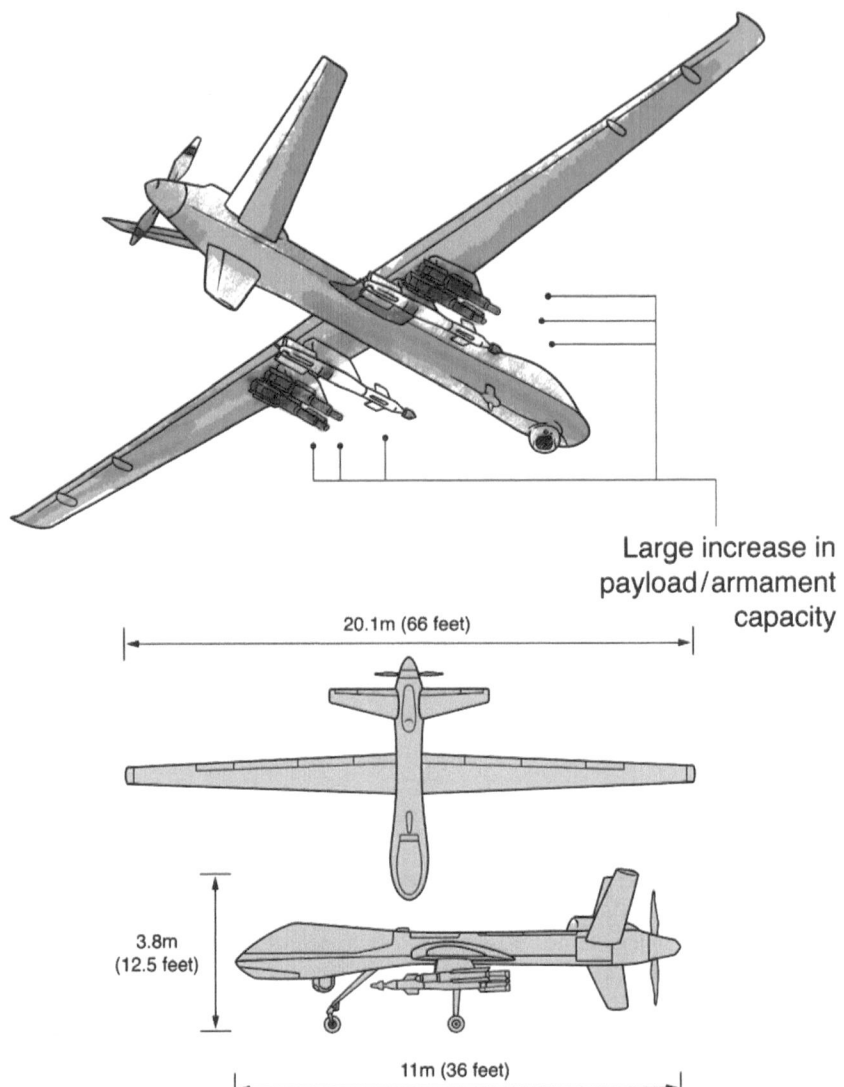

Large increase in
payload/armament
capacity

20.1m (66 feet)

3.8m
(12.5 feet)

11m (36 feet)

General Characteristics of the MQ-9 Reaper[9]

Primary Function: Intelligence collection in support of strike, coordination, and reconnaissance missions
Contractor: General Atomics Aeronautical Systems, Inc.
Power Plant: Honeywell TPE331–10GD turboprop engine
Thrust: 900 shaft horsepower maximum
Wingspan: 66 feet (20.1 meters)
Length: 36 feet (11 meters)
Height: 12.5 feet (3.8 meters)
Weight: 4,900 pounds (2,223 kilograms) empty
Maximum takeoff weight: 10,500 pounds (4,760 kilograms)
Fuel Capacity: 4,000 pounds (602 gallons)
Payload: 3,750 pounds (1,701 kilograms)
Speed: Cruise speed around 230 miles per hour (200 knots)
Range: 1,150 miles (1,000 nautical miles)
Ceiling: Up to 50,000 feet (15,240 meters)
Armament: Combination of AGM-114 Hellfire missiles, GBU-12 Paveway II and GBU-38 Joint Direct Attack Munitions
Crew (remote): Two (pilot and sensor operator)
Unit Cost: US$56.5 million (includes four aircraft with sensors, ground control station and Predator Primary satellite link) (fiscal 2011 dollars)
Initial operating capability: October 2007

In the first decade of twenty-first century, unmanned devices have both quickly developed and become much more common. Although in 2004 up to 40 states had some capabilities of this kind at their disposal (developed indigenously or, much more often, imported), by 2011 that number reached 76.[10] In the majority of cases, however, these capabilities are limited to the "micro-drones" or tactical UAVs at best. They are usually very small and light, sometimes so light that they are even launched by throwing them in the air directly by the operator. In addition, they have a short range (up to 200 km at best), low altitude and endurance, and are strictly for reconnaissance purposes. Nevertheless, the number of countries with more advanced and complex UAV technologies, including MALE/HALE reconnaissance drones or even UCAVs at their disposal, is increasing quickly and quite substantially.[11]

---

9   United States Air Force, MQ-9 Reaper, General Characteristics, Published Aug 18, 2010, available at <http://www.af.mil/AboutUs/FactSheets/Display/tabid/224/Article/104470/mq-9-reaper.aspx> accessed Dec 2014.

10   Sarah Kreps, Mikah Zenko "The Next Drone Wars: Preparing for Proliferation" [2014] Foreign Affairs 68, at 72.

11   MALE—*Medium Altitude, Long Endurance* category of UAVs consists of vehicles with ranges from 200 to 500 kilometers and altitude up to 9 kilometers. HALE (High Altitude, Long Endurance) UAVs are those with the maximum altitude above 9 kilometers and a range longer than 500 kilometers. Cf. CRS Report (n 6) at 8.

The United Kingdom army already uses combat drones purchased from the US. Some other countries—including, among others, France, Canada, Italy, and Turkey, but also Iraq—have already issued requests to the US government for the export of UCAVs technologies. Many others are planning such buys. Moreover, some countries like France, the UK, Russia, China, Republic of South Africa, and Iran (and even relatively poor and unstable countries like Pakistan) are trying to develop such advanced UAV/UCAV technologies indigenously. For example, Chinese and Iranian authorities claim that they have already constructed indigenously developed multi-role UAVs, capable of carrying weapons and performing lethal missions.[12] The drone market is expected to grow from USD5.2 billion in 2013 to USD8.35 billion by 2018.[13] Thus even though the US and Israel are still currently leaders in this segment of military capabilities, UCAVs particularly, the present pace of technological advancement in this field could soon bring about a change.

There are at least two additional developments worth mentioning in this regard. Firstly, in the wake of the "Charlie Hebdo" attacks in Paris on the morning of January 7, 2015, it is of interest to present the burgeoning UCAV capacities of France. This is certainly not to suggest that France is on the cusp of launching its own drone campaign across international borders in the name of counterterrorism. Rather, the intention is to highlight the fact that the expanding military capabilities in combat drone technology are widespread and that more and more governments will likely be facing pressure to deploy such means in the future if there are no shared constraints and uniform standards.

Specifically, under a program partnership of six European countries (France, Italy, Sweden, Spain, Greece, and Switzerland) the French aircraft manufacturer of military, regional, and business jets—Dassault Aviation—has produced the nEUROn demonstrator whose maiden flight took place in December 2012. Additionally, in March 2014, the nEUROn took part in the world's first flight operation with a combat drone flying in formation with other aircraft—a Rafale fighter and a Falcon 7X business jet.[14] The UCAV will have a stealth airframe design to penetrate undetected and the capacity to launch laser-guided bombs from an internal weapons bay.

---

12   Mikah Zenko, Sarah Kreps, "Limiting Armed Drone Proliferation" (2014), CFR Special Report No. 69,6; DF Holman, "The Future of Drones in Canada: Perspectives from a Former RCAF Fighter Pilot" (2013) Strategic Studies Working Group Papers, available at <http://www.cdfai.org/PDF/The%20Future%20of%20Drones%20in%20Canada.pdf> accessed July 2014.

13   Ibid, Zenko and Kreps at 7.

14   Dassault Aviation, External Relations and Corporate Communication, "UCAV: A World First for Dassault Aviation" (April 12, 2014) available at <http://www.dassault-aviation.com/en/dassault-aviation/press/press-kits/ucav-world-first-dassault-aviation/> accessed Dec 2014.

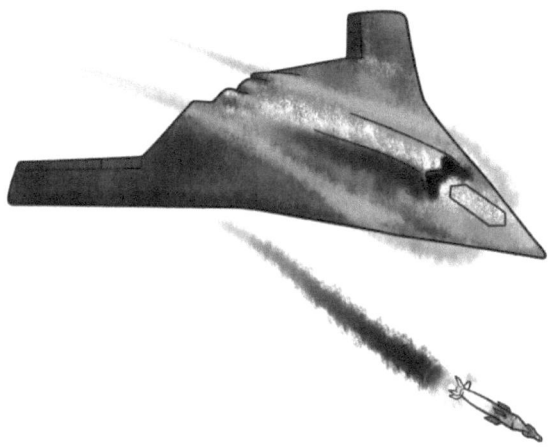

Figure 10.3   Dassault nEUROn

The second issue worth noting is that the Obama administration issued new rules for the international export of UCAVs in February 2015.[15] While the US State Department said that the new policy sets strict standards for the sale of combat drones—that is, the export rules will not permit buyers to conduct "unlawful" operations—this is a highly contentious issue as can be seen throughout this volume. Perhaps most importantly, and as was keenly expressed by the RMA specialist Peter W. Singer, "whether it's an F-16, an armed drone or a billy club, once you sell it to another country, you lose control over how it's used."[16] It is for this very reason that formulating international legal provisions that maintain the highest possible level of transparency as this technology develops and proliferates worldwide is advocated for in the conclusion of this chapter.

---

15   Missy Ryan, "Obama Administration to Allow Sales of Armed Drones to Allies" *Washington Post* (Feb 17, 2015) available at < http://www.washingtonpost.com/world/national-security/us-cracks-open-door-to-the-export-of-armed-drones-to-allied-nations/2015/02/17/c5595988-b6b2–11e4–9423-f3d0a1ec335c_story.html> accessed Feb 2015.

16   Cited in Scott Shane, "New Rules Set on Armed Drone Exports" *New York Times* (Feb 17, 2015) available at <http://www.nytimes.com/2015/02/18/world/new-rules-set-on-armed-drone-exports.html?_r=0> accessed Feb 2015.

III. UCAVs in the Fight against Terrorism

As mentioned, the acceleration of work on drone technologies has largely coincided with (or in fact was spurred by) the proclamation of the "war on terror" by the US in response to the 9/11 attacks. However, drones were already used in the fight against terrorism even earlier, for example to conduct surveillance and observation of terrorist training camps and bases in Afghanistan in the late 90s.[17] Nonetheless the decision has stimulated development of international cooperation in fighting to reduce or even eliminate threats posed by terrorism (primarily of radical Islamic fundamentalist groups) and simultaneously led to the increase of importance of military means and methods of counterterrorism, as well as the frequency of unilateral actions. Such a defined "war" on terror was also an important impulse for the development of UAVs technologies. Drones, for various reasons, started to be viewed as a promising tool of fighting terrorism.

On a technical level, drones' main advantage in comparison to traditional military means of fighting terrorism (like special units raid-type operations or conventional aircraft strikes) was their ability to conduct on regular basis missions of continuous observation-surveillance on a given area or target over long distances. This is mainly a result of their relatively long endurance. For example, the Predator drone could operate in the air for more than 24 hours, loitering over the target area for some 10 to 20 hours, depending on the distance from the landing site. The Reaper has similar capabilities—it could stay in the air for 28 hours, even with full payload (1,400 kilos, up to 14 Hellfire missiles). That ability of staying longer in the area of operation is primarily the consequence of the lack of the crew, which for obvious reasons cannot conduct its operational activities indefinitely, as it would be exhausted after some time and requires significant number of additional equipment on board (life support systems, etc.).[18]

Moreover, UAVs nowadays are still in great majority relatively cheap. However, their other characteristics—speed, payload, maximum altitude, maneuvering—seems currently less attractive than in case of conventional aircraft. Nevertheless, the pace of technological advancements in UAV technologies suggests that at least some of these characteristics, in particular the maximum payload or altitude, could be significantly and rather quickly upgraded (also because designers are not restricted by the crew safety requirements). Improving in other areas, particularly speed, would be much more demanding and could be conditioned by the increase in autonomization and automatization of their functioning (remote control over the UAVs—as in all current

---

17   Peter Bergen, Katherine Tiedemann, "Washington's Phantom War. The Effects of the US Drone Program in Pakistan" [2011] For Aff available at <http://www.cfr.org/wars-and-warfare/washingtons-phantom-war/p25381> accessed July 2014.

18   Singer (n 4) at 116–120.

vehicles of that kind—means that there are some delays between, for example, a decision about the change of the direction of flight and its execution by the drone).[19]

This extended loitering time makes it easier to identify correctly potential targets, to recognize daily practices of such persons, and to decide on the proper time and circumstances for a potential strike. This also offers an important psychological impact. Because persons under seemingly permanent watch by the vehicles, which they can often see or even hear, could feel cornered and under an inescapable pressure, there would conceivably be a significant mental consequence. Finally, the attack would be conducted by a remotely controlled machine; that is, "the enemy" would be "beyond of the reach" of the person attacked.

Additionally important are the political advantages of the use of UCAVs in fighting terrorism, and such advantages were probably decisive for the US decision to intensify drone operations as a central element of their counterterrorism actions, particularly during the first Obama term. Because of ambiguities in the legal status of such attacks (real or invented, as it will be discussed in next paragraphs), along with a lack of direct presence of the US personnel in the area of operation, many of the key facts for evaluation are, as practices of the initial years of the drone "war" on terrorism have shown, generally unknown to wider public and kept in secret. Therefore, they could largely escape from clear assessment in context of their legality under the laws of war, not to mention the ethical or moral aspect of the issue. (For further discussion of the strategic implications of this secrecy see Chapter 11 by Steven J. Barela.)

All this helps to understand why, soon after 9/11, actions with UAVs started to be executed on regular basis within the counterterrorism framework. Initially, as it was mentioned, their functions were almost entirely limited to surveillance and intelligence tasks, particularly during the US-led international intervention (Operation Enduring Freedom) in Afghanistan to oust the Taliban regime and then to counter remnants of their forces, allies from Al-Qaeda, and other terrorist groups.

However, as early as in November 2002, a UCAV was used for the first time in the fight against terrorism as a means of so-called targeted killing[20] outside the one field of combat in Afghan territory at that moment. In Yemen, a Hellfire

---

19 Marcel Dickow and Hilmar Linneskamp, "Combat Drones-Killing Drones" (2013) SWP Comments, available at <http://www.swp-berlin.org/fileadmin/contents/products/comments/2013C04_dkw_lnk.pdf> accessed July 2014.

20 The term "targeted killing" is a category neither clearly defined nor universally recognized in international law. According to Nils Melzer, by targeted killing we could understand an action characterized by five criteria: the use of lethal force; the killing is intentional, premeditated and deliberate; the target is an individually selected person (unlike unspecified collective targets, ie enemy soldiers); there has been no physical custody or due process before the killing; it was carried out by a state, or an actor that is subject to international law. Targeted killing defined in this way could be committed even in cases when the attacked is not posing a direct, imminent threat. Therefore, decisive factor for labeling some specific action as targeted killing is it predefined purpose—elimination of the given person. Cited in Nathalie van Raemdonck, "Vested Interest or Moral Indecisiveness? Explaining the EU's

rocket missile fired from a Predator UCAV (operated by Central Intelligence Agency (CIA), not US military staff) killed Qaed Senyan al-Harithi, one of the alleged leaders of the Yemeni branch of Al-Qaeda, responsible—inter alia—for the organization of successful attack on the *USS Cole* destroyer in 2000.[21]

This success, however, did not result in immediate and rapid growth in similar strikes in other regions where the counterterrorist operations were ongoing, including Afghanistan and Pakistan. UCAVs missions with characteristics of targeted killing on the borderland areas of these two states were only initiated in 2004. However, until 2008 such attacks were relatively rare (see Table 10.1). Only in the final year of the George W. Bush second term did the number of anti-terrorist UCAVs strikes start to grow significantly. This was mainly because of rapid destabilization in the Pakistani northern provinces (also due to the increase in violent activities of Tehrik Taliban Pakistan, an anti-American, but primarily anti-governmental group of "Pakistani Talibans" which emerged in December 2007); ineffectiveness (or rather idleness) of Pakistani security forces in fighting local terrorist structures; inability of US forces to directly intervene there due to opposition of the authorities in Islamabad to any direct American military presence in northern Pakistan; and a lack of an appropriate international—that is, UN—mandate for such an endeavor.

Counterterrorism operations with the use of drones were dramatically intensified in 2009, under the new US president Barack Obama. There were several reasons for this. Definitely significant was the interest of US authorities in improving the effectiveness of American military operations in Afghanistan, where, despite increasing spending and growth in numbers of troops from the United States (as well as other NATO and non-NATO participants in the International Security Assistance Force (ISAF)), the security situation was steadily worsening. One of the most important reasons for this was the existence of safe bases and supply routes of the Taliban in northern Pakistan, which meant that they were outside the reach of the international forces deployed in Afghanistan.

Undoubtedly, the intention of Obama's administration was also to increase an overall effectiveness of the global fight against terrorism. The idea was that this could be achieved by intensified and more precise attacks primarily on leaders of the most dangerous groups with headquarters located in northern Pakistan (but also active in other parts of the world), not their rank and file members. However, at least equally important was the desire to shape the global conflict, in particular the fighting in Afghanistan (highly criticized within Western societies already in 2009), in a much less controversial way for the sake of public opinion. The turn to the use of UCAVs as a "surgical" and "safe" weapon (primarily for the operators) was meant to help improve the public image of the US manner of fighting terrorism,

---

Silence on the US Targeted Killing Policy in Pakistan," IAI Working Papers 12/05, 3 available at <http://www.iai.it/pdf/DocIAI/iaiwp1205.pdf> accessed July 2014.

21   Michael N Schmitt, "Precision Attack and International Humanitarian Law" [2005] Int'l Rev Red Cross 445, at 448.

particularly in context of humanitarian issues and respect for human rights. In other words, the new American administration wanted to both increase actual effectiveness of the fight against terrorists and Taliban insurgency and improve its public image among American, Western, as well as other states' societies. It should be remembered that this image had been profoundly weakened by the then-recent experience of Guantánamo, the invasion of Iraq, the ill treatment of Abu Ghraib prisoners, and press revelations of a CIA program of torture and secret prisons.

With no acceptance from Pakistan for open foreign (or simply US) military involvement in anti-terrorist and counterinsurgency operations on its territory, the somewhat more "discreet" drone missions seemed to be an almost perfect (or at least the best available) solution.[22] The use of drones offered a no-casualty risk, only minimal risk of loses in equipment, minimal attention—at least initially—by US citizens (and other citizens), the possibility of avoiding (or at least by-passing) Pakistani authorities' reservations, and was relatively cheap.[23] As a result, between 2009 and 2011 in the Federally Administered Tribal Areas (Pakistani province bordering with eastern Afghanistan) US forces conducted at least 249 attacks by drones, of which as many as 122 were done solely in 2010 (see Table 10.1). Later on, partially due to growing criticism from Pakistani government and society, as well as intensified coverage of US drone campaign in northern Pakistan in the media, UCAVs missions were conducted less frequently and in the first five months of 2014 were even stopped entirely. However, US forces resumed drone attacks there in June 2014.[24]

Importantly, since their introduction into service, the unmanned vehicles were commanded and operated by the US military personnel—both reconnaissance and combat—on the territory of Afghanistan (as a theater of widely recognized armed conflict). Actions over Pakistani territory, as counterterrorist strikes, were coordinated by the CIA (a civilian institution). This somewhat peculiar solution has complicated to some degree the legal assessment of these operations in the light of international law regulations. However, even more important were the political and practical consequences of such an arrangement. In the context of the appropriate procedures and modes of conduct of drone missions, target selection,

---

22   Hillel Ofek, "The Tortured Logic of Obama's Drone War" [2010] The New Atlantis 35.

23   However, despite public objections to any external intrusions, unofficially the Pakistani government has tolerated or even supported UCAVs strikes, exploiting them to some degree for its own purposes of suppressing Pakistani local fundamentalist rebel groups or even criminal gangs in exchange for influence on US decisions on targeting priorities by offering intelligence data about potential targets. See Daniel Byman, "Why Drones Work: The Case for Washington's Weapon of Choice," (2013) 92(4) For Aff 32, available at <www.brookings.edu/research/articles/2013/06/17-drones-obama-weapon-choice-us-counterterrorism-byman> accessed July 2014.

24   New America Foundation, International Security, "Drone wars Pakistan. Analysis" available at <http://securitydata.newamerica.net/drones/pakistan/analysis> accessed July 2014.

Table 10.1   UCAV strikes in Northern Pakistan, 2004–2014

Year	New America Foundation		The Bureau of Investigative Journalism	
	No. of attacks	No. of victims* (incl. assumed militants)	No. of attacks	No. of victims (incl. assumed civilians)
2004–2007	10	179 (57)	51**	410–595
2008	36	298 (223)		(167–332)**
2009	54	549 (387)	52	465–744 (100–210)
2010	122	849 (778)	128	751–1108 (84–196)
2011	73	517 (420)	75	363–666 (52–152)
2012	48	306 (268)	50	199–410 (13–63)
2013	27	153 (145)	27	108–195 (0–4)
2014	22	151 (151)	25	115–186 (0–2)

Note: * an average of high and low estimates, ** data for whole period 2004–2008
Source: "Drone wars Pakistan. Analysis" (New America Foundation—NAF) <http:// securitydata.newamerica.net/drones/Pakistan/analysis>; "Get the data: Drone wars" (*The Bureau of Investigative Journalism—BIJ*) <www.thebureauinvestigates.com/category/ projects/drones/drones-graphs/> accessed February 2015.

ordering of strikes, etc., the intelligence agencies did not necessarily feel bound by the elaborated rules of engagement as did the military forces. Decision-making procedures can be more flexible, and are of a higher level of secrecy, so the supervision over operations certainly appears to be less strict.

Apart from Pakistani-Afghani borderlands, counterterrorism missions with lethal use of drones and coordinated by CIA were executed in at least in two other countries: Yemen and Somalia.[25] Between 2010 and 2014 up to 101 attacks by drones took place in Yemen (only the one strike mentioned on Harithi in 2002 occurred in the previous eight years). Additionally, counterterrorist strikes by drones have been committed, although much more sporadically, in Somalia since 2007 (some five to eight cases until the end of 2014, with some 10 to 24 casualties).[26]

---

25   At least one anti-terrorist drone strike was likely conducted in the Philippines in 2006: see Zenko and Kreps (n 12) at 9. Another area of operation of American UCAVs are territories in Northern Iraq and parts of Syria controlled since June 2014 by Sunni extremists from Islamic State of Iraq and the Levant (ISIL, also known as Islamic State or Islamic State of Iraq and Syria).

26   The Bureau of Investigative Journalism, "Drone Warfare" available at <www. thebureauinvestigates.com/category/projects/drones/drones-war-drones/> accessed July

Table 10.2   UCAV strikes in Yemen, 2002–2014

Year	New America Foundation		The Bureau of Investigative Journalism	
	No. of attacks	No. of victims (incl. assumed militants)*	No. of attacks (solely confirmed)	No. of victims (incl. assumed civilians)
2002–2009	1	6 (6)	14–17**	53–109**
2010	1	7 (2)		(16–36)
2011	9	102 (81)		
2012	47	487 (451)	29–36	173–215 (7–12)
2013	25	133 (114)	16	63–99 (17–26)
2014	19	128 (124)	13–15	82–118(4–9)

Source: Drone Wars Yemen. Analysis, <http://securitydata.newamerica.net/drones/yemen/analysis>; Get the data: Drone wars, The Bureau of Investigative Journalism, <http://www.thebureauinvestigates.com/category/projects/drones/drones-graphs/> accessed February 2015.

Note: * an average of high and low estimates; total number of victims of US special operations, incl. altogether 15 airstrikes between 2007–2014, ** data for whole period 2002–2011

IV. Operational Aspects

It is rather commonly assumed in the public debates that the legal characterization of lethal UCAVs missions is somewhat ambiguous. However, in fact it is possible to delineate the applicable law quite precisely within the current legal framework of international humanitarian law (IHL) and human rights regulations. As pointed out, UCAVs are simply a weapons platform subject to existing law. Existing doubts on the issue are in fact caused not by the specificity and unique characteristics of combat uses of UCAVs as such, but rather by possibilities of various interpretations (or, more precisely, claims that such possibilities exists) in the legal regime in which they are conducted. In fact, like in the case of other forms of targeted killing (special forces' operations, snipers, etc.), the legality of drone strikes depends primarily on whether they are conducted within the legal framework of armed conflict (inter- or intrastate) or not.

2014; see also Leila Hudson, Colin S Owens, Matt Flannes, "Drone Warfare: Blowback from the New American Way of War" [2011] Middle East Policy, available at <www.mepc.org/journal/middle-east-policy-archives/drone-warfare-blowback-new-american-way-war> accessed July 2014; and Sudarsan Raghavan, "In Yemen, US airstrikes breed anger, and sympathy for al Qaeda" *The Washington Post* (May 30, 2012).

The key problem here is to establish if the drone operations (combat in particular) in northern Pakistan, Yemen, Somalia, or elsewhere constitute an element of some kind of armed conflict (of any type) and a form of self-defense or not, as well as to decide to what extent (if any) conducting such actions by the US (or other) forces on the territory of third states constitutes a breach of these countries' sovereignty.

Although promoted rather consistently by US authorities, Washington's position on the legality of their drone strikes seems to be built on somewhat shaky fundaments. Firstly, it is based on the assumption that US fight by military means against terrorism, proclaimed by President George W. Bush after 9/11 as "war on terror" and continued (although not under such label) up until today, constitutes a specific, hybrid form of armed conflict—namely a transnational non-international armed conflict[27]—and the concept of such conflict is legally sound and universally recognized as in accordance with IHL and other international regulations. Only then can the regular use of lethal force with the intention to kill, especially on foreign territory outside the conflict-zone, be viewed as legal. This is, however, hardly the case because the very concept of such a hybrid, transnational non-international conflict as a distinct (in legal terms) category of armed conflict is disputable at best, if not incompatible with IHL regulations in their current common understanding. (For an in-depth discussion of the geographical scope of humanitarian law see Chapter 3 by Katja Schöberl.)

Highly disputable is also the second US assumption crucial for legality of drone counterterrorist strikes as a form of legitimate self-defense: that those who are the targets of US forcible activities within the framework of counterterrorism, often most referred as Al-Qaeda and its cooperatives, constitute the very same entity that attacked US on 9/11 and therefore, as internationally recognized (by the UN Security Counsel resolutions) aggressors, could be the targets of a US military response. Claims, particularly after more than a decade from New York City and Washington attacks, that those attacked by drones in Pakistan, Yemen, Somalia, or (potentially) elsewhere are members of the one big structure under unified command, with some permanent organizational linkages, and who communicate with each other on a more or less regular basis, are almost—if not totally—impossible to prove. This is especially the case if we take into account the level of decentralization of current terrorist organizations and depth and pace of changes within the phenomenon of terrorism starting from 2001.

---

27 "Transnational non-international armed conflict" describes conflict waged by a state against a non-state actor (actors) external to that state (ie who acts mainly or exclusively outside—or from the outside—the territory of that state), and therefore on territories or areas under jurisdiction of other states, who are not—as long as they are not supporting any party of the conflict—a participant of conflict. Cf. Philip Alston, "Report of Special Rapporteur on extrajudicial, summary or arbitrary executions, Addendum, Study on targeted killings" UN Doc A/HRC/14/24/Add.6 (May 28, 2010) at 17, para 53.

Moreover, even if the concept of transnational non-international conflicts were to be given a universal recognition,[28] there would still be a question as to which IHL rules—those applicable to intra- or rather inter-state struggles—should be treated as appropriate for such a hybrid category of armed conflicts, and customary law would still be applicable. Additionally, in any circumstances the legality of every form of use of lethal force, including drone strikes, would have to be assessed on the basis of their compatibility with the main principles of IHL, like proportionality and distinction. In the case of drone strikes particularly, compatibility with the latter principle could be difficult to achieve due to the fact that even use of civilians as human shields by terrorists (who function on a daily basis within societies), although it as such constitutes a war crime and therefore is often cited as an argument for admissibility of UCAV attacks, does not allow for terrorists' adversaries to use lethal force indiscriminately.[29] Therefore, any kind of civilian victims of attacks by UAVs could be a reason to question the legality of such actions as not sufficiently discriminating between combatants and non-combatants.

Thus the reason for the increasing reliance of the US on the use of UCAVs in fighting terrorism was not their exceptional legal status as a method of warfare, because there is hardly any specificity here. Hence, the decisive factors for that growth were rather of some technical and operational nature, which have led to some expectation of increasing both tactical and strategic efficacy of these actions as a method of counterterrorism.

Proponents of the use of drones in counterterrorism frequently point out the relatively high precision of their strikes and, therefore, effectiveness. At least two factors are decisive here. Firstly, crucial for their attractiveness—unachievable

---

28 The claim that US drone strikes outside Afghanistan and (until 2011) Iraq were executed as violent acts within the framework of the armed conflict, even if its hybrid, transnational form is somewhat discredited by the fact that these drones have been operated by personnel of CIA, a civilian (ie not a military) institution. It could mean that US authorities put—rather unnecessarily, taking into account scale and capabilities of US armed forces— the personnel of the agency in disadvantageous legal position. CIA agents engaged in drone strikes should be deemed as "directly involved in hostilities," which would result—with all legal consequences concerning legitimacy of targeting them—in loss of their civilian status (at least for the period of their participation in hostilities). "Targeting Operations with Drone Technology: Humanitarian Law Implications" Background Note for the American Society of International Law Annual Meeting, Human Rights Institute, Columbia Law School, New York (March 25, 2011) at 27–30. Seemingly due to their awareness of the problem, US officials started to suggest that the responsibility for drone operations will be moved fully from the CIA to the Pentagon. However, at the end of 2013 the transfer has been stalled for unknown reasons. See Gordon Lubold, Shane Harris "Exclusive: The CIA, Not The Pentagon, Will Keep Running Obama's Drone War" Foreign Policy (Nov 5, 2013) available at <http://complex.foreignpolicy.com/posts/2013/11/05/cia_pentagon_drone_war_control> accessed July 2014.

29 Michael W Lewis, Emily Crawford, "Drones and Distinction: How IHL Encouraged the Rise of Drones" [2013] GT J of Int'l Law 1127, at1149–1157.

for manned aircrafts, as already mentioned—is the ability of typical UCAV (at least of MALE category) to persist over a target for some 12 to 14 or maybe even more hours without the need to refuel or return to the base. Secondly, unlike the guided missiles fired from the larger distances, weapons fired from UCAVs can be diverted by the drone operator even at the very last moment if the situation in the targeting area changes (such as in cases when civilians or friendly forces are recognized within the probable explosion radius).[30]

However, exact assessment of actual tactical efficacy of drone attacks—regularity in eliminating correct targets, or the proportion between intended and unintended victims (collateral damage)—is extremely difficult. And it is even more difficult than in the case of other forms of targeted killing. Usually, drone attacks are executed in places distant from populated areas (at least those conducted by Americans, since this not entirely the case in context of Israeli operations which are performed mainly in urban environments). Such areas are hard to access (mountains, deserts, etc.), and collection and verification of whatever data there is about such operations is, for obvious reasons, really challenging. However, even more important are the possibilities for manipulation of information for propaganda (psy-ops) purposes, both by terrorists attacked and their sympathizers, as well as by US or other authorities. The current level of secrecy over US drone counterterrorist operations in northern Pakistan, Yemen, or elsewhere—even if recently decreasing (albeit slightly) thanks to the changes in position of the Obama administration[31]—is also a serious obstacle for collecting reliable data on the results of such strikes. In addition, UAV operations are constantly evolving in their missions, so their efficacy and accuracy have also been shifting in various periods and regions.

As a result, currently the efficacy of drones is assessed differently by various sources, at least in context of US campaigns in northern Pakistan, Yemen, and Somalia. Some sources, albeit primarily from Pakistan, claim that vast majority of victims of those attacks are civilians.[32] Quite to the contrary, US authorities, including President Obama and former CIA director and Secretary of Defense Leon Panetta, have repeatedly assessed the number of civilian casualties in drone operations as low, especially in comparison to alternative tactics and means.[33] Moreover, they stressed that the precision of drone strikes (understood

---

30   Zenko and Kreps (n 12) at 8–9.

31   For the critical assessment of the actual level of transparency of Obama's policy on drone strikes see Andrea Prasow, "The Year of Living More Dangerously: Obama's Drone Speech was a Sham," Human Rights Watch (May 24, 2014) available at <www.hrw.org/news/2014/05/24/year-living-more-dangerously-obamas-drone-speech-was-sham>, accessed July 2014.

32   Peter Bergen, Katherine Tiedemann, "The Hidden War," Foreign Policy (Dec 21, 2010), available at <www.foreignpolicy.com/articles/2010/12/21/the_hidden_war?page=0,5> accessed July 2014.

33   Cf. John O Brennan, Assistant to the President for Homeland Security and Counterterrorism "The Efficacy and Ethics of U.S. Counterterrorism Strategy" Woodrow Wilson International Center for Scholars, Washington DC (April 30, 2012) available at

as an ability to destroy solely intended targets), already rather high, is constantly improving.[34] Nevertheless, sources perceived as relatively reliable, for example the databases collected by New America Foundation and Bureau of Investigative Journalism, assess the share of civilian casualties in the total number of victims of the drone strikes as located between 10 percent and 25 percent.[35] Importantly, researchers (at least in the United States) generally support the US government's view that the precision and selectiveness of drone attacks are gradually, but constantly, improving (even if not sufficiently fast).[36]

What is more, even some quite reliable sources from other countries directly interested in the issue, namely Pakistan, suggest that accuracy of anti-terrorist UCAVs strikes is now reaching a quite impressive level. According to Pakistani newspapers, and based on a local government classified report of which fragments were leaked to the press at the end of 2013, the US drone strikes had killed some 2,160 militants and just 67 non-combatants. The same report states that the level of collateral damage by drone attacks had been reduced to virtually nil after 2012.[37] On the other hand, in his most recent speeches on drone operations, President Obama is slightly less firm on the issue of UCAVs strike precision, admitting regretfully that some of the US strikes have resulted in civilian casualties (although avoiding the provision of more detailed data and stressing simultaneously that the number of casualties is dwarfed by the potential number of victims of terrorist attacks anywhere in the world avoided, thanks to drone missions).[38]

<www.cfr.org/counterterrorism/brennans-speech-counterterrorism-april-2012/p28100> accessed July 2014. Chair of the US Senate Intelligence Committee Diane Feinstein has recently even stated that "the number of civilian casualties ... from such strikes each year has typically been in single digits." Jaclyn Tandler, "Known and Unknowns: President Obama's Lethal Drone Doctrine" (2013), FRS Note 07/13 at 2, available at <www.frstrategie.org/barreFRS/publications/notes/2013/201307.pdf> accessed July 2014.

34   Leon Panetta, "My Mission Has Always Been to Keep the Country Safe," Interview with NPR (Feb 3, 2013) available at <www.npr.org/templates/transcript/transcript.php?storyId=170970194> accessed July 25, 2014.

35   Andrew Callam, "Drone Wars: Armed Unmanned Aerial Vehicles" [2010], Int'l Aff Rev available at <www.iar-gwu.org/node/144> accessed July 2014.

36   According to NAF data, in 2010–2011, some 85 percent of victims of drone strikes in Pakistan were most probably terrorists and insurgents, while until 2009 that proportion was closer to 60 percent (although the number of strikes was lower, so the absolute numbers of civilian casualties could be lower as well): Bergen, Tiedemann (n 19). For an in-depth discussion of civilian deaths and proportionality see Chapter 7 by Avery Plaw and Carlos R. Colon in this volume.

37   Bakir Sajjad Syed, "2160 terrorists, 67 civilians killed by drones," *The Dawn* (Oct 31, 2013) available at <www.dawn.com/news/1053069> accessed July 2014.

38   Barack Obama, "Speech at National Defense University: The Future of our Fight against Terrorism," Washington DC (May 23, 2013) available at <www.cfr.org/counterterrorism/president-obamas-speech-national-defense-university-future-our-fight-against-terrorism-may-2013/p30771> accessed July 25, 2014.

Statistical data on casualties are relevant particularly for assessing the tactical efficacy of drone strikes—their ability to destroy precisely and exclusively those who were assumed to be valuable and justified targets. However, such efficacy depends largely on the value of intelligence on potential targets. Although some already mentioned characteristics of UAVs (especially their relatively long endurance) help to improve situational awareness, the identification of targets is always associated with some uncertainties and risks. Hence, targets could be identified as legitimate and valuable by mistake, on the basis of false or misinterpreted data, but nevertheless, thanks to the high precision and reliability of drones, will be most probably destroyed if attacked.

Such dangers of mistakes in target recognition are significantly increased by the US practice of "signature strikes" approved even if the personal identity of the target is unknown. These are authorized on the basis of evidence such as "pattern of life" and documented suspicious or hostile behavior of potential targets. This is assessed on the basis of surveillance/intelligence data collected by reconnaissance drones or other means, usually in the course of observation of other, already personally identified, potential targets. However, it is extremely difficult to estimate the exact scale of the problem posed by "signature strikes" from data currently available from open sources.[39] Therefore, taking into account the still high level of secrecy and restrictions on information about target selection procedures, it is nearly impossible to assess factual level of unjustified victims of UAVs. Of course, this is a real difficulty for all three spheres of legitimacy—legality, morality, and efficacy alike.

Irrespective of the actual proportion of collateral damage in drone strikes and remembering that their accuracy directly depends on the value of intelligence data on potential targets, one has to admit that some characteristics of the use of UAVs helps to improve their precision and effectiveness as a tool of fighting terrorism or, more precisely, terrorists. As it was mentioned, UAVs are capable of staying in particular region of operations for a long time, and conducting continuous reconnaissance activities, such as observation of suspected persons or places. That facilitates a collection of relatively comprehensive and verifiable data on persons under surveillance (which is essential for establishing their "patterns of life"). Additionally, it offers the opportunity to decide on the moment of actual strike in

---

39   The practice of "signature strikes" was allegedly given up by the CIA in mid-2013. However, it has just come to light in April 2015 that the practice was never abandoned with the administration's admission that a drone strike some three months earlier had killed two hostages who were unknown to be held at the targeted compound (M Mazzetti and E Schmidt, "First Evidence of a Blunder: 2 Extra Bodies" *New York Times* (April 23, 2015) at A1). Despite some effort in the US Congress, an official legal ban on such practices has not been introduced. Kathy Gannon, Sebastian Abbot, "Criticism Alters US Drone Program in Pakistan," Associated Press (July 25, 2013) available at <http://bigstory.ap.org/article/criticism-alters-us-drone-program-pakistan> accessed July 2014; Jason Ditz, "House Committee Refused to Ban Drone Strikes on Unidentified People," Antiwar.com (Dec 2, 2013) available at <http://news.antiwar.com/2013/12/02/house-committee-refused-to-ban-drone-strikes-on-unidentified-people/> accessed July 2014.

a way that minimizes the risk of collateral damage (for example by attacking the suspected person when they are alone).[40]

However, at the same time this relatively high precision of UCAVs could—quite paradoxically—lead to the growth in number of civilian victims of such strikes, at least in absolute numbers. Increasing accuracy of drones would stimulate their use and it is possible that the frequency of attacks would in such circumstances grow faster than their precision and the ability of operators to reduce collateral damage. As a result, the danger of growth in civilian casualties would increase, also because with the intensification of such operations they would no longer be treated as an extraordinary measure used to destroy only highly valuable targets, and become a rather routine method of counterterrorism, directed also against less important members of terrorist or insurgent groups (in context of which verification of their relevance and the level of threat posed by their activities would be, for obvious reasons, much more difficult).

Currently, US authorities claim that the relatively high accuracy of drone strikes—often presented simply as efficacy—is at least partially a result of careful, elaborated policy on the use of drones by Obama's administration. Such policy is based on several principles. Firstly, apart from the condition of legality of such operation under US law, drone strikes are assumed to be some kind of "last resort solution" in countering terrorism, used when capturing a suspected person—the most, at least officially, preferred option—is impossible or not feasible, because of terrain inaccessibility, serious risks for the US personnel engaged in any potential capturing operation, or possible political consequences of this specific form of military presence on foreign territory, etc.[41] Secondly, individuals attacked should constitute significant threat to the security of United States and its people; that is, a threat of a "continuing and imminent" nature, an explicit standard added to the policy's guidelines only in mid-2013.[42] Moreover, drone attacks are not intended to be a form of punishment, but should be intended exclusively to reduce the level

---

40   Cf. M Schmitt (n 23) at 446–447; Alston (n 29) at 24–25, para 81–82.

41   However, the actual practice of US counterterrorism actions, at least in recent years, seems to be quite different: Since September 2011 to the beginning of 2014 when US forces conducted some 200 drone strikes against alleged terrorists and made only three known capture attempts: Zenko and Kreps (n 12) at 10.

42   However, examples of threats "sufficiently significant" to pose a legitimate target for drone strikes offered by the US authorities in their official speeches suggest a rather broad interpretation of this term. For example, John Brennan (Assistant to the President for Homeland Security and Counterterrorism), in his remarks on April 30, 2012, listed in this context situations such as being an operational leader of Al-Qaeda, being in the midst of training or preparing to conduct an attack on the US or even merely possessing unique operational skills that could have a leverage in some planned attack. Hence, as was stated in the Presidential Political Guidance on drones of May 2013, the addition of the requirements of continuing and imminent nature of the threat posed by the potential drone's target seems to be a justified move, intended to elaborate current practice and limit the risk of unjustified use of lethal force by US institutions. Cf. Brennan (n 35). However, US authorities are

of threat of terrorism to the United States and international security. Thirdly, such attacks should be conducted only when there is a high level of confidence in identity of the target.[43] Lastly, except in "the rarest circumstances" drone strike should be ordered and conducted with "the near-certainty that no civilians will be harmed."[44] What is also very important is that US authorities stress the fact that drones are particularly effective in countering transnational terrorism in its current form. That is, it is fragmented and dispersed, often localized and made up of protean structures of numerous—but usually small—terrorist organizations functioning in inaccessible and distant areas not controlled fully by any government. Therefore, at least in principle, the rather frequent use of drones in counterterrorism seems to be perceived as a somewhat "temporary" practice, which could be modified or significantly limited (but unlikely abandoned entirely) once the threat posed by terrorist groups, primarily Al-Qaeda and its affiliates, is eliminated or transforms into other form.[45]

V. The Challenge of "Lawfare"

However, the issue of potentially higher precision of UCAVs strikes in relation to alternative methods of fighting terrorism (particularly those aimed at the physical elimination of terrorists) should not be discussed separately from other operational advantages and weaknesses. Therefore, I will briefly cross over into the concept of strategic efficacy to discuss the notion of "lawfare." Not only is there an inherent blurred line between these two types of efficacy but, as we will see, there is also an overlapping with the sphere of legality in the twenty-first century.

Use of armed UAVs is currently perceived as an option simply more economically efficient than other viable alternatives, such as permanent deployment of troops or even actions of special operation forces (SOF units). For rather obvious reasons (loitering at an altitude beyond the range of weapons of those who are attacked) drones are also more "politically" safe for politicians, helping to keep a large part of such activities in secrecy and to suppress the "body bag syndrome" caused by the soldiers' dead bodies returning home from the mission. Political attractiveness of drone strikes is augmented by the possibilities of various interpretations of their legality discussed in detail in the first section of this volume. This makes them a

---

continuously offering very scarce data concerning detailed mechanisms of assuring reliable assessment of the nature of the threat posed by prospected drone targets.

43   This high level of confidence was not equaled, however, with certainty. In other words, this is what led to the acceptance of the already mentioned—and now allegedly suspended—so-called signature strikes. As noted above in (n 39), this practice was never actually abandoned even though it was widely believed to be the case—a clear evidence of the problems created by secrecy for credible independent analysis.

44   Obama (n 40).

45   Ibid.

relatively less evident form of breach of other states' sovereignty or international norms than direct deployment of troops.

In the strictly operational dimension, there are also some important disadvantages of drone strikes. Apart from their possible low strategic effectiveness (due to the potential of causing public opposition and even outrage within societies experiencing the results of such campaigns, which could be exploited in psychological operations by terrorists and their supporters), liquidating terrorist or militants in that way means unavoidably also losing opportunities to collect additional intelligence data from valuable sources. Simply, "dead terrorists can't talk."[46] Questionable is also an assumption behind the very idea of current US drone counterterrorist campaigns that effective elimination of terrorist leaders will automatically lead to a decrease in intensity of terrorist attacks, as well as overall operational capabilities of terrorist groups.[47] (For a full discussion of the effectiveness of leadership removal, see Chapter 9 by J. Jordan.)

Even if such a goal can be achieved, it would most probably be a rather short-term solution. Those attacked would ultimately find some replacements and killing the leader of one group could have only minor influence on the behavior of commanders of other organizations. Moreover, apart from the risk of "martyr-creation," eliminating group leaders by drone attacks, being relatively easy to present as "treacherous," "inhuman," and "cowardly" methods of action, could simply backfire in the context of the psychological dimension of conflict, offering the enemies (terrorists) broad possibilities to wage propaganda war, including the relatively attractive option of so-called lawfare.[48]

Lawfare, the term introduced by the USAF colonel and lawyer Charles A. Dunlap in 2001, still remains quite controversial. It is used—and misused, sometimes intentionally[49]—in many different contexts and interpreted in various ways. Nevertheless, we could define it, as Dunlap did, as a "use of law as a weapon of war."[50] Or, as it has been more elaborately defined, "the strategy of using—or misusing—law as a substitute for traditional military means to achieve an

---

46 Charles Krauthammer, "Barack Obama: Drone Warrior" *The Washington Post* (June 1, 2012).

47 Some proponents of drone strikes stress that even the mere possibility of leaders being killed in drone strikes would lead to the decrease in activity of terrorist organization threatened in this way because their commanders and other valuable members would be turned into "permanent fugitives," who live in hiding and constant fear of being annihilated and therefore less engaged in group actions. However, such assumptions have not (yet) led to clear confirmation by empirical study. Cf. Byman (n 25).

48 Audrey K Cronin, "Why Drones Fail: When Tactics Drive Strategy" (2013) 92(4) For Aff 44 available at <http://www.foreignaffairs.com/articles/139454/audrey-kurth-cronin/why-drones-fail> accessed July 2014.

49 "Is Lawfare Worth Defining? Report of the Cleveland Experts Meetings" (Sept 11, 2010) Case W Res J Int'l Law, at 11.

50 Charles Dunlap, "Law and Military Interventions: Preserving Humanitarian Values in 21st Century Conflicts" [2009] Joint Forces Q 34.

operational objective."[51] Currently, however, it is most frequently used to describe the exploitation of vulnerabilities of the Western democratic states and their armed forces and secret services to allegations of violations of the law of armed conflict or human rights laws.[52] Alleged illegality of particular actions would weaken public support for such operations, making it more difficult for authorities to continue them. Likewise, international prestige and reputation could be subverted if the accusation of misconduct or breaches of law gained popularity.

All of this could lead to modification or even abandonment of methods or means of action in question—which would be beneficial to the enemy his operational goal. For obvious reasons, such mechanisms are most effective in the fight against democratic societies, where the impact of public opinion on the government policy is the strongest. Moreover, although lawfare is obviously easier to be executed in cases of evident breaches of particular international norms, even conformity with them when using actions that have come into question will not fully defend against accusations of their illegality. Thus it is the adoption of such lawfare techniques by the enemy that is most relevant since the key issue here is their propaganda prowess rather than the actual state of things.

In the context of current practice, particularly by the US, there is essentially—and quite paradoxically—substantial ground for such accusations that could constitute "an act of lawfare." As was already discussed, there are factors that could justify the allegations concerning their illegality, as well as significant opportunities for granting such views wider acceptance and recognition. Due to their technical specificity, drones are in fact more precise than majority of other weapons in use, and therefore offer greater chances to avoid unnecessary victims, damages, or losses. In other words, they could be more selective as weapons than their conventional alternatives and could also limit the overall number of victims of any violent action. That makes them potentially more "humanitarian" as weapons, because they offer relatively a high (maybe even the highest possible in current circumstances) level of conformity with the principles of distinction and proportionality. In fact, some scholars have even suggested that the rise of drones stems from their ability to increase conformity with humanitarian law standards.[53] However, and that is why it is somewhat paradoxical, that assumption is undermined by the serious doubts over the legality of US drone counterterrorist operations in their current form.

The problem here is obviously not their operational performance as such in relation to the principles of distinction and proportionality. The crucial issue is if the US has the right to use lethal force at all in their counterterrorist missions, such as those in northern Pakistan, Yemen, or Somalia, no matter what particular type of weapon is employed. In other words, it is an *ius ad bellum*, not *ius in bello*, question

---

51   Charles Dunlap "Lawfare Today: a Perspective" [2008] Yale J Int'l Aff 146, at 146.

52   Sibille Scheipers, "Is the Law of Armed Conflict Outdated?" [2013] Parameters 45, at 46.

53   See generally Lewis and Crawford (n 31).

(discussed thoroughly in Chapter 1 by Christian J. Tams and James G. Devaney). Such a right to use lethal force in these operations is disputable, irrespective of whether it is treated as a form of self-defense or not. Hence, the fact that use of drones in counterterrorism would be relatively more selective and proportional than their alternatives losses is somewhat insignificant. In such circumstances the use of drones to eliminate alleged terrorists or insurgents is only "a lesser evil" than other forms of using violence against such groups. Their tactical efficacy and possibly high proportionality and discrimination would probably not be sufficient reason to make an exception to the general rule of illegality of use of force outside the regular (that is, international or intrastate) armed conflict.

Moreover, building on the point made above, the higher precision of drones could lead to the temptation to attack more targets—if one has such a "lancet" and not a hammer, they could dare to plan and execute operations otherwise perceived as too risky, causing too much collateral damage, or politically and economically too costly. It seems logical, and has arguably been seen with the spike in drone attacks, that with higher number of missions the overall number of unnecessary (and otherwise possibly avoidable) victims will most probably grow as well, somewhat shadowing the fact of elimination of some important terrorists leaders or valuable operatives. Hence, even if this cannot be easily proven by those attacked by drones, the accusations of that kind could find popular acceptance or at least understanding rather easily, especially if supported by the skillful and intensive propaganda campaigns.

Distressingly, the unusual level of secrecy coupled with an unbounded territorial application opens the possibility of quite effective "lawfare" to the terrorists and their supporters. Thus the "tactical efficacy" of these unmanned engines should be put in doubt if they do no bring gains to the "desired end." In other words, simply creating more targets with dubious legal justifications, and then eliminating them, cannot be equated with tactical efficacy if their use is viewed as illegal—the aim of "lawfare"—and consequently undermines the final goal.

VI. Conclusion: The Future of UCAV Use for Counterterrorism

Despite all the weaknesses and doubts, current US practices in counterterrorism, as well as the direction of their evolution, seem to suggest that drones will be still perceived as attractive and viable option from the military, as well as economic and political perspectives.[54] The relatively high precision of drone strikes, and therefore the common understanding of the tactical efficacy of such operations as a method

---

54   As Leon Panetta, former US Defense Secretary and former director of the CIA said, in the context of combating foreign terrorist groups, particularly in remote, unlawful, and destabilized regions, where there are not clearly defined armed conflicts and it would be impossible or unwise to deploy the US troops, drone strikes remain in fact "the only game in town": Brief and Special Appendix for Plaintiffs-Appellants American Civil Liberties

of counterterrorism, is one of the key reasons for this reasoning. Moreover, the fact that the precision of these attacks improved in recent years—at least as far as US operations are concerned and when efficacy is understood as an ability to eliminate a selected target (chosen correctly or not) without causing collateral damage among civilians—should also be treated as relevant in context of potential future use of such weapon systems in fighting against terrorists or other irregular foes. Importantly, this improvement in precision of drone strikes is probably to a lesser extent the result of technological advancements than of modifications and elaborations of decision-making procedures concerning target selection for such attacks and their execution. Undoubtedly, these procedural changes were caused by the negative past experience— especially during the initial years of regular use of drones in counterterrorist operations by US the proportion of unintended victims among civilians to people actually targeted and deliberately eliminated was much higher than in current practice. Although technical means used then (UCAV types, weapons, and other support equipment) were in fact similar to those operating currently, with just minor upgrades.[55]

The rather mediocre initial record of drone strikes in context of the basic principles of humanitarian law, especially of proportionality and distinction, had led to a growth of interest in such operations among a wider audience in many countries. This resulted in a significant pressure on American authorities to at least—if stopping of such operations was judged by them as impossible or not feasible and counterproductive in the context of a fight against terrorism—amend their procedures of target selection and execution of UCAVs attacks. And, as above-mentioned data show, such changes proved to be effective in the reduction of collateral damage and improvement of drone strikes' tactical efficacy, even if not accompanied by major technological breakthroughs.

However, due to the constantly high level of secrecy on US drone strikes, it cannot be totally excluded that improvements in precision and therefore tactical efficacy of such operations signaled by publicly available data is first and foremost the result of effective public information management activities of the US government, who as a "perpetrator" of drone missions controls the access to majority of relevant information and sources.

Union (April 15, 2013) at 14, available at <https://www.aclu.org/national-security/anwar-al-aulaqi-foia-request-legal-documents> accessed Nov 2014.

55   As data in Table 10.1 show, in the initial years of the drone campaigns, under the Bush administration, the share of civilian victims in the overall number of victims of drone strikes in Northern Pakistan was higher than militants' share (respectively around 58 percent according to NAF and 55 percent according to BIJ). In the first year of Obama administration (2009) that share was still significant (some 28–30 percent of civilian victims accordingly to both sources), while currently, after changes in decision-making procedures and improvements in intelligence collection and cooperation with local authorities, this proportion was reduced to single-digit level (6 percent accordingly to NAF and 2 percent accordingly to BIJ data).

Obviously, advancements in efficacy on tactical level do not have to automatically lead to proportional (or any) improvement of strategic efficacy of UCAVs as a tool of counterterrorism. Higher precision of such strikes and an ability to reduce collateral damage, even on minimal level, does not necessarily mean that this smaller number of unintended casualties will be treated as justified and avoid public outrage or a new wave of recruits for terrorist groups or insurgents. The ability to destroy selected targets (and solely that target) with probability close to certainty does not guarantee that all relevant and influential terrorist leaders or operatives will be eliminated or that new leaders will not come to replace them. The precision of weapons used to counter terrorism, albeit important, is not the only factor decisive for ultimate results of such confrontation—hence strategic efficacy is to be explored in the following chapter by Steven J. Barela. Nevertheless, in light of the evolution of the US drone campaign against Al-Qaeda and its associates, as well as the growing interest it stimulates among other states in acquiring unmanned lethal technologies for their own irregular, asymmetric enemies or even transnational criminals,[56] a worldwide increase in use of UAVs in different forms of combat missions should be expected.

Therefore, the international community has to confront the issue of defining limits for legitimate missions of this kind, as well as principles of responsibility for potential misconduct. Current IHL and other relevant legal norms (especially in context of human rights regimes) are not as ambiguous as they have been assessed. Nevertheless, the rather common belief within US military and politicians in the high effectiveness of drone strikes as a counterterrorism and counterinsurgency tool suggests that in the near future such missions will not only be continued by the US (or at least treated as legally acceptable), but initiated by other countries too. Thus, efforts to secure complete fulfillment of IHL or HR norms in context of drone attacks could be futile in many cases. The result would undermine universal acceptance and adherence to the norms concerning use of force in international relations. Therefore, current regulations in this sphere require significant modifications to make them more coherent with the reality of current and future states' practice and the security environment. However, these changes should avoid automatic legalization of every form of drone use in combating terrorism or other non-state entities.

As a good starting point for such reforms would be the introduction of an obligation to maintain the highest achievable level in transparency on the use of UCAVs in counterterrorism, including decision-making processes concerning target selection, timing, and legal requirements of any given drone attack. The process of formulating such international legal provisions will undoubtedly not germinate effortlessly, and will probably not be quickly accomplished. However,

---

56 Turkey, for example, had plans to employ UCAVs in its fight against Kurdish rebels and organized trafficking syndicates in the eastern part of the country, close to the border with Iraq: Adam Entous, "U.S. Plans to Arm Italy's Drones," *The Wall Street Journal* (May 29, 2012).

it could also be instrumental for serious improvements in the ability to verify of the claim of allegedly high efficacy of drone attacks as counterterrorist weapon. Such a claim of "drones work!", although based more on common sense and a priori assumptions than empirical evidences, seems to constitute the central argument for the proponents of extensive use of drones in fighting terrorism. Having in mind the accelerating pace of development of unmanned military technologies, the risks associated with the lethal use of UCAV for international legal order could soon be simply too great to allow the international community to neglect or ignore this problem and accept discussions on so important issue based on mainly intuitive arguments.

Chapter 11

Strategic Efficacy: The Opinion of Security and a Dearth of Data

Steven J. Barela

The drones have done their job remarkably well: by killing key leaders and denying terrorists sanctuaries in Pakistan, Yemen, and, to a lesser degree, Somalia, drones have devastated Al-Qaeda and associated anti-American militant groups.

Daniel Byman, "Why Drones Work"[1]

The sometimes contradictory demands of the American people ... have fueled the technology-driven, tactical approach of drone warfare. But it is never wise to let either gadgets or fear determine strategy.

Audrey Kurth Cronin, "Why Drones Fail"[2]

I. Introduction

The overall strategic objective of counterterrorism is clearly to improve security for the targeted population. Yet it should be remembered that the concept of security was spotlighted in the eighteenth century as a subjective notion by "the celebrated Montesquieu."[3] This legal and political philosopher sensibly asserted, "[p]olitical liberty consists in security, or, at least, in the opinion that we enjoy security."[4] Specifically, Montesquieu was discussing the issue of the relationship between security and political liberty, that is, protection from injustice by the hands of government. Nonetheless, when safety was discussed he was careful to reference its subjective nature as an opinion,[5] and this insight into the idiosyncratic

---

1 Daniel Byman, "Why Drones Work: The Case for Washington's Weapon of Choice" (2013) 92,4 Foreign Affairs 32, at 33.

2 Audrey Kurth Cronin, "Why Drones Fail: When Tactics Drive Strategy" ibid 44 at 54.

3 Montesquieu was twice referred to in the Federalist Papers using this laudatory formulation (four times overall): Alexander Hamilton, No 78, "The Judiciary Department," Reprinted in *Federalist & Anti-Federalist Papers* (BN Publishing 2006) at 286; and James Madison, No 47, "The Particular Structure of the New Government and the Distribution of Power Among Its Different Parts," ibid at 184.

4 Montesquieu [1748] *The Spirit of the Laws*, Anne Cohler et al (trans and ed) (Cambridge Univ Press 1988) at vol XII (1).

5 Ibid at vol XI (6). In Montesquieu's quest to define a "free state" he drew a distinction between *objective* and *subjective* liberty (see eg David W Carrithers, "Montesquieu's

facet of security reveals the immense difficulty for putting forward a scientific measurement for the strategic efficacy of drones.

Additionally, the current usage of UCAVs (unmanned combat aerial vehicles) for cross-border counterterrorism creates its own particular stumbling blocks. To begin, the conventional model of threat assessment was built for identifying dangers posed by other states and hence using this same model for detecting those emanating from menacing non-state actors is decidedly problematic. Next, the US drone program has been cloaked in official secrecy leaving many gaps in the knowledge of analysts and the public alike. The fact that the program has chiefly taken place in some of the world's most remote regions—where there is little or no independent journalism for fact-checking—has made the available data questionable at best. Further, the government has used terms of indeterminate breadth when defining the enemy, thus making measurement of progress toward an endgame unattainable. Together these issues have led to two primary compounded problems for the determination of strategic efficacy of increasing security by using drone strikes: an unsuitable model for objective threat assessment; and a constellation of data that is thus far impossible to collate systematically or completely.

By analyzing the difficulties for methodically assessing this specific type of overall efficacy, this chapter will illuminate these obstacles as currently insurmountable impediments to definitive conclusions on this important question. There is little doubt that a conclusive answer to the question of whether armed UAVs (unmanned aerial vehicles) help to secure a nation from destructive and deadly threats from non-state actors is at the front of many citizens' minds when they assess the legitimacy of counterterrorism policies. However, this chapter unfortunately does not provide the answer. Rather, it explores the different (missing) datasets, along with the parameters themselves, that would be necessary to construct an empirical response to this query. It is thereby possible to glean other, equally important, aspects concerning the efficacy of the Obama administration's drone program—a policy that is setting precedents that other states could certainly follow as UCAV technology spreads.

Two other points should be mentioned. Firstly, the frequency with which the Obama administration has glided between arguments of efficacy, legality, and morality to describe its drone program requires a limited number of disciplinary crossovers in this chapter. However, such crossovers are made explicit, and are meant to expose particular points of overlap that exist between the spheres of legitimacy. Secondly, this chapter is meant to be complimentary to the work found in Chapter 9 by Jenna Jordan. Her empirical research is applied to the historical data available on the removal of terrorist group leaders, and what results ensued

---

Philosophy of Punishment" (1998) XIX History of Political Thought 213; Till Hanisch, *Justice et puissance de juger chez Montesquieu* [Classique Garnier, forthcoming 2015] esp the chapter *Modération, juste régime et liberté*). Although this will not be the precise tack followed in this chapter, we will find this same important division drawn in security studies.

for that organization, whereas this chapter investigates the questions that must be posed, and credibly answered, for strategic efficacy to be assessed.

This chapter is divided into five sections. Section II outlines the complications involved for being able to make general pronouncements on the effectiveness of targeted killing with drones by investigating the objective side of security: threat measurement. Section III then explores the difficulties raised by government secrecy and selective disclosures. Next, section IV addresses tracking the consequences of strikes in one of the deadliest countries in the world for practicing independent journalism. Section V investigates the challenges to systematizing the advancements made against a group or groups that is/are ill defined and largely bound together only by their agreement upon who they deem to be their enemy. Finally, section VI concludes and suggests that these enormously difficult problems force one to reexamine the morality of a society exercising deadly force abroad when its strategic benefits are hidden and/or unknowable.

II. National Security: General Pronouncements on Efficacy

The two epigraphs to this chapter capture the current state of the dispute over the efficacy of drones. In 2013, *Foreign Affairs* magazine published two investigations coming to directly opposing conclusions on the effectiveness of UCAVs for counterterrorism and promoted this debate on their front cover. One might expect that determining whether drones work would be a straightforward examination leading to a clear conclusion. But as this scholarly debate based on the publicly available evidence shows, the question of efficacy is many sided and deceptively simple.[6] This section of the book has a variety of chapters on efficacy so as to avoid some of the common amalgamating of the subject. Nonetheless, we will find in this chapter that, even when treated alone, strategic efficacy remains a tremendously complex question.

As a starting place, the concept of security has come to saturate political language and culture in the Western world. As one scholar has put it, "[n]early all political disputes and disagreements appear to centre on the conception of security,

---

6 For additional studies discussing the effectiveness of drone strikes see Patrick B Johnston and Anoop Sarbahi, "The Impact of U.S. Drone Strikes on Terrorism in Pakistan" Berlfer Center for Science and International Affairs, Kennedy School of Government, Harvard University (2011); Megan Smith and James Igoe Walsh, "Do Drone Strikes Degrade al Qaeda? Evidence from Propaganda Output" (2013) 25 Terrorism and Political Violence 311; see also the very recently published report by the CIA itself revealed by Wikileaks: CIA Office of Transnational Issues; Conflict, Governance, and Society Group, "Best Practices in Counterinsurgency: Making High-Value Targeting Operations an Effective Counterinsurgency Tool" (July 7, 2009) WikiLeaks release of Dec 18, 2014, available at <http://wikileaks.org/Conflans-Sainte-Honorine/WikiLeaks_Secret_CIA_review_of_HVT_Operations.pdf> accessed Dec 2014.

and nothing seems to advance a policy claim more than to be offered in the discourse of security."[7] Yet no matter how central to policy debate, or perhaps precisely because of this centrality, security claims remain open to political manipulation. Thus it is necessary to investigate why this wide margin of maneuver is possible.

In Arnold Wolfers' classic essay, "'National Security' as an Ambiguous Symbol," this difficulty is explored. Though its title might suggest otherwise, Wolfers does not argue that the concept of security is unclear or useless as a guide for research inquiries. Rather he tries to flesh out the confusions that are often unnoticed, and in doing so he provides a useful division between two basic portions of security. He explains, "security, in an objective sense, measures the absence of *threats* to acquired values, in a subjective sense, [it is] the absence of fear that such values will be attacked."[8] Hence, since the objective portion of security pivots on *threat assessment*, this is where our focus should turn in an attempt to avoid the "opinion" underscored by Montesquieu. In looking into the measuring of threats we will find that in the context of contemporary dangers this task has become particularly challenging.

For decades US intelligence agencies have had a remarkably consistent and enduring understanding of the concept of threat (the objective side of security); it has nearly always consisted of *capabilities* and *intent*. J. David Singer publically defined it in his 1958 seminal paper as "the quasi-mathematical form: *Threat Perception = Estimated Capability x Estimated Intent*."[9] In an adept historical and critical analysis entitled "Rethinking Threat," Charles Vandepeer investigated declassified intelligence reports, government publications, and intelligence literature and found they reveal an almost exclusive reliance upon this conventional model.[10]

However, in this model there are intrinsic assumptions leading to tangible limitations, which become acutely exposed as new threats develop. While the prototype worked relatively well during the Cold War when the primary threat to the United States came from the Soviet Union, it has real inadequacies when it comes to measuring the threat from non-state actors. In other words, it was a model built specifically for intelligence agencies tasked with assessing threats from other states. Former Central Intelligence Agency (CIA) Director James Woolsey vividly described this shift in threat assessment in his 1998 testimony before the US House Committee on National Security:

---

7   Mark Neocleous, *Critique of Security* (Edinburgh University Press, 2008) at 2.

8   Arnold Wolfers, "'National Security' as an Ambiguous Symbol" (1952) LXVII, 4 Pol Sci Q 481, at 485 (my emphasis).

9   J David Singer, "Threat-Perception and the Armament-Tension Dilemma" (1958) 2, 1 J of Conflict Res 90, at 94.

10   Charles Vandepeer, "Rethinking Threat: Intelligence Analysis, Intentions, Capabilities, and the Challenge of Non-State Actors" Doctoral Thesis, The University of Adelaide, School of History and Politics, (Oct 2011) at 11 available at <http://digital. library.adelaide.edu.au/dspace/bitstream/2440/70732/1/02whole.pdf> accessed July 2014.

[I]t is as if we were struggling with a large dragon for 45 years, killed it, and then found ourselves in a jungle full of poisonous snakes—and the snakes are much harder to keep track of than the dragon ever was. Snakes mean that we have to assess what might happen, not just what is likely.[11]

One problem is that as major threats have largely shifted, the debate over an analytical framework for a new context of non-state menaces has failed to progress sufficiently. What has primarily occurred is a discussion over whether to focus more on the intent or the capabilities of non-state actors, but not a reassessment of the model itself.[12] However, it must be recognized that this dominant model gives no consideration to the central problem of identifying the specific actors posing a threat. Identification is presumed. But, unlike states, non-state actors do not present a bounded or well-understood category of threat-actors because they are not confined by territory, government, or bureaucracy. Hence the exact non-state actors who pose a real threat need to be discovered by the analyst. Vandepeer described this problem as:

> The conclusion here is a subtle, and yet critical, one: the actor-based approach described by Singer was not designed for identifying covert non-state threats, but for assessing overt state-based threats. An actor-based approach assumes knowledge and understanding of the actor being assessed. Whilst there are no unknown states, there are unknown non-state actors. ... Indeed, if the existence, nature or characteristics of non-state actors are unable to be accurately identified or understood, any assessment of that group's intentions and capabilities will be partial, and perhaps even potentially misleading or entirely incorrect.[13]

Regrettably, there are obvious ways in which this difficulty of identifying the correct non-state threat-actors can be demonstrated in the "war on terror" with quite ghastly results. For example, there are a number of individuals who were tortured and/or ill-treated by agents of the US government because of their presumed status within the Al-Qaeda organization. However, the public information available reveals that in several different instances these tortured persons were either innocent or demonstrably unknowledgeable of future attacks.[14]

---

11 James Woolsey, CIA Director, Hearing of US House of Reps Committee on National Security (Feb 12, 1998) available at <http://www.globalsecurity.org/intell/library/congress/1998_hr/h980212w.htm> accessed July 2014.

12 Vandepeer (n 10) at 68–73.

13 Ibid at 74.

14 See Chapter 5 "Through the Lens of Efficacy: Torture on Suspicion" in Steven J Barela, *International Law, New Diplomacy and Counterterrorism* (Oxon, Routledge, 2014). See also the very recently released Senate Select Committee on Intelligence, "Study of the CIA's Detention and Interrogation Program, Findings and Conclusions, Executive Summary" (2014) Approved December 13, 2012, Updated for Release April 3, 2014,

Abu Zubaydah, who was captured in March 2002 and became the pivotal figure in the development of the torture program, is an egregious illustration.[15] Although a key "torture memo" identified him as "one of the highest ranking members of the Al-Qaeda terrorist organization" and "involved in every major terrorist operation carried out by Al-Qaeda,"[16] the government later recognized that these suspicions of his importance were incorrect.[17] As a result of this error, he was subjected to torturous waterboarding 83 times some four months after his capture (already indicating a stale intelligence value for imminent threats), but "not a single significant plot was foiled."[18]

However, Zubaydah falsely implicated at least one other person.[19] This wrongly identified captive was extraordinarily renditioned to Morocco, tortured with beatings and scalpel cuts all over his body (including his penis), transferred to the Guantánamo detention center, before finally being returned to the UK where he was released without charges within hours after a nearly seven-year tortuous saga.[20]

Declassification Revisions December 3, 2014, available at <http://www.intelligence.senate.gov/study2014 /sscistudy1.pdf> accessed Dec 2014.

15   The European Court of Human Rights recently handed down judgment in Zubaydah's case against Poland and found that the state had violated the applicant's rights by enabling the United States to secretly detain and torture the applicant on Polish soil, conducting an inadequate investigation into the acts of torture and ill treatment, and by allowing the applicant's transfer to Guantánamo despite the real risk he would be tortured and could be subjected to unfair trial and the death penalty by the United States, *Case of Husayn (Abu Zubaydah) v Poland* App no 7511/13 (ECHR, July 24, 2014).

16   Signed by Jay Bybee and authored by John Yoo "Memorandum for John Rizzo, Acting General Counsel of the Central Intelligence Agency: Interrogation of al Qaeda Operative," at 1 available at <http://www.justice.gov/sites/default/files/olc/legacy/2010/08/05/memo-bybee2002.pdf> accessed Sept 2014.

17   US District Court for the District of Colombia, *Zayn al Abidin Muhammad Husayn v Gates*, Civil Action No 08-cv-1360 (RWR), Respondent's Memorandum of Points and Authorities in Opposition to Petitioner's Motion for Discovery and Petitioner's Motion for Sanctions. See also, Brent Mickum "The Truth about Abu Zubaydah" *The Guardian* (London, March 30, 2009) <http://www.guardian.co.uk/commentisfree/cifamerica/2009/mar/30/ guantanamo-abu-zubaydah-torture> accessed Sept 2014.

18   Peter Finn and Joby Warrick, "Detainee's Harsh Treatment Foiled No Plots" *The Washington Post* (March 29, 2009) available at <http://www.washingtonpost.com/wp-dyn/content/article/2009/03/28/AR2009032802066. html?hpid=topnews> accessed Sept 2014.

19   Inspector General, Central Intelligence Agency "*Counterterrorism Detention and Interrogation Activities* (Sept 2001–Oct 2003)" (2003–7123-IG) (May 7 2004) available at the Homeland Security Digital Library: <http://www.hsdl.org/hslog/?q=node/5015> accessed Sept 2014: "information from Abu Zubaydah helped lead to the identification of Jose Padilla and Binyam Muhammed—operatives who had plans to detonate a uranium-topped dirty bomb in either Washington, D.C., or New York City" at 87.

20   This is a reference to the case of Binyam Mohammed al Habashi. For the most comprehensive and reliable information on his case see the decision by Judge Kessler who, based on this account of torture and cruelty, ruled that he could not serve as a credible

While this gruesome (yet enormously condensed) story certainly speaks to the (in) efficacy of torture, it also demonstrates the enormous problem of identifying the proper non-state threat-actors.

Additionally, we know that 780 detainees have passed through the detention facility in Guantánamo Bay, Cuba.[21] Following the Supreme Court's establishment of a right to judicial review in the 2008 *Boumediene* decision, there has been considerable legal action in the DC Court of Appeals to ascertain whether these persons have been rightfully detained as combatants (threat-actors) in the armed conflict.[22] At the time of this writing there has been a recent flurry of transfers out of the prison camp,[23] which helps bring some clarity to the numbers of those who are/were actually detained for directly participating in the non-international armed conflict.

While the Obama administration has designated 33 detainees for trial or commission (including those tried since January 2009), there are 35 who have been designated for indefinite detention without charge or trial.[24] This total of 68, with the vast majority not having stood trial, would already be a dismal success rate of 9.88 percent (including the nine who died in custody under the assumption they were indeed combatants). However, one former chief prosecutor for the Guantánamo military commissions presented an even lower count; he wrote that fewer than 4 percent ever sent there "have or will face charges."[25] Regardless of

---

witness to implicate another detainee at Guantánamo: US District Court, *Farhi Saeed Bin Mohammed et al.* v. *Obama*, 704 F. Supp. 2d 1. (DDC 2009).

21   The Guantánamo Docket, "Documents and research related to the 780 people who have been sent to the Guantánamo Bay prison since 2002" *The New York Times*, available at <http://projects.nytimes.com/guantanamo> accessed Jan 2015.

22   For an analysis of the US Supreme Court cases establishing the right to habeas corpus, and the trajectory of the litigation that followed, see Chapter 3 "Through the Lens of Legality: Detention without Judicial Review" of Barela (n 14).

23   Twenty-eight released in 2014 through mid-January of 2015: see Tara McKelvey, "Guantanamo Bay: What next for Cuba prison camp?" *BBC News, Guantanamo* (Jan 14, 2015) available at <http://www.bbc.com/news/world-us-canada-30820897> accessed Jan 2015.

24   Human Rights First, "Guantánamo by the Numbers" (last updated Jan 15, 2015) available at <http://www.humanrightsfirst.org/resource/guantanamo-numbers> accessed Jan 2015.

25   Col Morris "Moe" Davis, "Where Is Justice for the Men Still Abandoned in Guantánamo Bay?" *The Guardian* (London, Jan 16, 2015) available at <http://www.theguardian.com/commentisfree/2015/jan/16/justice-abandoned-guantanamo-bay#comment-46236175> accessed Jan 2015. In an email exchange with the author on Jan 17, 2015, Col Davis explained his calculations: "The less than 4% figure is based on current Chief Prosecutor Mark Martin saying some time ago that he anticipated perhaps 30 detainees standing trial, which would be 3.85% of the total detainee population. Since then, the DC Circuit said that providing material support for terrorism is not a legitimate law of war offense for pre-2006 conduct, so that will take away a few from Martin's total. I think the final tally will be about 20, which would be 2.57% of the 779 total population. There are

how one attempts to determine guilt, with so few trials carried out these numbers are distressingly low.

While there is evidence that indicates an increased intelligence capacity for identifying who is actually a member of Al-Qaeda, we must remember that when speaking about drones a portion of that identifying procedure is taking place with a camera in the sky. It is extremely difficult to guess how successful this method of identification, coupled with on the ground intelligence, has been for pinpointing the threat. Regardless, trusting an aerial identification process would often seem less reliable (other than viewing a direct participation in hostilities, of course) than actually having the suspected culprit in custody. Thus it seems fair to assume that identifying the precise threat-actors remains a real problem.

There are other complications beyond this fundamental difficulty of accurate threat identification. For example, while it is possible to state with a certain degree of confidence whether a state has the capability to attack another state, it is extremely difficult to claim that a particular group is incapable of carrying out a mass-casualty attack since there are such a variety of means and methods available for doing so. While weapons of mass destruction are the most dangerous and deadly, it is certainly not necessary for these to be obtained for a group to pose a veritable threat. Lest we forget that the attacks of 9/11 were initiated with only box-cutters in the hands of malicious individuals before they commandeered commercial aircraft. In such a context, measuring capability is of limited utility.

Additionally, it is important to note that terrorist groups have an acute interest in being recognized for both their capabilities and intent. Clearly, for a non-state organization to produce "terror" it must be believed to have both of these elements. Terrorism experts and politicians have long pointed out that such groups wish to bring as much attention as possible to their malevolent acts and objectives, which have been variously described as "theatre,"[26] "the oxygen of publicity,"[27] and "renown."[28] No doubt, the spectacular nature of terrorist acts is meant to grab the notice of the press and to generate as much media attention as possible. By doing so, such groups

---

6 cases completed, 2 where there are convictions but no final sentence, and 7 somewhere in the trial process. That makes 15 and there may be a couple of more that will face charges. I would also note that of the 6 cases completed, charges have been totally dismissed against 2 (Hamdan and Noor Uthman Mohammed) who were convicted of material support. David Hicks was convicted of the same and is currently challenging his case. Assuming the outcome is the same for Hicks, that means half the completed cases were detainees convicted in error."

26   Brian Jenkins "International Terrorism: A New Mode of Conflict," in D Carlton and C Schaerf (eds.) *International Terrorism and World Security* (Croom Helm 1975) at 16.

27   Margaret Thatcher, Speech delivered July 15, 1985 at Albert Hall, South Kensington, central London, to the American Bar Association: "And we must try to find ways to starve the terrorist and the hijacker of the oxygen of publicity on which they depend."

28   Louise Richardson, *What Terrorists Want* (Random House 2006) at 94–98.

intend to send out loud and clear signals of their intent and capabilities, thus throwing great confusion into the traditional model of objective threat assessment.

Of course, it is from this nature of terrorism and asymmetrical conflict that an enormous difficulty arises. There truly are non-state groups that have every intention of carrying out acts of violence; the threat to national security is very real. However, if we return to the Wolfers essay, we are reminded that the subjective side of threat is the "absence of fear."[29] Since terrorists unequivocally intend to cause fear with their malign intentions to do harm to unprotected and unsuspecting civilians, and less with their actual material threat, this circumstance largely forces the analysis into a realm that cannot be measured—"opinion."

One scholar who has identified many of the difficulties that arise when assessing general pronouncements on the strategic efficacy of drones is Stephanie Carvin in her article "The Trouble with Targeted Killing."[30] She pinpoints the following key problems: differing definitions of targeted killing;[31] a scarcity of sources from which to draw evidence;[32] comparisons across case studies that are not alike;[33] and the central fact that "there is no consensus as to what would actually constitute success."[34]

Carvin also cogently highlights an important point: All studies making general pronouncements on the efficacy of using armed drones for counterterrorism must engage in counterfactual history. She explains: "Essentially, it is challenging, if not impossible, to say what would or would not have happened if the policy of targeted killing had not been carried out or if a given situation would have ended up differently."[35] No doubt, this is an extremely complex question that has been pondered by historians, philosophers, and even scientists through discussions of causality, contingency, and imagination.[36] While it is not within the scope of this

---

29   Wolfers (n 8) at 485.

30   Stephanie Carvin, "The Trouble with Targeted Killing" (2012) 21 Security Studies 529.

31   Ibid at 543–46.

32   Ibid at 546.

33   Ibid at 547–48.

34   Ibid at 548–51. President Obama has explained the goal to be "to disrupt, dismantle, and defeat al Qaeda in Pakistan and Afghanistan, and to prevent their return to either country in the future," yet this still leaves many avenues for interpretation of at least the first two terms presented: "Remarks by the President on a New Strategy for Afghanistan and Pakistan" Dwight D Eisenhower Executive Office Building, White House Office of the Press Secretary (March 27, 2009) available at <http://www.whitehouse.gov/the_press_office/Remarks-by-the-President-on-a-New-Strategy-for-Afghanistan-and-Pakistan/> accessed Sept 2014.

35   Ibid at 551.

36   Martin Bunzl, "Counterfactual History: A User's Guide" (2004) 109, 3 The American Historical Rev 845; Randall Collins, "Turning Points, Bottlenecks, and the Fallacies of Counterfactual History" (2007) 22, 3 Sociological Forum 247; and Richard Ned Lebow, "Counterfactuals, History and Fiction" (2009) 34, 2 Historical Social Research

chapter to delve deeply into convoluted questions of what might have happened otherwise, it can be useful to remember that humans do not act in perfectly predictable ways. Thus, when making such counterfactual claims on an alternative history, it is wise to remain properly measured.

As this section shows, the line between objective and subjective security is disturbingly muddled since the conventional model focused on intent and capabilities is upended in the context of non-state-actor threat assessments. Consequently, general pronouncements on the strategic efficacy of drones face an enormously high hurdle. And even if we were to assume a working model for objectively assessing such threats, current circumstances present a dearth of data for doing so.

III. A Dearth of Data

A. Secrecy and Selective Disclosures

The subsequent difficulties that arise for drawing definitive conclusions on this question are ones that have been greatly exacerbated by the same government that wishes to be seen as efficacious in its use of armed drones. For the first years of the Obama presidency, while the administration conducted a counterterrorism program using UCAVs to target and kill suspected terrorists off the conventional battlefield in, at least, Pakistan, Somalia, Yemen, and the Philippines,[37] it refused to even acknowledge (officially) that it was doing so. This created quite an odd situation that can be described as surreal; although official responsibility can be denied, explosions, injury, and death are widely known. It is this great distance between the known and unknown, exacerbated by secrecy and selective disclosures that generates further problems for determining efficacy.

The drone program was initially cloaked in such official secrecy that it not only helped foster a public perception that the program was "illegal, unnecessary and out of control," but also contributed to internal dissent that eventually became public.[38] As a former Legal Advisor to the State Department (the country's top

---

57; for a more optimistic view of counterfactuals in quantum physics, science and everyday life cf. Osvaldo Pessoa Jr, "Counterfactual Histories: The Beginning of Quantum Physics" (2001) 68, 3 Philosophy of Science S519; and Roland Wenzlhuemer, "Counterfactual Thinking as a Scientific Method" (2009) 34, 2 Historical Social Research 27.

37   Most conspicuous in this list is the strike reported of 2006 in the Philippines: see Sarah Kreps, Mikah Zenko "The Next Drone Wars. Preparing for Proliferation" [March/April 2014] Foreign Affairs 68, at 71.

38   Harold Koh, "How to End the Forever War?" Speech at Oxford University (May 7, 2013); the so-called architect of the drone program, John O Brennan, was also reported to have fought both internally and publically for further disclosure of the program, see eg David A Graham "Meet John Brennan, Obama's Drone Czar and Nominee for CIA Director" The Atlantic (Jan 7, 2013) available at <http://www.theatlantic.com/politics/archive/2013/01/

international lawyer) expressed it in a speech after leaving the government: "The Administration must take responsibility for this failure, because its persistent and counterproductive lack of transparency has led to the release of necessary pieces of its public legal defense too little and too late."[39]

Even more pertinent to the argument here, however, is what this rigid adherence to secrecy has done to the possibility of conducting empirical studies on efficacy. As it has been pointed out, "[a] core problem with the current literature and a primary reason for discrepancy over the effectiveness of decapitation [leadership targeting] is a lack of solid empirical foundations."[40] And the fact that the Obama administration has been so opaque about the details of its killing program by remote control should indeed be seen as a chief cause of the difficulty here. To conduct a thorough study one would at least need to have full access to the pertinent data on the number of strikes (authorized, as well as considered and aborted missions so as to be able construct a view of what happened in the wake of negative decisions to strike), the intended target(s) and rate of successful strikes on them, the target's (suspected) position in the terrorist organization, the number of prevented attacks, the use of other tactics simultaneously or in replacement of drone strikes, along with changes in behavior by the other members of the group after assaults by drone.[41]

---

meet-john-brennan-obamas-drone-czar-and-nominee-for-cia-director/266884/> accessed Sept 2014.

39   Ibid.

40   Jenna Jordan, "When Heads Roll: Assessing the Effectiveness of Leadership Decapitation" (2009) 18 Security Studies 719, at 721.

41   For example, one scholar who has undertaken very similar research to analyze the effectiveness of targeted killing in Israel is Edward H Kaplan. When contacted to ask if he would be interested in undertaking a research project to analyze the effectiveness of US drone strikes he expressed analogous concerns. Kaplan communicated the need to systematically look at hit rate success, collateral damage, and downstream attacks after such strikes. Yet he also raised important doubts in an email exchange (Nov 1, 2013): "I don't even know if the available data from places like *The Long War Journal* or *Pakistan Body Count* are sufficiently accurate/credible, nor am I aware of a publicly available database that records Taliban insurgent attacks in Pakistan as relates to the drone strikes, nor am I aware of data on alternative counterinsurgency operations in the same locations and time periods (eg sweeps to arrest militants). Assembling such data would be the first step; creating an appropriate model that is able to squeeze information out of the data is a big second step." It is also interesting to note that such an empirical model would be based upon the theory that terrorist activity is the product of a "terror stock," a concept developed by Keohane and Zeckhauser ("The Ecology of Terror Defense" [2003] 26 The Journal of Risk and Uncertainty 201). This is a provocative theory in which they argue, "that the optimal counterterrorism policy ought to include efforts aimed at both reducing the flow of new capacity by terrorist organizations and directly reducing the stock of capacity that already exists" (at 202). Nonetheless, there is an inherent difficulty built into this model. There is insufficient recognition that "terror stock" cannot actually be known; terrorist groups are trying to exaggerate their presence and capacity while governments wish to trumpet any and all captures and eliminations of suspected members from the group. Kaplan himself

Compounding this issue of secrecy is the selective use of disclosed information in an attempt to shape the public opinion concerning the efficacy of drone strikes. As one journalist described it when noting a 68 percent spike in articles citing unnamed officials after a UCAV killed a key Taliban commander: "This is selective secrecy, and it inhibits the kind of reporting that would help Americans answer a very basic question: Is the drone war working?"[42]

During the first four years of the Obama administration some of these selective disclosures were carried out by unidentified government sources leaking information to news outlets, while still others were affected through open public statements. Regardless of how orchestrated all of these admissions have been, what we do know is that it is clearly within the direct interest of the president to have his program understood as legal, moral, and effective. Nearly all of the leaks pointed in this very direction, and as noted in the Introduction to this book, President Obama struck these very same three chords in his speech of May 2013.[43]

Even before that landmark speech, his administration began to address these questions in the run-up to his reelection. Most notably, the president's chief advisor on counterterrorism gave a speech in April of 2012, which was conspicuously entitled "The Ethics and Efficacy of the President's Counterterrorism Strategy." It was heralded as the most extensive presentation until that time of the UCAV program and was part of a charm offensive in view of the coming election. In addition to the blunt title, John O. Brennan stated unequivocally: "Yes, in full accordance with the law—and in order to prevent terrorist attacks on the United States and to save American lives—the United States government conducts targeted strikes against specific Al-Qaeda terrorists, sometimes using remotely piloted aircraft, often referred to publicly as drones."[44]

Much like the president's speech that was to come a year later, Brennan tried to focus the question of efficacy on tactical matters and mostly avoided strategic

---

recognizes this trouble in his work, explaining that "the size of the terror stock cannot be observed," which thus leads to a statement that conclusions from the study, "although not definitive, should be taken seriously" (Edward H Kaplan, Alex Mintz, Shaul Mishal and Claudio Samban, "What Happened to Suicide Bombings in Israel? Insights from a Terror Stock Model" [2005] 28 Studies in Conflict and Terrorism 225 at 227). It is agreed that results from such a study should indeed be taken seriously, but it is difficult to know just how seriously when the "terror stock" is neither observable nor measurable.

42   Tara McKelvey, "Covering Obama's Secret War" *Columbia Journalism Review* (May/June 2011) available at <http://www.cjr.org/feature/covering_obamas_secret_war.php?page=all> accessed Sept 2014.

43   See (n 23–25) and accompanying text in the Introduction to this volume: President Barack Obama, "The Future of Our Fight against Terrorism" National Defense University (May 23, 2013) available at <http://www.whitehouse.gov/the-press-office/2013/05/23/remarks-president-barack-obama> accessed Sept 2014.

44   John O Brennan, "The Ethics and Efficacy of the President's Counterterrorism Strategy" Woodrow Wilson International Center for Scholars, Washington DC (April 30, 2012) available at <http://www.lawfareblog.com/2012/04/brennanspeech/> accessed Sept 2014.

consequences other than to say that "[t]he death of bin Laden was our most strategic blow yet against al-Qa'ida."[45] However, as Jenna Jordan explores in Chapter 9 of this volume, the data on leadership decapitation is complex and does not automatically lead to the demise of a terrorist group.[46] In any case, Brennan focused on the achievements of drones in disrupting the communication between members of the group, reducing the number of places to train, a struggle to attract new recruits, low morale evidenced by fighters giving up and returning home, and even cited documents of Osama bin Laden found in his safe house which were encouraging his leaders to flee the tribal regions of Pakistan, "away from aircraft photography and bombardment."[47] Most noteworthy in this speech by the president's chief advisor on this issue, however, was the naming of specific leaders who had been targeted and killed by the armed drones:

> In Pakistan, al-Qa'ida's leadership ranks have continued to suffer heavy losses. This includes Ilyas Kashmiri, one of al-Qa'ida's top operational planners, killed a month after bin Laden. It includes Atiyah Abd al-Rahman, killed when he succeeded Ayman al-Zawahiri as al-Qa'ida's deputy leader.[48]

There is little doubt that this tool of rhetoric—providing specific names and accusations of a high-level status within the organization—is powerful for convincing a society of efficacy. Nevertheless, these are selective disclosures meant to provide evidence of a particular point.

It should also be pointed out that the obstinate insistence upon secrecy for the drone program has not only come from the executive branch. After President Obama's signaling of a push to open the program to greater transparency and accountability in May 2013, what followed was a similar drive from different branches and departments. However, there have been at least two examples of how these internal efforts, which would have allowed for better data gathering, have been stymied by the US Congress, and at times at the behest of actors within the executive branch.

At the end of 2013, the Senate Intelligence Committee passed a provision in its intelligence authorization act for the following year that would require annual reporting by the executive branch on "the total number of combatants killed or injured during the preceding year by the use of targeted lethal force outside the United States by remotely piloted aircraft," as well as the provision of exact figures

---

45   Ibid.

46   However, in this speech Brennan did explicitly say, "[w]e've always been clear that the end of bin Laden would neither mark the end of al-Qa'ida, nor our resolve to destroy it," ibid.

47   Ibid.

48   Ibid.

for the numbers of civilians killed or injured.[49] Cleary, such reporting on the exact number of people killed and injured away from the conventional battlefield would be an important step forward for data collection and public scrutiny.

However, in April of 2014, this provision was quietly stripped out of the legislation following the request of US Director of National Intelligence, James Clapper, who had expressed concern that such reporting would "undermine the effectiveness of the program."[50] Despite this blanket claim of effectiveness, it is important to note that the US government has "yet to engage in a systematic cost-benefit analysis of targeted UAV strikes as a routine counterterrorism tool"; this is a critical point observed in a bipartisan report released in 2014 that called for credible data to be able to engage in "strategic analysis."[51] In other words, it does not appear that the government itself has done its own full analysis of drone efficacy.[52]

Additionally, at the beginning of 2014, an omnibus spending bill was passed that included a provision in a classified annex that would block the president's stated goal of moving drone operations out of the hands of the CIA and into the Defense Department.[53] Not only are there important impacts on humanitarian law (opening the door to civilian operators being legally targeted),[54] there are also significant consequences for this issue of secrecy and data collection. CIA operations are considered covert, which in practical terms means that it is "extremely difficult for journalists and outside researchers to obtain data from the CIA about its drone operations. And they are still briefed to Congress as covert operations, so relatively few lawmakers and congressional staff know about them."[55]

49　Intelligence Authorization Act for Fiscal Year 2014, Section 312 (version of Nov 2013) available at <http://www.lawfareblog.com/2013/11/ssci-approves-intelligence-authorization-bill-for-fy2014/> accessed Sept 2014. It is worth noting that there was an exception for reporting on such lethal force prior to the end of combat operations in Afghanistan and in the territory of countries where there is a declared war in the future.

50　Mark Mazzetti, "Senate Drops Bid to Report on Drone Use" *The New York Times* (April 28, 2014) A10.

51　Gen John Abizaid (US Army, Ret), Rosa Brooks "Recommendations and Report of the Task Force on US Drone Policy" Stimson Report (June 2014) at 31 and 11 respectively.

52　However, a very recently released CIA document suggests that the agency did produce some materials in this regard. See CIA "Best Practices … " (n 6).

53　President Obama (n 40).

54　For a breakdown of the different possible statuses in IHL of drone operators (much turning the enforcement of the laws of war by the CIA's chain of command) see Michael Lewis, Emily Crawford "Drones and Distinction: How IHL Encouraged the Rise of Drones" (2013) Georgetown J of Int'l Law 1127, at 1161–1162; see also Chapter 2 by Patrycja Grzebyk and Chapter 10 by Marek Madej in this volume.

55　Gordon Lubold, Shane Harris "Exclusive: The CIA, Not The Pentagon, Will Keep Running Obama's Drone War" Foreign Policy Magazine (Nov 5, 2013) available at <http://complex.foreignpolicy.com/posts/2013/11/05/cia_pentagon_drone_war_control> accessed Sept 2014.

Nonetheless, despite this classified congressional move at the beginning of 2014, the issue is not entirely settled. The US military came out publicly in favor of the transfer of the drone program into their control in May of that year, explaining that it was necessary to counter criticism at home and abroad: "long-standing U.S. secrecy surrounding drone operations has bolstered support for Al-Qaeda in places like Yemen."[56] Thus the final result of this debate and struggle over secrecy is worthy of attention by analysts who wish for fuller access to the facts and figures.[57]

Another way in which we can gain insight into these carefully chosen disclosures of facts practiced by the Obama administration is by crossing over into the sphere of legality, where it arguably overlaps with efficacy. A legal brief treating this very question was prepared by the American Civil Liberties Union (ACLU) during its three-year litigation to bring transparency to the targeted killing program.[58] The proceedings eventually resulted in an April 2014 order for the administration to publicly release an important internal memo drafted in 2010, but in a redacted form.[59] It was a key piece of the legal puzzle long sought after by the ACLU, reporters from the *New York Times* newspaper, legal scholars, and concerned citizens alike. Prepared by the Office of Legal Counsel (OLC), it provided the legal justifications for a targeted killing of US citizen Anwar Al-Aulaqi that was successfully carried out in September of 2011.

To argue its case for the public release of this document the ACLU put together an extensive recounting of the disclosures and admissions made by the Obama administration since ramping up its UCAV program for counterterrorism. The (successful) argument was that in trying to reassure the public of the legitimacy of drones, the administration's disclosures had been sufficiently detailed and substantial to fundamentally undermine the arguments being made in the litigation for continued concealment. In other words, what had already been released in

56   Julian Barnes, Siobhan Gorman "U.S. Military Pushes for More Disclosure on Drone Strikes: Effort Is to Counter Criticism of Program in U.S. and Abroad, Particularly Yemen" *Wall Street Journal* (May 22, 2014) available at <http://online.wsj.com/news/articles/SB10001424052702303749904579578443104419368> accessed Sept 2014.

57   At the time of this writing, legislation (similar to that stripped out by the Senate requiring reporting on the numbers killed by remotely piloted aircraft) had been introduced in the House of Representatives in April 2014. This bill differs from the earlier legislation in that it would require accounting for deaths and injuries caused in the previous six years. "H.R.4372, Targeted Lethal Force Transparency Act" now lingers in committee awaiting assignment for vote by the Speaker; text and status available at <https://beta.congress.gov/bill/113th-congress/house-bill/4372> accessed Sept 2014.

58   American Civil Liberties Union, "Request under Freedom of Information Act" (Oct 19, 2011) (submitted approx three weeks after the reported death of Anwar Al-Aulaqi via drone in Yemen) available at <https://www.aclu.org/national-security/anwar-al-awlaki-foia-request > accessed Sept 2014.

59   US Court of Appeals for the Second Circuit, *The New York Times Company et al v United States*, Case No 13–422-cv (April 21, 2014).

defense of the drone program left no further reason to keep secret the particular documents pursued.

Documenting the trail of what had trickled out over the years was meant to show that "[t]he government's disclosures about the program have been selective and self-serving, and they leave the public record about the targeted killing program incomplete in crucial respects."[60] For example, the beginning was a statement in May 2009 from Leon Panetta, who was the CIA director, explaining that "these operations have been very *effective* because they have been very precise in terms of the targeting and it involved a minimum of collateral damage."[61] Nearly a year later, Harold Koh, then legal adviser to the Department of State, explained to the American Society of International Law that the unmanned vehicle strikes "*comply with all applicable law.*"[62] Right from the start, the government was sending out explicit signals that these operations were legal and effective.

It is not a surprising conclusion from the ACLU that that cherry-picked information the Obama administration divulged was essentially to paint a flattering picture of the drone program—otherwise kept under very tight wraps. What is more interesting here, however, is that the District Court that first heard this case came to a very similar conclusion.[63] And it was one that the Court of Appeals went out of its way to substantiate: "we note initially the numerous statements of senior Government officials discussing the lawfulness of targeted killing of suspected terrorists, which the District Court characterized as 'an extensive public relations campaign.'"[64] Yet because that public relations (PR) campaign did not include a release of calculable data, it served only to distort the picture of efficacy.

The substance of this central OLC document that the court ordered released should be addressed since it is suggested here that this point represents an overlap of legality and efficacy.[65] Most importantly to our argument, all of the pertinent facts related to the specific case of Anwar Al-Aulaqi are redacted. Additionally, the author of the memo accepts all the facts presented by other departments of the executive branch as accurate. Thus the memo as it stands reads much more like an outline of the applicable law (much as we have seen presented in the first section of this book).[66] Yet, without the facts of the case, very little can be gained by

---

60   Brief and Special Appendix for Plaintiffs-Appellants American Civil Liberties Union (April 15, 2013) at 10, available at <https://www.aclu.org/national-security/anwar-al-aulaqi-foia-request-legal-documents> accessed Nov 2014.

61   Ibid at 14 (my emphasis).

62   Ibid at 11 (my emphasis).

63   US District Court, *The New York Times Company et al v United States Department of Justice*, 915 F Supp 2d at 524.

64   US Court of Appeals (n 59) at 37.

65   David Barron, Acting Assistant Attorney General, "Re: Applicability of Federal Criminal Laws and the Constitution to Contemplated Lethal Operations Against Shaykh Anwar al-Aulaqi" (July 16, 2010).

66   It should be noted that this is an important step forward from the Bush administration who repeatedly interpreted away the applicability of international law. For a full discussion

reading this memo. In effect, the memo essentially provides a framework within which analysts, or even an intelligence source on the ground, might "discover" facts to fit what would be construed as legal targeted killing, knowing that there is no reliable check to follow.

For example, the memo does not provide any further explanation of the remarkable and novel expansion of the definition of "imminence" that the administration has been using.[67] Additionally, the memo does not discuss any of the criteria used to determine when capture is "infeasible," thus triggering a consideration of targeted killing in its stead.[68] In each of these circumstances the author of the memo simply defers to the facts represented by the agents, who are the same individuals requesting the authorization for a drone killing.

When there is no oversight or transparency on the authority to exercise lethal force it is undoubtedly an awesome power, and one that can be abused. Without a check whether that power is in reality misused or not, those suspecting the very worst of intentions can never be disproven. This creates an environment in which matters of life and death, for citizens and non-citizens alike, rest entirely on the assumption of good faith by (often foreign) government officials. This consequence creates legal problems for the rule of law, particularly because it avoids any judicial review or outside check; put another way, legality is only an idea treated and tested in the minds of those making the policy.[69]

At the same time, this generates difficulties for strategic efficacy in two regards. Firstly, the belief that the facts of a drone killing can be held tightly in secret and will not later become exposed creates a political incentive to disclose only the data that colors the program in a favorable light; to wit, there is a motivation to hide inconvenient facts necessary for empirical testing. Secondly, it undermines a faith in the legitimacy of the government that exercises such a breathtaking authority without oversight, review or even a cursory check. This is one important reason to recognize the idea of a Drone Court, as set forth in Chapter 13 by Guiora and Brand in this volume, as representing a veritable step forward by creating such a check.

---

of the specific manner in which this was done see generally Barela (n 14).

67   Barron (n 65) at 39: "on the facts represented to us, a decision-maker could reasonably decide that the threat posed by al-Aulaqi's activities to United States persons is 'continued' and 'imminent.'" For a historical discussion of this standard of "imminence" see Chapter 6 in this volume.

68   Ibid at 40: "[i]n addition to the nature of the threat posed by al-Aulaqi's activities, both agencies here have represented that they intend to capture rather than target al-Aulaqi if feasible; yet we also understand that an operation by either agency to capture al-Aulaqi in Yemen would be infeasible at this time."

69   For an analysis of "Law versus the Rule of Law" see Stimson Report (n 51) at 12: "changing technologies and events have made it increasingly difficult to apply the law of armed conflict and the international law relating to the use of force in a consistent and principled manner, leading to increasing divergence between "the law" and core rule of law principles that traditionally have animated US policy." See also Chapter 14 of this volume by Farer and Bernard.

A government without legitimacy, or even one suffering a legitimacy deficit, can certainly not be in the strategic interests of citizens or the government itself.

B. Independent Journalism

The next difficulty to be addressed is that of credible and independent information gathering. We have discussed the fact that one of the primary sources for material on drone strikes is the very government that has: (1) all of the information on numbers, decision-making processes, and a good view on the consequences; and (2) a vested interest in how the program is perceived. Since the great bulk of the UCAV strikes have taken place in the Federally Administered Tribal Areas (FATA) of Pakistan (see Tables 10.1 and 10.2), our attention will be focused on the viability of neutral data collection in this region as an illustration of the problem. Namely, we will look into the fact that investigative journalists have become an endangered species in this geographical area.

Although one should use prudence and not attribute too much importance to every *fatwā* issued (an interpretation of the Qur'an and the Prophet Muhammed's teachings delivered by an Islamic law specialist), one such declaration helps illuminate our point on the threat to media freedom in Pakistan. In October 2013 a standing *fatwā* was reissued via a post on Twitter by a group that supports an outlawed coalition, Tehreek-i-Taliban Pakistan, and identified certain media organizations and specific journalists as "enemies of the Mujahideen."[70] It named several leading local media sources and the BBC and "included the photos of two nationally-known journalists—Hamid Mir, host of a TV program, and Hasan Nisar, a columnist and commentator."[71]

The fact that Hamid Mir was shot six times in an assassination attempt only months later in April of 2014 is telling of the imperiled independence of the profession. This is even more so since the media outlet he works for blamed the feared Inter-Services Intelligence (ISI) agency for the attack,[72] since violence against reporters comes from both non-state and state actors. Such patently intolerant and violent threats, at times followed by deadly attacks, have indeed had an impact on the impartial reporting coming out of the FATA region. This is

---

70   The Newspaper's Correspondent, "Fatwa 'threatens media freedom' in Pakistan" *Dawn.com* (Islamabad, Pakistan, Oct 26, 2013) available at <http://www.dawn.com/news/1051960/fatwa-threatens-media-freedom-in-pakistan> accessed Sept 2014; see also Huma Yusuf, "Intimidating Pakistan's Press" *The New York Times* (Oct 31, 2013) available at <http://latitude.blogs.nytimes.com/2013/10/31/intimidating-pakistans-press/?_php=true&_type=blogs&emc=eta1&_r=1> accessed Sept 2014.

71   Ibid.

72   Mohammed Hanif, "The Hamid Mir case: 'In Pakistan, they used to censor journalists—now they shoot us'" *The Guardian* (London, April 23, 2014) available at <http://www.theguardian.com/world/2014/apr/23/hamid-mir-pakistan-journalist-shooting-isi-mohammed-hanif> accessed Sept 2014.

why Pakistan has been recognized as one of the deadliest countries in the world for the practice of journalism.[73]

This point is demonstrated by two recent non-governmental organization reports that were compiled to increase attention to this disturbing proclivity in the specific region where drone strikes off the "hot battlefields" have been most prevalent. This targeted violence on journalists is not highlighted here to simply draw attention to even further bloodshed in the conflicts, but rather to identify another major problem for the ability to draw conclusions on the question of drone efficacy.

In 2012, a report entitled "Reporting from the Frontlines" was published by Intermedia Pakistan and focused specifically on media coverage in the FATA region where the government has forbidden reporters to travel independently. The overarching conclusion of the report was that journalists working along the northwest border with Afghanistan lived in constant fear (74 percent of respondents had received threats during the prior 12 months)[74] because "reporting on human rights violations from both militants and government security agents exposes them to threats from both sides."[75] As a result, a large number have been forced to relocate and many "report on their region while staying outside of the tribal agencies in FATA."[76] So along with the fact that some (53 percent) of these journalists are working without any salary, commission, financial compensation, or any kind of medical and life insurance from their organization (86 percent), an alarming number (73 percent) said that they count on the political administration as "one of their primary sources of news."[77] This last figure is of importance because the independence of such coverage should be questioned when free access to the area in question is so tightly controlled.

Next, Amnesty International compiled an updated report in 2014 with research that identified many of the same problems, including the fact that the perpetrators of violence "continue to operate in a general climate of impunity. In only one

---

73   "Murder of Journalist Illustrates Dangers, Again" *Radio Free Europe, Radio Liberty* (Oct 15, 2013) available at <http://www.rferl.org/content/journalists-in-trouble-murder-of-journalist-illustrates-dangers/25137170.html> accessed Sept 2014: "The US-based Committee to Protect Journalists (CPJ) has for several years running designated Pakistan among the most dangerous countries for journalists in the world, citing the number of reporters killed and the impunity that attaches to such crimes." For further details on all of the journalists killed in Pakistan since 1992 (with motives confirmed and unconfirmed), visit the CPJ website <http://cpj.org/killed/asia/pakistan/> accessed Sept 2014.

74   Sadaf Baig "Reporting from the Frontlines: The Tough Working Conditions of Journalists in Pakistan's Tribal Areas" *Intermedia Pakistan* (2012) at 41 <http://journalistsafety.files.wordpress.com/2012/09/reporting-from-the-frontline-final.pdf> accessed Sept 2014.

75   Ibid at 5.

76   Ibid at 1.

77   Ibid at 2.

of the 73 cases investigated ... have the perpetrators been brought to justice."[78] Additionally, the report underscores the fact that the expanding public violence, coupled with explicit threats, creates a range of precarious "red lines," or unspoken boundaries enforced by both state and non-state actors. As it was chillingly explained, "[t]his creates a nearly impossible course for journalists to navigate because appeasing one perpetrator to avoid the risk of abuse almost inevitably increases the risk of abuse from others."[79]

These reports represent some of the research that has been conducted in an attempt to gather an overall picture of the situation in Pakistan, and in the specific region of concern. However, sometimes a local resident, with invaluable experience on the ground, can capture in a few words the major problem we identify here for those wishing to compile a true picture of drones' efficacy. Pir Zubair Shah, a Pulitzer Prize (shared) winning journalist and Fellow of Journalism at Harvard University,[80] concludes that the story of alliances and enemies in North Waziristan is far more complicated than has been presented by the administration. This native to one of the world's most inaccessible regions wrote:

> Although the drone campaign has become the linchpin of the Obama administration's counterterrorism strategy in Central Asia—and one it is increasingly exporting to places such as Yemen and the Horn of Africa—*we know virtually nothing about it.* I spent more than half a decade tracking this most secret of wars across northern Pakistan, taking late-night calls from intelligence agents, sorting through missile fragments at attack sites, counting bodies and graves, interviewing militants and victims. I dodged bullets and, once, an improvised explosive device. At various times I found myself imprisoned by the Taliban and detained by the Pakistani military. *Yet even I can say very little for certain about what has happened.*[81]

In light of such reporting, there is little doubt that the reliability of the facts available should raise great skepticism. In addition to this, it is even questionable whether the Obama administration itself has more dependable sources for explaining what is actually happening on the ground in Pakistan and, by extension, whether the drone killing is in fact efficacious in targeting the right people.

------

78   Amnesty International, "'A Bullet Has Been Chosen for You' Attacks on Journalists in Pakistan" (London, April 30, 2014) available at <http://www.amnesty.org/en/news/pakistan-journalists-under-siege-threats-violence-and-killings-2014-04-30> accessed Oct 2014.

79   Ibid at 8.

80   Corydon Ireland, "A Pakistani journalist explains the dangers of his trade" *Harvard Gazette* (Dec 15, 2011) available at <http://news.harvard.edu/gazette/story/2011/12/dateline-classroom/> accessed Oct 2014.

81   Pir Zubair Shah "My Drone War" [March/April 2012] Foreign Policy (my emphasis) available at <www.
foreignpolicy.com/articles/2012/02/27/my_drone_war> accessed Oct 2014.

C. Defining the Enemy

In 2003, Secretary of Defense Donald Rumsfeld wrote an internal memo that recognized a central problem and posed an important question: "Today, we lack metrics to know if we are winning or losing the global war on terror. Are we capturing, killing or deterring and dissuading more terrorists every day than the madrassas and the radical clerics are recruiting, training and deploying against us?"[82] Although this probing only occurred after the launching of two wars, the detaining of hundreds without judicial review in an off-site prison, and introducing a program of torture, the question is certainly essential. In this context, attributing "madrassas and radical clerics" as the only source of new recruits certainly seems incomplete. Nonetheless, the idea of ascertaining and defining who exactly the enemy is in this conflict should actually precede the idea of reducing the "terror stock" (quantified capability).[83]

It is important to note that the issue of knowing precisely who your adversaries are can be more directly related to strategic, rather than tactical, efficacy.[84] Since the widely accepted purpose of hostilities "is not to *kill* the enemy but to *defeat* him,"[85] then you must know who to accurately count as your foes in order to achieve the strategic goal. This is one place where avoiding the mixing of tactical and strategic objectives becomes particularly important. Put another way, killing many of the wrong people to eventually remove an important leader can still be counted as a tactical success, but whether this translates into an overall victory is not obvious, clear or certain (see Chapter 9 of this volume by Jenna Jordan).

One way in which this matter has been treated in the literature (similar to the Rumsfeld memo) is by looking at the creation of new recruits.[86] For instance, it has

---

82   Donald Rumsfeld, Secretary of Defense, "SUBJECT: Global War on Terrorism" (Oct 16, 2013) available at <http://usatoday30.usatoday.com/news/washington/executive/rumsfeld-memo.htm> accessed Oct 2014. It was also here where he famously referred to the wars in Afghanistan and Iraq war as a "long, hard slog."

83   This concept developed by Keohane and Zeckhauser is discussed above in (n 41).

84   See eg Katherine Zimmerman, "The al Qaeda Network: A New Framework for Defining the Enemy" American Enterprise Institute's Critical Threats Project (Sept 10, 2013): "[u]nderstanding precisely which groups contribute to the al Qaeda network and how they operate within that network will better enable American policymakers and decision makers to develop a comprehensive strategy to defeat al Qaeda. Absent that understanding, the United States will continue to engage in a tactical battle that promises only occasional battleground victories, but no real prospect of winning the war" at 1.

85   Nills Melzer, *Targeted Killing in International Law* (OUP 2008) (original emphasis) at 427.

86   For an extensive treatment of this question see International Human Rights and Conflict Resolution Clinic (Stanford Law School) and Global Justice Clinic (NYU School of Law), "Living Under Drones: Death Injury and Trauma to Civilians from US Drone Practices in Pakistan" (September 2012) at 131–7, available at <http://livingunderdrones.org/> accessed Oct 2014; see also Amnesty International, "Will I Be Next?: US Drone

been pointed out that the confessed and convicted New York City Times Square bomber, Faisal Shahzad, told a US District Court judge that his attempted terrorist attack was in part in retaliation for "the drone strikes in Somalia and Yemen and in Pakistan."[87] Yet even though there are a good number of policy analysts, officials and independent observers who argue that this incident is not simply an outlier in the collection of facts, the quantitative data proving a connection between drone strikes and future participation in hostilities is very limited. For example, one study that found it "probable that drone strikes provide motivation for retaliation" also admitted, in the very same sentence, that "it is impossible to prove direct causality from data analysis alone."[88]

On the other hand, one scholar who has frequented Pakistan for research published a critical newspaper piece in October 2014.[89] Christine Fair's article dug deeply into the methodological practices of three referenced reports in this chapter and concluded that the results pointing toward stopping the use of armed drones raised serious ethical issues. Her critique identified problems concerning conflicts of interest, sampling troubles, ignoring inconvenient voices, and the use of children as witnesses. In her conclusion, Fair vehemently charged that these reports should be denounced since their "approaches are extremely misleading and raise serious ethical questions about the intent of the studies in the first place."[90]

This strident criticism is puzzling because Fair published a piece in 2010, emphasizing that some "FATA residents are strong proponents of the drones" and

---

Strikes in Pakistan" (2013) available at < http://www.amnestyusa.org/research/reports/will-i-be-next-us-drone-strikes-in-pakistan> accessed Oct 2014; Colombia University Law School and the Center for Civilians in Conflict "The Civilian Impact of Drone Strikes: Unexamined Costs, Unanswered Questions" (2012) available at <http://web.law.columbia.edu/human-rights-institute/counterterrorism/drone-strikes/civilian-impact-drone-strikes-unexamined-costs-unanswered-questions> accessed Oct 2014; Human Rights Watch, "Between a Drone and Al Qaeda: The Civilian Cost of US Targeted Killing in Yemen" (October 2013) available at <http://www.hrw.org/reports/2013/10/22/between-drone-and-al-qaeda-0> accessed Oct 2014.

87   Chris Dolmetsch, "Times Square Bomber Vows Revenge in Al-Arabiya Video" *Washington Post* (July 14, 2010) <http://www.washingtonpost.com/wp-dyn/content/article/2010/07/14/AR2010071404860.html> accessed Oct 2014; see also Jo Becker and Scott Shane, "Secret 'Kill List' Proves a Test of Obama's Principles and Will," *New York Times*, (May 29, 2012), where it was argued "drones have replaced Guantánamo as the recruiting tool of choice for militants; in his 2010 guilty plea, Faisal Shahzad, who had tried to set off a car bomb in Times Square, justified targeting civilians by telling the judge, 'When the drones hit, they don't see children.'"

88   Leila Hudson, Colin Owens & Matt Flannes "Drone Warfare: Blowback from the New American Way of War" (2011) XVIII, 3 Middle East Policy Council 122, at 126.

89   Christine Fair, "Ethical and Methodological Issues in Assessing Drones' Civilian Impacts in Pakistan" *Washington Post* (Oct 6, 2014) available at <http://www.washingtonpost.com/blogs/monkey-cage/wp/2014/10/06/ethical-and-methodological-issues-in-assessing-drones-civilian-impacts-in-pakistan/> accessed Oct 2014.

90   Ibid.

"when children hear the buzz of the drones, they go [sic] their roofs to watch the spectacle of precision."[91] Not only does Fair indulge in the same rhetorical tool of illuminating voices that agree with her own point of view, the argument of children rushing to rooftops when they "hear the buzz of the drones" is directly undermined by her 2014 charge that the reports are flawed for any discussion of the psychological impact of constant buzzing since drones "are mostly flying at altitudes that tend to be inaudible."[92] Nonetheless, there are valid concerns raised by Fair and the inaccuracies of both sides are due to the general difficulty of knowing with complete certainty what actions will cause what future results.

In addition to this uncertain tack of tracking enemies that have not (yet) joined the fight, the difficulties in defining the enemy can also be demonstrated by the imprecise notion of "associated forces." One can grasp the significant problems created by the Obama administration's inclusion of this term in its definition of the enemy by tracing its origins into an overlap with the realm of legality. In a court brief filed in March of 2009, the executive branch framed its detention authority as pertaining not only to those responsible for the attacks of 9/11, but stemming from the fact that the president "also has the authority to detain persons who were part of, or substantially supported, Taliban or al-Qaida forces or *associated forces*."[93]

While the "substantial support" prong of this definition raises similar concerns, particular attention is directed to the inclusion of "associated forces" because of how the administration has used this same interpretation of the law for targeting killing. Specifically, in Attorney General Eric Holder's letter to Congress in May 2013, it was clarified that targeting decisions for the use of lethal force would be based (even after a constriction of the drone program) on whether a person is "a senior operational leader of al Qa'ida or its *associated forces*."[94] Hence, this broad term takes on significance beyond detention authority and its subjectivity for targeting decisions is equally problematic.

From the perspective of domestic law, membership in Al-Qaeda or the Taliban plainly falls within the authority granted by Congress in the 2001 Authorization to Use Military Force (AUMF).[95] But the National Defense Authorization Act for Fiscal Year 2012 (NDAA) included a provision that was meant to delineate

---

91 Christine Fair, "Drones Over Pakistan—Menace or Best Viable Option?" *The World Post* (Aug 2, 2010) available at <http://www.huffingtonpost.com/c-christine-fair/drones-over-pakistan----m_b_666721.html> accessed Nov 2014.

92 Ibid.

93 Respondents' Memorandum, Regarding the Government's Detention Authority Relative to Detainees Held at Guantánamo Bay, Guantánamo Bay, Detainee Litigation No 08–0442 (DDC), (March 13, 2009) (my emphasis).

94 Eric Holder, Attorney General, "Letter to Chairman of the Judiciary Committee, US Senate" (May 22, 2013) (my emphasis).

95 Authorization for Use of Military Force, Public Law 107–40, 115 Stat. 224,(September 18, 2001).

the categories of persons who could be indefinitely detained.[96] However, this legislation moved beyond those directly responsible for the original attacks with the insertion of the words "*associated forces*,"[97] closely following the Obama administration brief discussed above.

As an indication of the problem created by introducing a category of indeterminate breath, it is worth looking at the 2012 District Court ruling in the *Hedges v Obama* litigation, which included a permanent injunction enjoining enforcement of the provision containing this language.[98] Most pertinent to our point here is the vagueness of the terms used by the administration to define their authority—charting who can be classified as an enemy—which indeed occupies an important portion of the District Court opinion. It was declared that "'*associated forces*' is an undefined, moving target, subject to change and subjective judgment."[99] Additionally, in a footnote, the judge dissected one of the pre-trial government memoranda, concluding that the "argument is carefully crafted and does not exclude the concept of *associated forces* constituting groups the executive branch 'believes' may be tied to Al-Qaeda or the Taliban."[100]

The Obama administration has defined the enemy broadly, using terms and parameters that cannot be fenced in, so as to make strategic efficacy itself dependent on the administration's own moving definition. The original 2009 brief strikingly argued that:

> [i]t is neither possible nor advisable, however, to attempt to identify, in the abstract, the precise nature and degree of "substantial support," or the precise characteristics of "*associated forces*," that are or would be sufficient to bring persons and organizations within the foregoing framework.[101]

This subjective margin of maneuver continues to be something the administration is intent on maintaining. At the time of this writing in early 2015, there has been a novel application of this 2001 AUMF law that directly demonstrates the vast problems for strategic efficacy created by such a wide interpretation of authority. On the day before the 13th anniversary of 9/11, President Obama stepped before the

---

96   National Defense Authorization Act for Fiscal Year 2012, Public Law No 112–81, 125 Stat 1298, (December 2011).

97   Ibid at §1021(b)(2) (my emphasis).

98   US District Court, *Hedges v Obama*, No 12-CV-331, 2012 WL 1721124 (SDNY 2012) (slip opinion). The presiding judge rejected the government's argument that the statute simply affirmed what was already in the AUMF exactly because of the difference in language between the two. Namely, the statute no longer required a link to 9/11, and so something novel was indeed introduced into this legislation.

99   Ibid at 107 (my emphasis).

100   Ibid at 105–106 (my emphasis).

101   Respondents' Memorandum (n 93) at 2 (my emphasis).

US citizenry to announce: "I have the authority to address the threat from ISIL."[102] He was referring to the extension of hostilities against the group variously known as the Islamic State of Iraq and the Levant (ISIL), the Islamic State of Iraq and Syria (ISIS), or Dawlat al-Islamiyah f'al-Iraq wa al-Sham (Daesh).[103] Regardless of the nomenclature that eventually sticks to this group, the idea that President Obama had the authority to direct attacks against them raises real difficulties for defining the enemy.

It should be clarified that the exact legal interpretation being used by the Obama administration is not clear at this point in time. The night of the speech, a senior administration official spoke specifically of the 2001 AUMF, and the interpretation discussed above, including *"associated forces."*[104] As Al-Qaeda has publically disowned ISIL, and they currently compete for allegiances, the assertion of a connection between them is suspect.[105] There were two predominant interpretations posited in legal circles at the time of that speech in September 2014: (1) that ISIL is still a force "associated" with Al-Qaeda; or (2) ISIL was previously known as "al-Qaeda in Iraq" and thus a part of Al-Qaeda proper, leading to the conclusion that the later split between the two groups is irrelevant.[106]

---

102   President Barack Obama, "Statement by the President on ISIL" State Floor, White House (Sept 10, 2014) available at <http://www.whitehouse.gov/the-press-office/2014/09/10/statement-president-isil-1> accessed Oct 2014. It should be noted that the president also "welcome[d] congressional support for this effort."

103   Lizzie Dearden, "ISIS vs Islamic State vs ISIL vs Daesh: What do the different names mean—and why does it matter?" *The Independent* (UK, Sept 23, 2014) <http://www.independent.co.uk/news/world/middle-east/isis-vs-islamic-state-vs-isil-vs-daesh-what-do-the-different-names-mean-9750629.html> accessed Oct 2014.

104   Statement reported and posted at Just Security <http://justsecurity.org/14799/legal-theory-presidents-military-initiative-isil/> accessed Oct 2014; however, an official explanation has not been released.

105   See Barak Mendelsohn, "After Disowning ISIS, Al Qaeda Is Back On Top" Foreign Affairs (Feb 13, 2014) available at <http://www.foreignaffairs.com/articles/140786/barak-mendelsohn/after-disowning-isis-al-qaeda-is-back-on-top> accessed Dec 2014; Liz Sly, "Al-Qaeda disavows any ties with radical Islamist ISIS group in Syria, Iraq" *Washington Post* (Feb 3, 2014) available at <http://www.washingtonpost.com/world/middle_east/al-qaeda-disavows-any-ties-with-radical-islamist-isis-group-in-syria-iraq/2014/02/03/2c9afc3a-8cef-11e3-98ab-fe5228217bd1_story.html> accessed Dec 2014; Rezaul H Laskar, "IS announces expansion into AfPak, parts of India" *Hindustan Times* (New Delhi, Jan 29, 2015) available at <http://www.hindustantimes.com/india-news/is-announces-expansion-into-afpak-parts-of-india/article1-1311533.aspx> accessed Jan 2015.

106   See eg Jennifer Daskal, "Democracy's Failure" Just Security blog, (Sept 11, 2014) <http://justsecurity.org/14820/democracys-failure/> accessed Oct 2014; Jens David Ohlin, "The 9/11 AUMF does not cover ISIS" Opinion Juris blog, available at <http://opiniojuris.org/2014/09/11/911-aumf-cover-isis/> accessed Oct 2014; and Marty Lederman, "Tentative first reactions to the 2001 AUMF theory" available at <http://justsecurity.org/14804/first-reactions-2001-aumf-theory/> accessed Oct 2014. While the first two blogs are quite critical

By the time this work is in print it is possible that Congress will have weighed in and circumscribed both the authorization to use force against the ISIL group and the president's applicable authority.[107] However, regardless of whether this happens or not, the dangers, difficulties, and follies of the confused interpretations of who is the enemy has made the judgments of strategic efficacy immensely difficult, if not impossible. If we do not know precisely who the enemy is, how do we know when they have been defeated?

VI. Conclusion

Without a method for measuring strategic progress on security the public are forced to be spectators of a game with no view of the scoreboard, and they do not even have the tools to keep score themselves. Notably, it is the spectators whose safety is the primary goal in that game, regardless of whether that security is based on objective threats or simply an "opinion." Secrecy of the drone program has rendered the necessary data for full assessment unattainable, and selective disclosures have only distorted an already fragmented picture. On top of this, those who might be able to provide facts and figures to fill in the many gaps—that is, investigative journalists working in the regions where the armed drones are operating—are themselves targeted and threatened on all sides, throwing credible data collection into a chaotic state. Lastly, the fact that the enemy in this conflict is ill defined, and even changing as new violent groups materialize, makes the ambition to provide conclusive answers on the strategic efficacy of UCAVs beyond our current reach.

Nonetheless, this certainly does not mean that nothing of consequence can be said about the efficacy of using armed drones to kill individuals within the territory of another state. Most importantly, one must certainly question the reasoning behind killing suspected individuals if the benefits are not publicly known, or perhaps not even possible to know. However, this would largely become a moral question outside the scope of this chapter, but would be well worth further pursuit. Additionally, another moral question arises as to whether it is acceptable to kill if it effectively increases the subjective "opinion" of security.[108] Hence it is possible

---

of the possible interpretations and applications of the 2001 AUMF, the third blog posits that the Obama administration chose this more restrictive route specifically because using the president's Article II commander-in-chief power or the War Powers Resolution would have stretched the president's authority to use force even further.

107   President Obama introduced an authorization request for the use of force against ISIL in February 2015. However, his previous interpretation of the 2001 AUMF to include "associated forces" still raises important questions over how and if the enemy will in fact be delineated in any limited way: see eg Bruce Ackerman, "Obama Still Believes in Unlimited War" *New York Times* (Feb 12, 2015) op.-ed, A29.

108   Although the conventional wisdom is that the US public largely supports the use of drones, the mainstream public opinion polls often simply ask whether or not respondents support the use of drones to kill terrorists in a foreign territory. However, a

to conclude that the absence of definitive answers over whether armed drones effectively increase objective security reframes important questions of morality, and thus offers other insights into the legitimacy of their use for cross-border counterterrorism.

We have already seen an empirical investigation into the historical data on leadership removal in Chapter 9 (Jenna Jordan), and an exploration of the tactical efficacy of UCAVs as a weapons platform in Chapter 10 (Marek Madej). To further examine another essential component of efficacy the following chapter (Robert Kolb) will delve into what the use of lethal force across borders means to the system of international law in place today. In other words, one must also ask the question: even if tactical and strategic efficacy were to be assumed, would extending the current use of armed drones to all members (as must be supposed), still leave the international system functioning?

---

thorough pursuit of the above research question would need to dig more deeply into poll results. For example, one research has recently suggested, "there are now good reasons to question the conventional wisdom on public support for armed drones": Ryan Goodman, "Social Science Data on Public Reactions to Drone Strikes and Civilian Casualties" Just Security website (July 3, 2014) available at <http://justsecurity.org/12556/social-science-data-public-reactions-drone-strikes-civilian-casualties/> accessed Dec 2014.

Chapter 12

Systemic Efficacy: "Potentially Shattering Consequences for International Law"

Robert Kolb

I. Introduction

The subject matter of the present volume is the use of armed drones by the US and other countries in the fight against terrorism. As seen in this volume, such practice raises various questions of a political, moral, and legal nature. One of the problems posed by the use of drones lies in the overall effects of their use on the system of public international law. This system is based on a certain number of parameters, general assumptions, and principles. Varying practices can have a lesser or greater impact on a specific subset of rules of international law, which may end up prodding modification. However, such practices can also have important impacts on general parameters and principles upholding the system as a whole. In this short contribution, it is proposed to look squarely into this issue.

Such concerns are not entirely new; they have existed before. But the magnitude and extent into which they are now cast and intertwined is of an unprecedented nature. Thus, we may be on the shifting lines, where quantity tends to become quality. The point of the present contribution is not to discuss all these issues at length, but to provide a provisional mapping. The focus will lie on the tectonic aspects of these questions, looking to their potential consequences on the functioning of the international legal system. The rationale behind this approach is to take account of these potentially adverse consequences in order to feed them into the overall assessment of the efficacy of drone warfare. This efficacy cannot be measured only by reference to the tactical military gains, or even a strategic success brought about by such engines. It must also take into account the collateral losses on the legal system itself, so vital in our days for a minimum global order.

It is also useful to point out that this is not the first time that such concerns have been harbored in the context of counterterrorism. International jurist Antonio Cassese identified "potentially shattering consequences for international law"[1] in the United Nations Security Council (UNSC) Resolutions passed just in the wake of the attacks of 9/11, and within the former shadows of the Twin Towers in New

---

1 Antonio Cassese, "Terrorism Is Also Disrupting Some Crucial Legal Categories of International Law" (2001) 12,5 Eur J Intl Law 993.

York.[2] His worry was that they introduced a disruption of crucial legal categories, and the contention here is that this disquiet becomes amplified with the use of drones employing lethal force across international borders. This chapter will address this strain and the danger it causes to the global structure as we know it.

The first section of this volume delves deeply into the legal challenges raised by the use of UCAVs (unmanned combat aerial vehicles) for cross-border counterterrorism, and there is certainly no need to replicate the in-depth juristic work that has been competently carried out there. Thus this chapter will build off of this solid foundation and present more of a bird's-eye view on the relevant international rules so that we can appreciate the collisions occurring between competing or overlapping bodies of law created by present interpretations thereof. In order to be in a position to elaborate on these systemic consequences, it is first considered useful to provide the reader with some basic knowledge on the character and on the evolution of international law. Thus this brief historical picture will be illustrated in section II. Next, this chapter will depict in section III a raised perspective on the international system by delineating the specific challenges presented by the current use of drones, that is, United Nations Charter law, humanitarian law, and human rights law. Finally, section IV will conclude by offering a plea for an attentiveness to the gradually developing law of peace constituting the foundation of the international community so as to keep emerging technologies from razing the painstakingly constructed edifice thus far constructed of an international system.

II. International Legal Order: Nature, Evolution, and Pillars

International law is principally the legal system governing certain relations among states, and secondarily with or among other subjects of the law.[3] In other words, international law contains legally binding rules for the intercourse of states and some other subjects, generally (customary law) or specially (treaty law) accepted as addressees of international law. States, like any other person, have regular or irregular interaction. The latter may be relations of cooperation, treaty relations, schemes for dispute settlement, responsibility for breach of the law, hostilities and warfare, foreign policy, etc.

For all these relations, there must be some legal rules providing at least a minimum degree of certainty and attunement. In the complex and manifold relations of today, the capacity to function in legal terms for some questions and subject matters is a precious asset of global governance. These questions are thus extracted from the maelstrom of the general relations between states and stabilized through shared legitimate expectations of a legal nature.

---

2   UN Security Council, Resolution 1368, UN Doc S/RES/1368 (2001); Resolution 1373, UN Doc S/RES/1373 (2001).

3   For a definition of international law, see eg L Oppenheim (ed) in RY Jennings, and A Watts, *International Law* (9th ed, vol I, London, 1992) at 4ff.

Two terms of the definition need some further elaboration: "certain relations" and "states." The term "certain relations," used above, tends to emphasize that not all the contacts and exchanges between the subjects are extensively regulated by law, as little as all inter-human relations are cast in the realm of law. Thus, a love affair between two individuals has certain legal dimensions, for example, in regard to mutual gifts or marriage. But the essence of it remains extra-legal. The same is true for states. Their international policy remains free—"discretionary" as the lawyer would say—in many regards. But it is limited by the applicable treaties or by other applicable rules of international law. Thus, for most states, choosing a policy sympathetic to the United States, or to Russia, or to some other state, is a matter of unfettered discretion, a matter not regulated by international law. There is no lacuna here, since international law is not expected to regulate this question. However, a certain policy benefiting this or that other state may be incompatible with an applicable treaty, say a treaty of alliance. In such a case, the foreign policy of the concerned state is bound legally, at least as long as the treaty of alliance has not been denounced or suspended.

As to the other prong of the above-mentioned definition, it continues to stand to reason that public international law is in the first place the legal order applicable among states. However, neither in the past, nor all the more in the twentieth or twenty-first century, international law has been or is exclusively the law of inter-state society. There are a series of other legal persons participating in the dealings of international law, whose importance varies according to subject matter and context, for example, international organizations, insurgents, movements of national liberation, governments in exile, the ICRC (International Committee of the Red Cross), the Holy See, the individual, or even multinational corporations. All these persons perform some acts which, to a varying degree, are relevant to public international law: Questions of immunity of international organizations, the law of armed conflict applicable to armed groups, ICRC powers to conclude agreements with states, human rights for individuals, etc.

The so-called international law of the classical period (roughly the nineteenth century up to 1919) was essentially based on the protection of the sovereignty of the existing "civilized" states.[4] Its regulations were essentially geared at bilateral relationships of states. The three main pillars of the law were (1) diplomacy, (2) transactions, and (3) war. Diplomacy (with a well-elaborated and age-honored diplomatic law setting) was a means of foreign policy; transactions was a way of cooperating when necessary, to solve disputes, to conclude treaties creating new

---

4   On this period and the evolutions since then, see G Abi-Saab, "Cours general de droit international public," *RCADI*, vol 207, 1987-VII, at 45ff; B Simma, "From Bilateralism to Community Interest in International Law," *RCADI*, vol. 250, 1994-VI, at 217ss; see also W Friedmann, *The Changing Structure of International Law*, (Columbia University Press 1964). In the context of international responsibility, see S Villalpando, *L'émergence de la communauté international dans la responsabilité des Etats* (Presses Universitaires de France 2005).

law; and war was a means to settle disputes and to sanction violations of the law, as well as a political tool for power policies. The interests taken account of by the law were essentially those of the states themselves, as a sovereign and supreme entity each one in itself (*uti singulus*). Thus, for example, self-preservation was writ large, since it allowed the states to protect themselves, or through alliances, from external insecurity; fundamental rights of states were insisted upon, since they allowed the state to possess the legal tools for the most varying policies; territorial integrity was legally developed, since it allowed each state to enjoy an appreciable degree of self-government, without intervention in its internal affairs; and so on. At the same time, only some "civilized" states were recognized the privileged status of a state under international law. Hence, a law based on formal (equal) sovereignty of each of these "civilized" states became compatible with the greatest degree of inequality among peoples. Many of them were refused statehood and kept within the relative legal limbo of "barbarian" or "savage" peoples.[5]

Modern international law, born in 1919 with the creation of the League of Nations and consolidated in 1945 with the UN, is based on a quite different set of parameters. It has witnessed a dramatic shift from the old law, at least in the following three aspects to be discussed below. However, this must be stressed immediately, the layers of the old law did not altogether disappear. Most of them continued to apply, such as diplomatic law and treaties, while others have been largely relinquished. The power to resort to war as a means to settle international disputes as an instrument of national policy is no longer available, and is the paramount shift seen.[6] The crucial point is that some new realities cast the old law into a fresh environment and most importantly that some new layers added themselves to the old ones. Which are, then, the three main aspects on which modern international law departs from the old law?

---

5   This distinction of circles of civilized, barbarian and savage collectivities was quite common at that time. See eg H Bonfils, *Manuel de droit international public* (3rd ed, A. Rousseau, 1901) at 82.

6   This is a primary reason why the International Court of Justice has repeatedly referred to the ban on the use of force between states as a "cornerstone" of international law; see *Armed Activities on the Territory of the Congo* (*Democratic Republic of the Congo v Uganda*), [2005] ICJ Rep 201, at para 148: "[t]he prohibition against the use of force is a cornerstone of the United Nations Charter"; *Military and Paramilitary Activities in and against Nicaragua* (*Nicaragua v United States*) (Nicaragua case) [1986] ICJ Rep 14, separate opinion of President Singh, at 153: "the very cornerstone of the human effort to promote peace in a world torn by strife"; *Case concerning Oil Platforms* (*Islamic Republic of Iran v United States of America*) (Oil Platforms case), Judgment of Nov 6, 2003, [2003] ICJ Rep 161, dissenting opinion of Judge Elaraby, at 291: "The principle of the prohibition of the use of force in international relations ... is, no doubt, the most important principle in contemporary international law to govern inter-State conduct; it is indeed the cornerstone of the Charter"; ibid, separate opinion of Judge Simma, at 328; C Joyner, *International Law for the 21st Century* (Rowman & Littlefield 2005), at 165.

A. Expansion of the Material Scope of International Law

The first is a quantitative point. Classical international law contained a rather small amount of rules; and most of these rules were bilateral (multilateral treaties were unknown up to 1856). Conversely, in the twentieth century, international law has witnessed a material expansion of huge dimensions. Today there is hardly any question on which there is no international legal regulation: commerce (e.g. World Trade Organization (WTO)), labor (e.g. International Labour Organization (ILO)), intellectual property, patents, telecommunication, cultural property, energy, cooperation to fight crime, etc. For most subjects, the legal regulation consists of a constantly shifting mix of national law and international law, different for each state according to its internal laws and regulations, as well as according to the international conventions it has or has not ratified. International law is thus virtually everywhere. It is complexly blending with municipal law. That makes it impossible to uphold the old rigid distinction between internal and international questions (domestic jurisdiction). The merger between the two bodies of law is too intimate.

B. Expansion of the Personal Scope of International Law

In the second place, concomitantly, international legal personality has considerably spread. In a somewhat exaggerated version, it is often affirmed that classical international law was purely inter-state law, whereas modern international law is a sort of transnational law encompassing the most different of legal persons. However, it is not true that classical international law was limited to states; conversely, it is definitely true that modern international law is a legal order applying to a much broader international society than the one based only on states. Especially in the economic and social area, a great number of actors have entered onto the scene. There are also actors in the political arena (international organizations) or else in the humanitarian area (the individual). The number of objective (sources) and subjective (legal entitlements) relationships has thus also dramatically increased. Moreover, not only were many more types of subjects entering the legal system as recognized actors and addressees of international law. The number of states has also exploded. Decolonization brought the number of states from 50 to 150 in a short number of years, and the new states were of a much more varied political, cultural, and ideological background.

C. The Emergence of Community Interests

Third, the structure of international law has shifted by the crystallization of collective or common interests of the "international community," or more soberly of a "community of states." Many legal regimes are now at least in part thought of in the perspective of the common weal of states. The catastrophes of the twentieth century (war, famine, diseases, common threats, etc.) have developed a sense for the need of "cooperation," be it through international organizations, or other settings, to achieve some common interests of states and their peoples.

Thus, for example, Article 11 of the League of Nations Covenant announced new law when it affirmed peremptorily that a "war," wherever it erupted and even if it concerned non-members of the League, was a matter of concern for the League, which should take immediate action. Peace thus became the first common value, which all states should observe and toward which all states should cooperate.

Other common interests would follow, such as human rights, environmental protection, the fight against epidemics, etc. International law developed legal norms in order to express and to cope with such "common concerns": the multilateral treaty, especially "integral" treaties or "common-interest treaties";[7] a strengthened universal customary international law with rules applicable to all states (a "common international law"); the creation of a series of international institutions, which would work to realize some common aims in the most differing subject matters; and the emergence of categories such as *jus cogens*, *erga omnes* obligations, international crimes, international universal jurisdiction, and the like.

In such modern international law, gaps in the law were seen as an anomaly and as a danger. They would allow the stronger to act as it sees fit (power politics), in absence of any regulation. Soon, great attention was thus spent on the so-called international law system,[8] that is, on its completeness, coherence, and its needs. Any action of single states was scrutinized as against the background of its impact on the interrelated system of the law of cooperation (and the mutual confidence on which it necessarily rests). The states composing the international community now criticized unilateral action in defiance of the system, or with potentially devastating effects on the system (e.g., by danger of emulation). The need for some order in the modern quick-paced world, with its delicate equilibria and economic underpinnings, has been of such an overwhelming importance that the watch has been fierce to avoid as far as possible "destabilization," or at least to manage it multilaterally when it occurs. The decolonization process is an example of such a practice. One may think also of the atomic bomb; or today the challenge of international terrorism.

It must be moreover noted that the Western States, and especially the Great Powers of the West, notably the US, have had a decisive impact on the configuration and the functioning of the modern law of nations. The United States largely initiated the shift from classical law to modern law through the establishment of

---

7   See eg the *Reservations to the Genocide Convention*, Advisory Opinion, ICJ, *Reports* [1951], at 23.

8   C Focarelli, *International Law as a Social Construct*, (OUP 2012), at 255ff.

the League of Nations. Later, European Powers were the backbone in the League. In the first 15 years, the UN was under Western domination.[9]

Today the picture has clearly changed. However, the impact of Western behavior on the existence and working of international law norms still remains of great importance.[10] The post 9/11 world, with its fight against terrorism, heeds an agenda reflecting mainly (albeit not exclusively) Western concerns and priorities. Western practice is among the most publicized and hence among the most considered when it comes to constructing norms of customary international law.

Drones have up to the present day remained largely a Western tool, albeit this will probably soon change. The pace is set for a race to automated warfare. (For further discussion of this point see Chapter 8 by Hayim.) In short, the implications of Western attitudes for the system of international law remain significantly more important than those of other states of the world. This situation of fact may not be entirely compatible with the "sovereign equality" of states under Article 2(1) of the UN Charter; on the other hand, the Latin refrain would seem applicable here *quod fieri non debet, factum valet* (What ought not to be done, when done, becomes a fact).

Nonetheless, this short history is directly relevant for our present quest. As the intercourse between states, organizations, and individuals (not to mention infections and digital devices) has become an everyday occurrence, the need for predictability and liability grows in tandem. There has been a concerted effort, particularly during the twentieth century, to construct a system of law that can begin to offer some of that needed regularity. Consequently, when investigating the efficacy of combating international terrorism, it is appropriate to take stock of how such efforts might impact this scrupulously constructed international system.

III. Specific Challenges by Drones to the International Legal Order

Considering the significant advancements succinctly reviewed above, it is important that the use of drones not disrupt the international legal system. When looking into their current use, we see that they can indeed challenge that system.[11]

---

9 See E Luard, *A History of the United Nations*, vol. I, *The Years of Western Domination, 1945–1955*, (Palgrave Macmillan 1982).

10 See M Byers, and G Nolte (eds), *United States Hegemony and the Foundations of International Law*, (Cambridge University Press 2003).

11 On the issue, see eg S Ahmad, "A Legal Assessment of the US Drone Strikes in Pakistan," (2013) 13 Int'l Crim Law Rev 917; DI Ahmed, "Rethinking Anti-Drone Legal Strategies: Questioning Pakistani and Yemeni 'consent'," (2013) Yale J Int'l Aff 1; R Barnidge, "A Qualified Defense of American Drone Attacks in Northwest Pakistan under International Humanitarian Law," (2012) 30 Boston U Int'l Law J 409; L de Beer, *Unmanned Aircraft Systems (Drones) and Law* (Wolf Legal Publishers 2011); G Boutherin, "Les systèmes de drones au coeur des conflits modernes: de la nécessité d'une clarification," (2011) 12 Annuaire français de relations internationales 756; I Henderson,

Under the "law of peace," the use of drones puts stress onto the conception of *jus ad bellum*/self-defense as well as onto the rules relating to territorial integrity/ consent to foreign military intervention (state sovereignty). Under international humanitarian law (IHL) or the law of armed conflict, a series of issues arise, in particular the question as to the threshold of applicability of that law, the definition of the combatant and of civilians directly participating in hostilities, and targeting (protection of civilians and proportionality issues). Finally, under both of these branches of law, there are problems of human rights law (HRL), in particular with regard to targeted killings/extra-judicial killings (right to life). The main three problems can thus be summarized as follows: A right of self-defense against "armed attacks" by individuals; a battlefield without borders; and a use of lethal force without due process of law.

A. Jus ad Bellum, Self-Defense and Consent

At the level of *jus ad bellum* (that is, who has the right to use force, when, and under what conditions?), the first issue relates to *self-defense*. Under Article 51 of the UN Charter, self-defense is conditioned upon the existence of an armed attack. This rule in principle excludes preventive action. Moreover, the drafting

---

"Civilian Intelligence Agencies and the Use of Armed Drones," (2010) 13 YB of Int'l Humanitarian Law 133; MW Lewis, "Drones and the Boundaries of the Battlefield," (2012) 47 Texas Int'l Law J 293; N Lubell, "The War (?) against Al-Qaeda," in E Wilmshurst (ed), *International Law and the Classification of Conflicts* (OUP 2012) at 421ff; N Lubell and N Derejko, "A Global Battlefield? Drones and the Geographical Scope of Armed Conflict," (2013) 11 J of Int'l Crim Justice 66; M McNab and M Matthews, "Clarifying the Law Relating to Unmanned Drones and the Use of Force: the Relationships between Human Rights, Self-Defense, Armed Conflict and International Humanitarian Law," (2011) 39 Denver J of Int'l Law and Pol 661; ME O'Connell, "Remarks: The Resort to Drones under International Law," (2011) 39 Denver J of Int'l Law and Pol 585; ME O'Connell, "Unlawful Killing with Combat Drones: A Case-Study of Pakistan," in S Bronitt, M Gani and S Hufnagel (eds), *Shooting to Kill: Socio-Legal Perspectives on the Use of Lethal Force* (Hart Publishing 2012); AC Orr, "Unmanned, Unprecedented and Unresolved: The Status of American Drone Strikes in Pakistan under International Law," (2011) 44 Cornell Int'l Law J 729; JJ Paust, "Self-Defense Targetings of Non-State Actors and Permissibility of US Use of Drones in Pakistan," (2010) 19 J of Transnat'l Law and Pol 237; JJ Paust, "Propriety of Self-Defense Targeting of Members of Al Qaeda and Applicable Principles of Distinction and Proportionality," (2012) 18 ILSA J of Int'l and Comp Law 565; M Sterio, "The United States' Use of Drones in the War on Terror: the Legality of Targeted Killings under International Law," (2012) 45 Case Western Reserve J of Int'l Law 197; MV Vlasic, "Assassination and Targeted Killing," (2012) 43 Georgetown J of Int'l Law 259; D Whetham, "Drones and Targeted Killing: Angels or Assassins?," in BJ Strawser (ed), *Killing by Remote Control: The Ethics of an Unmanned Military* (OUP 2013) at 69ff; J Yoo, "Assassination and Targeted Killings after 9/11," (2012) 56 NY Law School Law Rev 57; DH Dunn, "Drones: Disembodied Aerial Warfare and the Unarticulated Threat," (2013) 89 Int'l Aff 1237.

of Article 51 was at least implicitly predicated upon an armed reaction of a state against an armed aggression of another state. The fight against terrorism has questioned these assumptions in many regards.

First, there is the timing of self-defense. The point is frequently made that to wait for a terrorist attack in order to react thereto is to have waited in excess. The legal consequence drawn is that self-defense must be conceded for anticipatory (action to thwart future threats) or at least for preemption (action to thwart imminent threats) purposes.[12] Both notions allow for a certain degree of subjective appreciation; even imminence is a question allowing various interpretations. The existence of a terrorist cell as such can be seen as an imminent threat, since the activities of such a group are geared toward the commission of acts of violence. It is interesting to note that while the great majority of states have condemned preventive self-defense, action against imminent threats seems now quite widely accepted[13]—while it was a very controversial issue before 9/11![14] The attacks of drones are situated in this gray area between a reaction to past "attacks" and the forestalling of future violent action. The timing of action, and therefore the question of what is to be seen as a reaction to what, therefore becomes blurred. The net result is that it is difficult to draw a clear line between aggression and self-defense.[15] Allowing armed action outside the trigger of an armed attack having already occurred inevitably leads at least to some extent to merge the two notions into one another. Indeed, the general rule (but it is not without exception) that the one who uses force in the first place is the aggressor is here frontally challenged. The issue is all the more serious in view of the existence of International Criminal Court (ICC) jurisdiction over aggression, which will become operational in 2017, albeit in a restricted fashion.[16] (For a historical discussion on this question see Chapter 6 by Steven J. Barela.)

---

12   Although there are differing views on which precise terms carry which definition, this terminology is used to reflect the Department of Defense dictionary definitions discussed in Chapter 6 by Steven J. Barela.

13   See eg UN Doc A/59/565 (2004), § 188; see also MN Schmitt (ed), *Tallinn Manual on the International Law Applicable to Cyber Warfare* (Cambridge University Press 2013) at 63. On the problems posed by this notion, see eg O Corten, *Le droit contre la guerre*, (A Pedone 2008) at 617. See also generally N Lubell, *Extraterritorial Use of Force Against Non-State Actors* (OUP 2010) at 55ff.

14   R Kolb, *Ius contra bellum, Le droit international relatif au maintien de la paix*, (2nd ed, Helbing & Lichtenhahn 2009) at 278.

15   See eg J Kunz, "Individual and Collective Self-Defense in Article 51 of the Charter of the United Nations," (1947) 41 AmJIL 877.

16   Article 8bis ICC Statute. See G Della Morte, "La Conferenza di revisione dello Statuto della Corte penale internazionale ed il crimine di aggressione," (2010) 93 Rivista di diritto internazionale 697.

Second, there is the issue of the subjective scope of self-defense. Traditionally, Article 51 self-defense was limited to relations between states.[17] This is still the reading and the position of the International Court of Justice (ICJ).[18] However, a significant part of state practice and of doctrinal opinion tends to erode this rule.[19] Arguments of necessity or of enlarged self-defense are made to encompass in Article 51 the cases of "attacks" by terrorist groups, at least when these attacks display an intensity making them appear a war-like act. This intensity can also be reached by the aggregate of a series of smaller actions, such as a series of transboundary attacks by rebel groups. This is a significant broadening of the scope of self-defense. One of its consequences is to privilege *armed* responses to terrorist action (with all the danger of escalation and collateral damages) over purportedly less effective police action and criminal law cooperation. In turn, one of the problems of this approach is that every state views differently who is a terrorist and who is a freedom fighter, or who is a legitimate and who is an illegitimate group. A quite broad consensus could be reached on the nature of this or that group such as Al-Qaeda, but beyond this there is much uncertainty, with the danger of a serial spread of armed action and armed reaction.

Third, if non-state actors can be included in the target list of self-defense, there arises the delicate problem of the "double strike." It would not be all too problematic to concede self-defense against such terrorist groups if they could be hit in isolation. There are a series of other self-defenses under international law, such as individual self-defense of the soldier or unit self-defense in times of armed conflict; each follows its own rules. The common point of these allowed self-defense reactions is that the use of force hits only the culprit. No third party is illegitimately suffering from that use of force. This may be so, for example because there is an armed conflict on the territory at stake, and hence armed violence is generally allowed there. Therefore, if the terrorist group were gathered on a ship on the high seas hoisting a flag of Al-Qaeda, it would not be unlawful to sink that ship in the exercise of the right of self-defense.

However, since such terrorist groups operate from the territories of a great number of states, the issue arises whether armed action on the territory of all these states is allowed, even when it is directed to hit the terrorist infrastructure or personnel. The armed strike inevitably hits also the state, its territory, and its

---

17   This does not exclude today a broader teleological interpretation: see eg C Tomuschat, "Der 11. September und seine rechtlichen Konsequenzen," (2001) 28 Europäische Grundrechte-Zeitschrift 540.

18   *Armed Activities* case, ICJ, Reports, 2005, §141ff; *Wall* Opinion, ICJ, *Reports*, 2004-I, § 139. On this latter finding, see the criticism by CJ Tams, "Light Treatment of a Complex Problem: The Law of self-defense in the Wall case," (2005) 16 EurJIL 963.

19   On the issue, see eg N Lubell, *Extraterritorial ...* (n 13) at 25ff; E Cannizzaro, "Entités non-étatiques et régime international de l'emploi de la force," (2007) 111 RGDIP 333; A Tancredi, "Il problema della legittima difesa nei confronti di milizie non statali alla luce dell'ultima crisi tra Israele e Libano," (2007) 90 Rivista di diritto internazionale 969.

objects. There is thus the necessity to justify such armed action against the so-called territorial states,[20] since self-defense is in principle a right limited in its exercise to the subject having conducted the armed attack; it cannot by itself extend to third parties, since it is not a general right of necessity.

In this context, new standards of attribution are devised, holding for instance that a state being "unwilling or unable" to halt the use of its territory by the terrorists has to accept that the attacked third states intervene militarily on its territory to protect their rights.[21] However, this statement still raises many questions. First, there is the danger that self-defense might quickly escape from its traditional setting of just two actors and become a sort of general right *erga omnes*, that is, that it could be directed against an undefined number of states in an undefined number of situations, according to the discretion of the state acting under the color of self-defense.[22] Second, self-defense thus risks loosing its status as a narrowly construed exception to the non-use of force rule. For an exception to gain the status of a widely applicable power *erga omnes* is not really a progress for the legal system; it is rather a danger multiplied.

In short, the use of force could spread beyond any legal control. This would jeopardize the main tenet of modern international law, namely the non-use of force rule enshrined in Article 2(4) of the UN Charter, and the rule for the peaceful settlement of disputes, in Article 2(3). There would be a sort of open discretion to engage in one avenue (use of force) or the other (cooperation against terrorism) as each saw fit. Whether that is to be seen as progress—because it gives more latitude for action to the victim state—remains to be seen. If many states claim such rights (e.g. Russia, China, and others), the situation may quickly become destabilized. Finally, it may be noted that certain conditions for the exercise of self-defense can only with greatest difficulties be adapted to such widely applicable *erga omnes* defenses, namely the criteria of necessity and proportionality.[23]

In the context of the law of peace, difficult questions arise in the protection of *territorial integrity*. Assuming that the argument of self-defense is not made or appears inapplicable, there remains the possibility that armed action against terrorists is covered by an authorization of the UN Security Council (as happened in Mali, 2013)[24] or by the consent of the local sovereign. The problem in this latter

---

20   See eg J Delbrück, "The Fight Against Global Terrorism: Self-Defense or Collective Security as International Police Action?," (2001) 44 German YB of Int'l Law 18.

21   See eg R van Steenberghe, *La légitime défense en droit international public*, (Larcier 2012) at 268ff; a criticism against this enlargement can be found in O Corten (n 13) at 669ff. On the criterion, see also UN Doc A/68/389, *Promotion and Protection of Human Rights and Fundamental Freedoms while Countering Terrorism*, at 15–16.

22   See eg Cassese (n 1) at 997.

23   On these notions, see eg J Gardam, *Necessity, Proportionality and the Use of Force by States* (Cambridge University Press 2004).

24   UNSC Res 1973 (n 2). See on this issue G Bartolini, "L'operazione 'Unified Protector' e la condotta delle ostilità in Libia," (2012) 95 Rivista di diritto internazionale 1012.

avenue is a flagrant lack of transparency. The local governments have sometimes consented to US action on their territory, as seems to have been the case of Pakistan, at least in a first phase.[25] But it is often impossible for these governments to openly admit such consent, for obvious internal policy reasons (their population being violently adverse to such US strikes). The problem is thus that it is impossible to know from the outside whether drone action is lawful or not. The law is silenced or put in a limbo, simply because the facts are not clear. This again marginalizes international law. It tends to give the impression that "anything goes, anyway."

It may briefly be recalled that the consent of the local government would eliminate the unlawfulness of the armed action on the territory at stake. Nevertheless, it would not cure violations in the context of humanitarian or human rights law since in this context they are absolute obligations, which cannot be renounced by a state or even by the concerned protected persons themselves.[26] It must also be noted that if consent is finally withheld, the United States cannot any more rely on this title to justify the use of force. The only legal argument remaining available, in the absence of UN Security Council authorization, is then self-defense.[27]

B. International Humanitarian Law

With regard to IHL, the problems arise on different planes. And when such difficulties arise, they tend to weaken or hamper the functioning of some fundamental rules of this body of law.

1. The existence of an "armed conflict"
In order to be able to invoke the bulk of the rules of IHL—allowing a more generous recourse to killing than the rules applicable in peacetime—there must be an armed conflict occurring. Such a conflict supposes belligerent parties, that is, states or armed groups, both identifiable and placed under a responsible command. Shadowy terrorist networks are not usually a belligerent party; they have no rights and should not have the rights of a belligerent; they are and remain a criminal organization.[28] The fight of a state against a non-state group pertains at best to

---

25 At least up to 2008: UN Doc (n 21) at 15. Pakistan did not protest against these attacks in this phase.

26 See eg Articles 6/6/6/7 and 7/7/7/8 Geneva Conventions I-IV of 1949.

27 UN Doc (n 25) at 15.

28 On the issue of the war against terror, see eg L Condorelli and Y Naqvi, "The War against Terrorism and Jus in Bello: Are the Geneva Conventions Out of Date?," in A Bianchi (ed), *Enforcing International Law Norms Against Terrorism*, (OUP 2004) at 25ff; C Emanuelli, "Faut-il parler d'une 'guerre' contre le terrorisme?," (2008) 46 Canadian YB of Int'l Law 415; DE Graham, "The Law of Armed Conflict and the War on Terrorism," in R B Jacques (ed) *Issues of International Law and Military Operations* of *International Law Studies*, vol. 80, 2006, at 331ff; C Greenwood, "International Law and the 'War against Terrorism'," (2002) 78 Int'l Aff 301; K Kress, "Some Reflections on the International Legal

the realm of non-international armed conflicts (NIAC); the opposing parties are not states, but a state and a non-state armed group. However, a NIAC exists only in the context of fighting taking place against groups having a certain degree of organization and possessing certain intensity.[29]

If it is accepted that varying terrorist groups can or even do fulfill such criteria, the consequence is that the battlefield (and the application of IHL) becomes spatially unlimited. There would an "armed conflict" between the US and an unlimited number of groups, involving an unlimited number of states. We would arguably already find ourselves in the third world war, the US being entitled to use belligerent lethal force on the territory of an undefined number of states. Moreover, the terrorist groups would have been upgraded into belligerents. True, their members are not vested with the combatant privilege, that is, the right to use force against military targets. They had no right to use force, and they are still subjected to various criminal codes, that of the states on whose territory they operate, and that of the victim states. But there remains the fact that they would be entitled to claim some undefined degree of "equality" between the belligerent parties and to negotiate on their claims.

Taking the case of the drones used in Pakistan, the United States considers itself as a party to a NIAC against the Taliban and their associates.[30] It is true that the US has flown a great number of attacks on Taliban positions, but the number of attacks of against the US and its allies by terrorist groups has remained extremely low. Would such a low-intensity conflict suffice to qualify as a NIAC under Article 3 common of the Geneva Conventions, or even to Additional Protocol II of 1977? Is Al-Qaeda sufficiently organized, or is it just a loose network, a sum of loose affiliations? The International Criminal Tribunal for the former Yugoslavia (ICTY) has interpreted the criteria for the existence of an NIAC in an extensive

---

Framework Governing Transnational Armed Conflicts," (2010) 15 J of Conflict and Security Law 245; MW Lewis, *The War on Terror and the Laws of War: A Military Perspective* (OUP 2009); ME O'Connell, *International Law and the 'Global War on Terror,'* (Editions-Pedone 2007); M Sassòli, "La 'guerre contre le terrorisme,' le droit international humanitaire et le statut de prisonniers de guerre," (2001) 39 Canadian YB Int'l Law 211; M Sassòli, "Use and Abuse of the Laws of War in the 'War Against Terrorism'," (2004) 22 Law and Inequality: J of Theory and Practice 195. See also N Lubell (n 13) at 112ff.

29   See eg the jurisprudence of the ICTY: *Limaj* (Trial Chamber, 2005), § 83ff; *Mrksic* (Trial Chamber, 2007), § 405ff; *Haradinaj* (Trial Chamber, 2008), § 37ff; *Boskoski* (Trial Chamber, 2008), § 175ff. This is a traditional IHL criterion, developed initially by a commission of experts under the auspices of the ICRC. See R Kolb and R Hyde, *An Introduction to the International Law of Armed Conflicts*, (Hart Publishing 2008) at 78. See also D Schindler, "The Different Types of Armed Conflicts According to the Geneva Conventions and Protocols," (1979) 163 RCADI 153.

30   See section III(A).

way in the context of events in Macedonia (the former Yugoslav Republic of Macedonia or FYROM) in 2001,[31] but there remain important doubts.[32]

The situation could also be considered from another perspective. There indeed remains the possibility that the US acts (on invitation?) in the context of another NIAC between the Pakistani forces and the Taliban groups on its territory. That conflict would certainly fulfill the conditions of organization and intensity; it would even satisfy the condition of a territorial control of the rebels, as required by Article 1(1) of Additional Protocol II (AP II) for the applicability of that Protocol.

Overall, there is manifestly a lack of certainty. Traditionally, the law of armed conflicts applies in a set of well-defined and well-tailored situations of belligerency, between well-determined belligerent parties. The "war on terror" implies a significant loss of such focus. There is thus a pressure for the development of new rules of IHL, tailored to the situations of such fuzzy military engagements, a new law straddling between law enforcement and punctual conduct of hostilities. This is the gist of the US argument that the Geneva Conventions are out of date. Indeed, not all rights and duties of belligerents under Article 3 common and AP II, or other customary rules, would fit such situations of low intensity and of low organization of the "terrorist" forces. Nor would IHL really fit the "criminal law" background of most of these situations. This is precisely the reason why new rules would have to be devised. However, this would split up IHL into a potentially high number of context-related rules: here for terrorism-fighting (drones); there for drug trafficking-fighting (Mexico); still elsewhere, territories not actually occupied but only controlled from the air or the outside, etc.

IHL is already quite complex in its taxonomy.[33] To add up such further categories will overstretch it and lead to its increasing ineffectiveness. It rather seems that IHL is perfectly equipped to deal with situations of true belligerency— but that the war against terror is largely not an issue of IHL at all. Thus we can see great tension created by trying to force fit legal rules that were in fact drafted for different circumstances of belligerency; when this happens we can indeed see cracks appearing in the edifice of the international system.

2. Distinction

Second, there is the issue of distinction in the targeting. The problems posed here are not inherent problems of unmanned vehicles, but rather concern the specific use of drones by the United States. It is generally recognized that as an instrument among others drones are neutral. They can be used to increase the accuracy of military strikes, eliminate some physical stress of the pilot on the spot, wait for the

---

31   ICTY, *Boskoski* case (Trial Chamber, 2008), § 173ff.

32   Lubell, "The War(?) ... " (n 11) at 436. See also Lubell (n 13) at 112ff.

33   M Milanovic and V Hadzi-Vidanovic, "A Taxonomy of Armed Conflict," in ND White and C Henderson (eds) *Research Handbook on International Conflict and Security Law: Jus ad Bellum, Jus in Bello and Jus post Bellum*, (Cheltenham 2013) at 256ff.

best moment of attacking, and take the necessary time to verify all the parameters due to the long flying capacity of the engines.[34] (For discussion of the operational benefits of drones see Chapter 10 by Marek Madej.)

Conversely, drones can also be used in such a way as to pose problems with regard to targeting, especially if the targeting process does not live up with the targeting conditions posed under IHL (especially under Articles 50–58 of AP I, or Article 13 AP II and related customary international law). The danger is to loosen the central pillars of IHL and to prepare the terrain for still less stringent requirements in targeting. Thus, it is known that fully autonomous unmanned vehicles (without any guiding by a human being) are now being developed. Such engines, it is said, will in certain situations not have the necessary time to comply with a full proportionality assessment, which is an assessment required by Article 51(5)(b) of AP I, in the context of international armed conflicts; but the proportionality rule is applicable also in NIAC as a customary rule for all types of armed conflict.[35] The result would be that technical reasons could lead to a dropping or to a significant restriction of a central requirement of IHL. Drones open the door to such developments.

In Pakistan, the greatest problem is that of the "signature strikes." Drone attacks by the US have been classified in two categories: personality strikes and signature strikes. The former target specified persons "known" to be affiliated to a terror or insurgent group, such as Al-Qaeda or the Taliban. Conversely, signature strikes target persons on the basis of an analysis of their displacements on the territory and further on the basis of a bundle of suspect behaviors, such as the bearing of arms. It has been reported that roughly 90 percent of drone attacks are signature strikes.[36] It stands to reason that the criteria used for signature strikes are inherently imprecise. To be traveling in an area controlled by the Taliban bearing weapons (a signature used by the US)[37] can hardly be squared with the targeting requirements of IHL. It ignores contextual and cultural realities, for example, the fact that in the tribal Afghan societies bearing arms is not necessarily a sign of terrorist or insurgent action. The same can be said for some forms of traditional use of arms in wedding ceremonies (and there has been more than one US attack in the context of such ceremonies).[38]

---

34   See Lewis (n 11) at 297. See also UN Doc (n 21) at 23; see also Chapter 10 of this volume by Marek Madej.

35   See JM Henckaerts and L Doswald-Beck, ICRC, *Customary International Humanitarian Law*, vol I, (Cambridge University Press 2005) at 46ff.

36   KJ Heller, "One Hell of a Killing Machine," (2013) 11 J of International Crim Justice 89, at 90.

37   Ibid at 98.

38   See eg Robert Worth, "Drone Strike in Yemen Hits Wedding Convoy, Killing 11" *New York Times* (Dec 12, 2013) available at <http://www.nytimes.com/2013/12/13/world/middleeast/drone-strike-in-yemen-hits-wedding-convoy-killing-11.html?_r=0> accessed Oct 2014.

In a NIAC, the proper targeting under IHL must rest on the distinction between "fighters" and civilians, or, more precisely, between civilians participating directly in hostilities and civilians not participating in hostilities,[39] on the legal basis of Article 13, AP II. The issue of targeting turns on the notion of direct participation. Such a "direct participation" exists in two situations. First, in case of a "continuous combat function," an individual is "affiliated" to an armed group for as long as they remain a member of this group. He or she can then be targeted as a combatant without any temporal limitation (e.g. also when he or she is sleeping).

Alternatively, there can be a sporadic participation in the conflict on the basis of the "revolving door" principle. According to this principle, the person participating can be targeted for the time-span that he or she is effectively participating in hostilities by committing acts "likely to adversely affect the military operations or military capacity of a party to an armed conflict."[40] Apart from this threshold of harm, there must also be a direct causality link between the act and the harm likely to result (participation must be direct, not indirect). Additionally, there must be a belligerent nexus, that is, the act must be specifically designed to directly cause the required threshold of harm in support of a party to the conflict (we must be confronted with belligerent action, not to a criminal action on occasion of the armed conflict but not linked with it).

These criteria of threshold of harm, direct causation, and belligerent nexus are applicable to both categories of participation, continuous and sporadic, but will be regularly fulfilled by members of an armed group, whereas they have to be tested in a more precise way for sporadic participation in an armed conflict. The mentioned revolving door principle means that a person is for a short time-span directly participating, and can be targeted; then this person revolves into a civilian protected, when he or she has ceased to commit harmful acts for the adverse party, and cannot be militarily targeted any more.

The aim of this principle is to protect the civilian population from excessive attacks. The only way to ensure that no egregious abuses are committed is indeed to limit the military targeting of such sporadic participants to the phase when they are militarily active. It would be odd to allow a military targeting (e.g. by bombardment) while they are at home or in their civilian working position. This would heavily jeopardize innocent civilians and provoke significant collateral damages. And there would also be the danger to target persons whose participation in the conflict is not proved. In the phase of civil life of such sporadic fighters,

---

39 On the notion of direct participation in hostilities, see Nils Melzer, ICRC, "Interpretive Guidance on the Notion of 'Direct Participation in Hostilities'," *International Review of the Red Cross*, vol 90, 2008, at 991ff. On this notion in our context, see N Lubell (n 13) at 135ff. On this notion in general, see the further literature indicated in M Sassòli, A Bouvier and A Quintin, *How Does Law Protect in War* (3rd ed, vol. I, ICRC 2011) at 262–265.

40 Ibid, Melzer at 16.

what remains is the possibility to arrest and try for their unlawful participation in the armed conflict.

These are, in a nutshell, the principles emerging from the ICRC Interpretive Guidance on the issue of direct participation, a notion featuring in Articles 51(3), AP I (for international armed conflicts), in Article 13(3), AP II (for NIAC's covered by AP II), and in customary international law for all armed conflicts.

Of course, there are some problems with the categories created by the ICRC study.[41] For one, there remain persons who can be targeted throughout (continuous combat fighters and members of the armed forces), while others cannot always be targeted (sporadic fighters), thus creating some inequality among belligerents. But that is unavoidable and is based on sound reasoning; after all, only those who fight continuously should be targetable continuously. Moreover, the principle of equality of belligerents applies only in a very limited way to NIAC's. Second, there remains the problem of proof of "continuous combat function" and that of possible abuses in this context. But this problem is also unavoidable. International law, including IHL, must at the end of the day rely on a degree of good faith effort on the part of the involved states.

Now, as can be seen, drone strikes, especially those based on "signature," cannot be rendered compatible with these principles. Persons are hit on the sole basis of degrees of suspicion, and not on the basis of reliable evidence as to the "direct participation in hostilities." Therefore, purportedly applicable IHL categories are rendered inapplicable and even irrelevant.

Thus, once again, we find that the use of remotely piloted aircraft puts great strain on this already tenuous principle of distinction by drastically increasing the occasions when lethal force can be exercised. It must, however, be noted once more that the problem lies not in the action by drones themselves; it is rather the particular use of drones for signature strikes which has shattering consequences for the application of IHL. The odd result of such practices is the targeting, that is, the death and injury, of a series of persons who are not "fighters" but "innocent civilians"—precisely what IHL tries to avoid as much as feasible. It is not improbable that such action will trigger a great thirst for revenge and will increase enrolment in the terrorist groups. The strategy would then be self-defeating. There is indeed a body of persuasive argument contending that for every terrorist or civilian killed, the terrorist groups are able to enroll a still greater number of new volunteers.[42]

---

41   For the response to other critiques, see N Melzer, "Keeping the Balance between Military Necessity and Humanity: A Response to Four Critiques of the ICRC's Interpretive Guidance on the Notion of Direct Participation in Hostilities," (2010) 42 J of Int'l Law and Politics 831.

42   See eg International Human Rights and Conflict Resolution Clinic (Stanford Law School) and Global Justice Clinic (NYU School of Law), "Living Under Drones: Death Injury and Trauma to Civilians from US Drone Practices in Pakistan" (September 2012) at 131–7, available at <http://livingunderdrones.org/> accessed Oct 2014; Leila

3. Proportionality

There remain the principles linked to the preparation and carrying out of attacks, especially the principle of proportionality.[43] The rule is that an attack shall not be launched if it may be expected to cause incidental loss of civilian life, injury to civilians, damage to civilian objects, or a combination thereof, which would be excessive in relation to the concrete and direct military advantage anticipated. This rule is contained in Article 51(5)(b) of AP I for international armed conflicts, and is valid also for NIAC on the basis of customary international law, perhaps in the latter category slightly more loosely (as inferred from Article 13 of AP II and state practice). Contrary to the wording used, the rule is not really one of proportionality, but rather the prohibition of disproportion. The armed forces may attack only military objectives, but they must be free in principle to always attack. Otherwise, a belligerent could not attack anything anymore. A restriction on the ability to target military objectives is consequently provided only for the case of an "excess" in the collateral civilian losses, as compared to the military advantage gained.

In the context of drones, it is difficult to reach clear-cut opinions. The point at issue is again at best a particular use of drones, not the fact that drones are used. Signature strikes do not raise in themselves problems of proportionality, but rather problems of directly attacking persons who are not "fighters." The issue is therefore one of the proper selection of the military objective in the first place, not the collateral damage a lawful attack on a military objective may produce. However, a signature strike (or a personality strike) targeting a civilian not participating directly in hostilities or a "fighter" can also produce collateral effects on other "innocent" civilians. Then the issue of proportionality arises.

Conversely, as such, drones may be used for precise if not surgical targeting; in this sense, they could lead to an improvement in the implementation of the proportionality requirement. Precise information on the issue of proportionality is rather scarce. However, it seems that in Pakistan a number of "innocent" civilians have been killed or injured in the attack of drones. It is impossible from the outside

---

Hudson, Colin Owens & Matt Flannes "Drone Warfare: Blowback from the New American Way of War" (2011) XVIII(3) Middle East Policy Council 122; and Jo Becker and Scott Shane, "Secret 'Kill List' Proves a Test of Obama's Principles and Will,' *New York Times*, (May 29, 2012), where it was argued "Drones have replaced Guantánamo as the recruiting tool of choice for militants; in his 2010 guilty plea, Faisal Shahzad, who had tried to set off a car bomb in Times Square, justified targeting civilians by telling the judge, 'When the drones hit, they don't see children.'"; cf. Christine Fair "Drones Over Pakistan—Menace or Best Viable Option?" *The World Post* (Aug 2, 2010) <http://www.huffingtonpost.com/c-christine-fair/drones-over-pakistan----m_b_666721.html> accessed Oct 2014. See also Chapter 11 by S Barela at section III(C).

43   See generally in our context N Lubell (n 13) at 155ff. See also Orr (n 11) at 747. Generally on proportionality in the context of armed conflicts, see I Henderson, *The Contemporary Law of Targeting* (Brill 2009) at 197ff; and WH Boothby, *The Law of Targeting* (OUP 2012) at 71, 94–96.

to assess whether the principle has thereby been violated, since that would require a mastery of the particular facts: What was the military advantage pursued (which depends on the importance of the person targeted)? What have been the measures for a proper preparation of the attack in view to minimize collateral losses? How many persons could be reasonably expected to be collaterally hit, and how gravely? On the general plain, apart from a particular operation, what is the overall quota of collateral damages? Thus, in this regard, it certainly does not seem that drones have shattering consequences on the system of IHL—and in fact could even lead to improved proportionality.

C. Human Rights Law

Finally, there remain issues of HRL, especially of targeted killing or extra-judicial killings, and their relation to the right to life.[44] The tendency of the Unites States is to argue for a *lex specialis* approach because of its greater license to kill. The meaning of this approach is that in case of a conflict between a norm of the law of armed conflict and human rights law in the context of targeting persons in armed conflict, the norm of IHL should prevail over HRL. Further, the argument of the United States is that there is no extraterritorial application of HRL—but this narrow approach is rejected by a unanimous body of international jurisprudence and is indeed completely unsound.[45] To the extent that HRL is applicable—and in legal terms, it is—the use of drones may lead to an unprecedented weakening of one of the cornerstone rights under these instruments, namely the right to life. This right is not absolutely protected; exceptions are recognized, such as killing in wartime. But the right goes to the prohibition of arbitrary killing, and extra-judicial killings will often be qualified as such because these are killings without due process of law. They are involving the heaviest sanction possible (namely the death penalty), and this is exercised without any judicial review. To be sure, not every extra-judicial killing

---

44   On the issue of targeted killings, see mainly N Melzer, *Targeted Killings in International Law* (OUP 2008). In our context see also N Lubell (n 13) 169ff. For a general analysis under the right to life, see G Gaggioli, *L'influence mutuelle entre les droits de l'homme et le droit international humanitaire à la lumière du droit à la vie*, (A Pedone 2013). See also Y Shany, "Human Rights and Humanitarian Law as Competing Legal Paradigms for Fighting Terror" in O Ben-Naftaly (ed), *International Humanitarian Law and International Human Rights Law*, (OUP 2011) at 13ff.

45   See N Lubell (n 13) at 193ff. See also RK Goldman, "Extraterritorial Application of the Human Rights to Life and Personal Liberty, including Habeas Corpus, during Situations of Armed Conflict" in G Gaggioli and R Kolb (eds), *Research Handbook on Human Rights and Humanitarian Law* (Cheltenham 2013) at 104ff. For internal debate within the US government see also Harold H Koh, Legal Advisor to the US State Department, "Memorandum Opinion on the Geographical Scope of the International Covenant on Civil and Political Rights" (Oct 19, 2010).

would be unlawful, but most will be incompatible with the right to life.[46] In what situations could an extra-judicial killing be compatible with HRL?

In order to remain within the bounds of HRL, drone killings must be "in accordance with the law," not "arbitrary," and respect the principles of "necessity" and "proportionality." It is not always clear whether the drone attacks comply with domestic US law. The question remains whether there are some secret legal instructions covering such action, but these would hardly qualify as legal basis under HRL, if only because of the lack of public knowledge explaining the parameters for a fatal sanction. A lethal attack could avoid being arbitrary if it is shown that it took place against a fighter under IHL rules (*lex specialis*).[47] However, as many of these attacks do not comply with IHL requirements (e.g. the signature strikes discussed above), there is little chance that an international human rights monitoring body would accept such a plea, divesting it, moreover, of its own jurisdiction.

On top of this initial difficulty of legality, the greatest problems arise under the requirements of necessity and proportionality. Under HRL, these are quite strict criteria.[48] It must be shown that there was no lesser-force option available; that the arrest was prepared and attempted; that proper training and equipping of the acting forces in view of lesser options than death were organized. (The HRL parameters are fully discussed in Chapter 4 by Gaggioli.)

These are quite demanding requirements in extraterritorial counterinsurgency operations. At the end of the day, the requirements of necessity and proportionality risk being fulfilled only when there was an imminent threat by the targeted person. The easiest option to get out of this quagmire is to play the card of *lex specialis*, that is, of a priority of IHL in the context of a particular operation. This could be the case for drone attacks fully complying with the rules of IHL. Indeed, such a priority could be an incentive to comply with IHL rules. However, the incentive is quite remote in view of the fact that there is no international human rights court with compulsory jurisdiction and even the less with enforcement powers. Thus, we find that the current uses of armed drones again place great pressure on the system of international law—and in this case it is stress aimed at arguably the most cherished norm (the right to life).

IV. Conclusion: Consequences on the International Legal Order

Two remarks can be proffered at the end of this short study exploring the "effects" of drone attacks on the international legal system. First, it is too early to thoroughly

---

46   See the various reports by the UN Special Rapporteurs, among whom Philip Alston, Report of the Special Rapporteur on Extrajudicial, Summary or Arbitrary Executions [2006] UN Doc E/CN.4/2006/53.

47   On this issue, see N Lubell (n 13) at 236ff. See also M Milanovic, "Norm Conflicts, International Humanitarian Law, and Human Rights Law" in O Ben-Naftaly (n 44) at 95ff.

48   See the well-known *McCann v. UK* case of the ECtHR (1995), Series A, no. 324.

assess empirical data relating to the effects of drones on the legal system. The drone practice is still recent and only very few states use it extraterritorially for killing operations. Time has to pass before an assessment will be possible on an empirical rather than a prospective basis. However, this is not to say that a series of probable effects and developments of the use of drones cannot be analyzed today. The potential effects on the international legal order have been outlined here, and the tension that has been created is palpable. Thus they are well worth our attention for the assessments of systemic efficacy.

Second, the series of effects—negative or positive—span the different branches of public international law. Some of these effects are cross-related, as the complex relationships between IHL and HRL (including *lex specialis* issues) demonstrate. The most important of these effects have been addressed in the preceding pages. However, apart from these specific effects in this or that branch of international law, there are also some general effects on the system as a whole. These should now be shortly considered. It may also be emphasized once again that drones can also have some useful effects in certain contexts, and the present writer does not wish to condemn their use outright.

What are then these adverse general effects on the legal system as a whole? At least two such influences and impacts can be singled out. The first is a risk of excessive extension of the "law of war" or armed conflict, as opposed to the realm of the law of peace. At the time of Hugo Grotius and his renowned *De jure belli ac pacis* (1625),[49] the greatest portion of international law was devoted to the law of war; only a small fraction was concerned with the law of peace. Civilization has worked hard since then to invert this proportion and to give the limb of peaceful cooperation among states its place of pride; concomitantly confining the law of war to an exceptional state of diseased inter-state relations. The danger is of importing into the law of peace, elements of hostility and of the law of war—like a series of viruses.

This unhealthy tendency is prolonged by a general trend to make exceptions to the rule, for example, in the context of self-defense writ large, extended to "imminent" or even future threats. We would then live in a permanent state of exception. This will backfire on the workability of the law of peace. Too many "states of exception" and "states of necessity" bending the law to their particular needs may lead to an infection in its body and to a non-functioning of a number of its rules. In short, we now seem to live in a permanent state of exception and of necessity. Hard cases make bad law; and there seem now to be more hard cases than easy ones. The utter difficulty to conclude important multilateral conventions since 1998 (year of conclusion of the ICC Rome Statute) is just an illustration of that situation.

This brings us to the second, related, point. International law can work and prosper only on the basis of some minimum confidence among the actors it binds. Generalized distrust tends to destroy it, since it attacks the root of the

---

49 Hugo Grotius, *De Jure Belli ac Pacis*, James Brown Scott (ed) *Classics of International Law*, (first published 1625, Clarendon Press 1925).

principles pacta sunt servanda (agreements must be kept) and good faith. Times of hazard, of uncertainty, and crisis are never times where the law features well. It is then sacrificed to political aims; but society as a whole is then as ill as its politics. International law, to function and to create the new norms international life needs, must be accompanied by confidence-building measures. The recent crisis of multilateralism, and the parallel return to insolent forms of unilateralism, are expressions of this deplorable state of affairs.[50] If we further aggravate it, a future commentator will note a perhaps even more dramatic decline of public international law in world affairs than one contemporary commentator has found in past decades.[51] Such a decline would not be good for the world; an even greater anarchical distribution of power and a series of uncoordinated power politics will hardly be able to address the numerous common and vital challenges lying ahead.

As a consequence, an important caveant consules is warranted: The prospective systemic effects of lethal force exercised across international borders must be considered in the complex web of an overall assessment and balancing of the interests at stake in counterterrorism.

---

50   As was noted already in 1999 by C Tomuschat, *International Law: Ensuring the Survival of Mankind on the Eve of a New Century*, (Martinus Nijhoff Publlishers 1999) at 51-53. See also I Brownlie, "International Law in the Context of the Changing World Order", in N Jasentuliyana (ed), *Perspectives on International Law*, (Kluwer 1995) at 49ff.

51   M Koskenniemi, *The Gentle Civilizer of Nations, The Rise and Fall of International Law 1870–1960*, (Cambridge University Press 2001).

SECTION IV
Creating a Drone Court—
Integration via a Policy Proposal

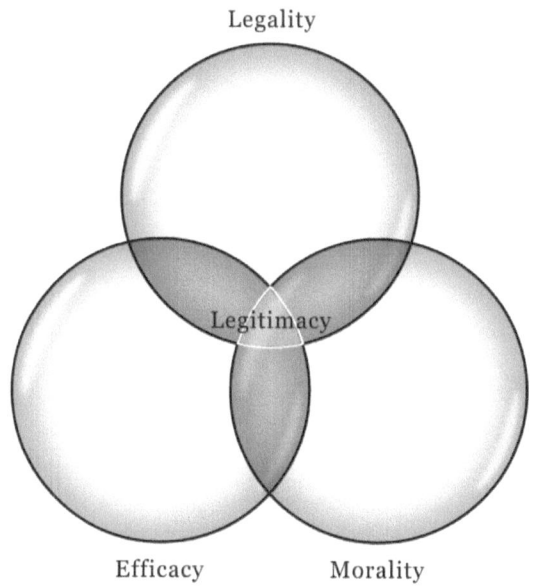

Chapter 13

Establishment of a Drone Court: A Necessary Restraint on Executive Power

Amos N. Guiora and Jeffrey S. Brand[1]

I. Introduction

On April 20 and 21, 2014, the United States—perhaps in concert with Yemini forces—attacked individuals and targets reportedly affiliated with the terrorist organization, Al-Qaeda in the Arabian Peninsula (AQAP). According to press reports, over 30 individuals were killed or "successfully targeted" as the government reported, including those who allegedly bore responsibility for attacks against the United States. The primary weapons used during the US attack were drones.

Immediately after the attacks, President Obama's former national security advisor, Thomas Donilon took to CNN's airwaves and affirmed the obvious: The use of drones is the go-to weapon in the Obama administration's arsenal for its war on terror and is the future of modern warfare. Donilon's exchange with CNN's Wolf Blitzer clarifies this:

> BLITZER: And the president remains, as far as you know, as determined as ever to continue these drone strikes? I asked the question of Peter Bergen, our national security analyst here at CNN, an expert on terrorism. He's called this President Obama's war, this drone war, as it is.
>
> DONILON: I think it's fair to say … that the United States remains committed to defending the country. It remains committed to degrading and defeating Al-Qaeda and those groups around Al-Qaeda that would do harm to the United States. And it has been, from the outset of this administration, a determined

1 The authors wish to acknowledge the following individuals for their invaluable input: Dean Erwin Chemerinsky and Professor of First Amendment Law Raymond Pryke, at the University of California Irvine School of Law; Mark P Denbeaux, Professor of Law, Seton Hall School of Law; Kevin Govern, Professor of Law, Ave Maria School of Law; Monica Hakimi, Professor of Law, University of Michigan Law School; Peter Honigsberg, Professor of Law, University of San Francisco School of Law; David Luban, Professor of Law, Georgetown Law Center.

and focused effort against those individuals and groups who pose a threat to the United States and would do harm to the United States ... *because we're at war*.[2]

That exchange followed on the heels of this question and answer:

> BLITZER: So he would authorize it even if there were a possibility some innocent civilians, what they call collateral damage, could be struck in the process? He knows the risks?

> DONILON: There are very strict rules that have been laid out here, and it's very important, I think, to go through this. And with respect to the death or harm to non-combatants, civilians, it's very, very high thresholds. And indeed, the threshold has been articulated with near certainty that you're not going to harm a non-combatant or a civilian in one of these efforts.

The Obama administration's claims voiced by Donilon motivate what follows in this chapter. We have no doubt that the use of drones will continue and that drone warfare is, indeed, here, and here to stay. However, we have serious doubts that "high thresholds" exist to authorize an attack or that "near certainty" exists about its outcome. The reality is that the administration conducts its drone warfare unencumbered by any external restraints or accountability, thereby raising grave legal (both constitutional and international), policy, and moral questions. In sum, it is our belief that the certainty of drone warfare combined with the uncertainty that pervades the drone strike decision-making process mandates the establishment of a Drone Court.[3]

The authors come to this endeavor from distinct backgrounds. Professor Guiora served for 20 years in the Israel Defense Forces where he was involved in targeted

---

2   CNN: The Situation Room with Wolf Blitzer, "Source: 'Unprecedented' Attack Targets al Qaeda" (April 21, 2014) (emphasis added) available at <http://transcripts.cnn.com/TRANSCRIPTS/1404/21/sitroom.01.html> accessed Dec 2014.

3   The authors use the term Drone Court as a means of focusing on drone warfare and its consequences. Despite this narrow moniker and focus, we also believe that the notions of separation of powers under the United States Constitution, and the role of the judiciary that it demands, make the arguments set forth here equally applicable to other executive decisions made in the name of national security, most notably with regard to surveillance. The authors have written extensively about these other arenas and made some of the same arguments that are set forth here. See eg Professor Brand's, "Eavesdropping on Our Founding Fathers: How a Return to America's Core Values Can Help Resolve America's Surveillance Crisis" (2015) 6 Harvard Nat'l Security J 1. Thus, the notion that a broader National Security court rather than a narrowly named Drone Court might, in the end, be the most efficacious to preserve the checks and balances envisioned by the framers of the United States Constitution.

killing decision-making.[4] Professor Brand is a long-time academic who has studied, written, and passionately defended individual rights and social justice.[5] Despite the differences in our backgrounds and perspectives, we share a common concern: We fear, in the words of Justice Jackson's opinion in *Youngstown Sheet and Tube v. Sawyer*,[6] that "unfettered executive" discretion, devoid of external restraints and accountability, lies at the heart of America's drone policy and that it must be checked if our democracy is to thrive. We both recognize the urgency of constructing a jurisprudential architecture to ensure that any drone policy is conducted in accordance with the rule of law.

As such, we propose the establishment of a Drone Court to review executive branch drone decisions *prior* to their execution. That said, we harken back to Chief Justice Earl Warren's wise words in *Miranda v. Arizona*[7] in which he called on, if not challenged, Congress and others to recommend alternative means to protect those subject to police interrogations. In the spirit of Chief Justice Warren's entreaty, we are not wedded to the establishment of a Drone Court. Moreover, we do not claim to have all of the answers. We are, however, wedded to the creation of a process that will ensure that US counterterrorism be conducted in accordance with the rule of law and the principles of morality in armed conflict which can effectively and concretely be monitored and measured. In other words, we advocate for the use of drones falling into alignment with the precepts of legitimacy explored in this volume.

We present our proposal in the context of two realities that must be noted for the record. The first reality is that since the horrific attacks on September 11, 2001, the United States has struggled, as it never has in its history, with finding the appropriate balance between the protection of individual rights and the demands of national security. Repeatedly, as we note below, particularly in the area of surveillance, national security has trumped individual rights leading to abuses that have shocked the nation and led to calls for reform.[8] Those abuses have occurred

---

4   AN Guiora, *Legitimate Target: A Criteria-Based Approach to Targeted Killing* (OUP 2013); AN Guiora, *Targeted Killing and the Law* (OUP 2013); AN Guiora, "Drone Policy: A Proposal Moving Forward," JURIST.com, Academic Commentary (March 4, 2013) available at <http://jurist.org/forum/2013/03/amos-guiora-drone-policy.php> accessed Dec 2014; AN Guiora, "Targeted Killing as Active Self-Defense" (2004) 36 Case W Res J Int'l Law 319.

5   Jeffrey S Brand, "Striking a Blow for Democracy," Op.-ed, *Washington Post* (1997); Jeffrey S Brand, "The Supreme Court, Equal Protection, and Jury Selection: Denying that Race Still Matters" (1994) Wisconsin Law Rev 511; Jeffrey S Brand, "The Second Front in the Fight for Civil Rights: The Supreme Court, Congress and Statutory Fees" (1990) 69 Texas Law Rev 291.

6   *Youngstown Sheet & Tube Co v Sawyer*, 343 US 579 (1952).

7   *Miranda v Arizona*, 384 US 436 (1966).

8   WW Burke-White, "Human Rights and National Security: The Strategic Correlation," (2004) 17 Harvard Human Rights J 249.

despite the formal existence of institutions intended to maintain the balance of power among the branches of government.

Thus, the Foreign Intelligence Surveillance Court (FISC) was established pursuant to the Foreign Intelligence Surveillance Act (FISA) supposedly to insure that executive branch decisions with respect to surveillance were carried out in a way to protect the rights of those surveilled. The reality, however, is that fundamental flaws in the structure of the FISC, detailed more fully below, have turned the court more into a rubber stamp of executive decision-making than a constitutional buffer between citizens and executive branch abuses that FISA sought to eliminate. Indeed, the mere existence of FISC has been cleverly used by the executive branch to bolster claims that the massive surveillance programs disclosed since 9/11 pass constitutional muster—after all, goes the argument, the FISC said it was OK!

We are under no illusion that the same process might not occur were a Drone Court to be established. It too, if not constructed properly with safeguards to maintain its independence, could be hijacked and used as a shield for abusive executive decision-making rather than as a sword to insure appropriate checks on the executive branch.

The task will not be easy. The lessons learned from past abuses must be incorporated into the Drone Court structure and the Court must have sufficient power to insure that the executive responds in good faith to the requirements imposed on it. As we make our proposal, we recognize that such a structure has yet to be created in America's 250-year struggle to find the appropriate balance between individual rights and national security. We believe, however, that if we keep that struggle in mind, along with past dismal failures, we may have the best hope of succeeding in the future. It is with that mindset that this proposal is offered, knowing the perils that may exist, including the misuse by the executive branch of a judicial order sanctioning its decision-making in the name of national security.

We also present our proposal with reservations rooted in a second reality—the need for executive power and the use of force. Our proposal, at its core, reflects deep concern regarding executive power, particularly when not subject to external review or restraint. We believe that post 9/11 US foreign policy and national security decision-making has reflected the ills of a unitary executive; the reality of the past 13 years is that neither Congress nor the courts have forcefully interjected themselves while consecutive administrations have conducted complicated, complex, and controversial operational counterterrorism throughout the world.

While we understand the need for executive power and the use of force we, simultaneously, are deeply concerned about its unrestrained nature. That deep concern drives this project. However, we also understand that certain circumstances justify executive decision-making under extreme, time-limited conditions. That is a reality. One of us has been involved in similar dilemmas. For that reason, and with a sense of discomfort, our proposal for the establishment of a Drone Court is tempered by several caveats and compromises, one of which is particularly fraught: We propose, as we detail below, that under very unique and limited circumstances of a narrowly defined "imminent threat" to the national security, there might be an exemption to the requirement of prior judicial approval of lethal drone strikes. Despite our discomfort with this aspect of our proposal, we also understand that operational realities justify this problematic "wiggle room." However, to ensure the Court's effectiveness and legitimacy the executive branch will be required to explain—retroactively—why the exceptional circumstances justified the exemption in such a limited case.

The Court's power in these limited circumstances will be twofold: the possibility of holding the president in contempt if the case is insufficiently made; and, ingraining deep skepticism into any statement by the executive branch that "imminent threats" and "national security" justify unilateral drone strikes that have not received requisite judicial review before they are launched. As the former President (Chief Justice) of the Israel Supreme Court, Aharon Barak, famously wrote "national security is not a magical phrase."

In devising our model for the Drone Court, two core principles guide our thinking: first, the recognition that the target, no matter how grave the accusations, has protected rights; second, the critical importance of the principle of checks and balances to rein in unfettered executive discretion and to sustain our constitutional democracy. We recognize that given the gravity of the threat posed by terrorism to the United States it is tempting to discard these principles in our zeal to insure the security of the nation. We reject that logic, believing that such expediency undermines the very principles we seek to preserve.

To address both the substantive and procedural aspects of the proposed Drone Court this chapter is divided into the following sections: section II: Underlying Principles; section III: Prior Models: Why Replicating the FISA Court Won't Work; section IV: How Will the Drone Court Work; section V: Meat to Bones—A Hypothetical Fact Scenario; and section VI: A Concluding Thought.

At the outset, it bears emphasis that our purpose is neither to cast aspersions nor to engage in unnecessary, divisive "finger-pointing." Rather our goal is to create an effective mechanism that bolsters the constitutional foundations of our democracy and protects individual rights, which are compromised by a unitary executive theory allowing for the unfettered exercise of executive power—a theory employed by both the Bush and Obama administrations.

Our emphasis on external restraints is practical and philosophical: practical because it is predicated on direct involvement in operational counterterrorism; and, philosophical owing to a deep belief that limits must be placed on executive power. We recognize that there is a risk in proposing restraints. We do so, however, recalling the wise counsel of Aharon Barak who urged that Israel apply a paradigm of "self-imposed restraints," reminding the nation that "sometimes a democracy must fight with one hand behind its back." Barak presciently opined that operational counterterrorism demands answers to difficult questions including the criteria required to make decisions to kill and the restrictions that should be imposed before such decisions are made. Our proposal seeks to achieve that proper balance.

As we envision it, the Drone Court will rarely engage in a post-attack review and assessment; rather, the executive branch will submit in advance to the Court relevant intelligence information and other supporting material that justifies a particular attack. The Drone Court will generally be asked to act before rather than after the fact.

Moreover, the Drone Court will review authorization requests with respect to US citizens and non-citizens alike. In contrast to the Department of Justice (DoJ) White Paper,[9] which articulated standards for drone attacks applicable solely to US citizens, the Court we propose would have a broader mandate. Indeed, our thinking is greatly influenced by the Department of Justice White Paper that defines the drone policy and which is discussed more fully below. If the broad, imprecise definitions of imminence, threat, and target found in the White Paper are applicable to US citizens we can but wonder how elastic, ungrounded, and problematic are the standards when non-US citizens are the intended targets of a drone attack.

In part, we base our recommendations on the line of United States Supreme Court cases decided after 9/11 that involved the rights of detainees at Guantánamo: *Rasul v. Bush, Hamdi v. Rumsfeld, Hamdan v. Rumsfeld*, and *Boumediene v. Bush*.[10] *Boumediene* is of particular importance and is discussed more fully below. There, the Court held that the Suspension Clause of the *Writ of Habeas Corpus* extended to *non*-US citizens who had been designated as enemy combatants by the Bush administration and were being held at Guantánamo. In a 5–4 opinion authored by Justice Kennedy, the Court held that the executive branch could not strip alien, non-US citizens of their right to meaningful review of their detention. Justice Kennedy wrote: "[t]o hold that the political branches may switch the Constitution on or off at will would lead to a regime in which they, not this Court, say, 'what the law is.'"[11]

We recognize that *Boumediene*, and the writings of many commenting on its reach, note that the Court's ruling was premised on the fact that Boumediene was being held at Guantánamo, which the Court found to be essentially located within the United States. Nonetheless, as other commentators have

---

9   Office of Legal Counsel, US Department of Justice, White Paper entitled "Lawfulness of a Lethal Operation Directed against a US Citizen Who Is a Senior Operational Leader of Al-Qa'ida or an Associated Force," draft copy, (Nov 8, 2011) (DoJ White Paper hereinafter) available at <http://msnbcmedia.msn.com/i/msnbc/sections/news/020413_DOJ_White_Paper.pdf> accessed Nov 2014. This White Paper sets forth the legal framework in which the US government could use lethal force in a foreign country against a US citizen who is a senior operational leader of Al-Qaeda or an associated force that is actively engaged in planning and carrying out operations to kill Americans.

10   *Rasul v Bush*, 542 U.S. 466 (2004); *Hamdi v Rumsfeld*, 542 US 507 (2004); *Hamdan v Rumsfeld* 548 US 557 (2006); *Boumediene v Bush*, 553 US 723 (2008).

11   Citing *Marbury v Madison*, 5 US 137 (1803); 1 Cranch 137, at 177.

noted,[12] *Boumediene* is rooted in a broader logic that seeks to protect the proper balance between the branches of government that the Framers envisioned. We agree that the philosophical and jurisprudential essence of *Boumediene* is that the foundational American constitutional value of a separation of powers and maintaining effective checks and balances on each branch of government is essential if we are to maintain individual rights.

From that assertion flows the necessary conclusion that the doctrine of the separation of powers requires a meaningful role for the judiciary. *Boumediene* reminds us that that principle applies to a naturalized citizen of Bosnia-Herzegovina as much as it does to a US citizen. *Boumediene* seeks to insure the judiciary "a seat at the table"[13] regardless of whether the detainee is a citizen or not. This chapter seeks to design the contours of that seat in the case of individuals targeted for execution by the executive branch.

Our analysis is also rooted in international norms and comparative international practices that suggest the need to protect the rights of those targeted for assassination in ways that current executive decision-making does not. As Jonathan A. Geltzer notes in his helpful analysis of the relationship of due process to the *Writ of Habeas Corpus*, "a number of recent Supreme Court decisions aim to place American human rights practices in line with those of comparable countries around the world."[14] In fact, the chief proponent of that developing line of cases may be Justice Kennedy,[15] the author of *Boumediene*'s majority opinion as well as the Court's decisions in *Roper v. Simmons*[16] and *Lawrence v. Texas*.[17] We suggest, as does Geltzer, that Justice Kennedy's views express an "urge to ensure that the

---

12 JA Geltzer, "Of Suspension, Due Process Guantánamo: The Reach of the Fifth Amendment After *Boumediene* and the Relationship between Habeas Corpus and Due Process" (2012) 14(3) J of Constitutional Law 761: "In contrast, a second understanding of *Boumediene* focuses less on the detainees themselves and more on the judiciary as an institution. This reading of the decision emphasizes its insistence that the courts not be silenced or sidelined by the political branches and instead, they defend the separation of powers among those branches ... From this perspective the real constitutional violation identified by the *Boumediene* Court was a structural overstepping by the political branches, not a deprivation of individual detainees' rights"; see also SI Vladeck, "The New Habeas Revisionism" (2011) 124 Harvard Law Rev 941; GL Neuman, "The Habeas Corpus Suspension Clause After *Boumediene v Bush*" (2010) 110 Columbia Law Rev 537; AL Tyler, "Suspension as an Emergency Power" (2009) 118 Yalw Law J 600.

13 Ibid, Geltzer at 768.

14 Ibid at 769.

15 Ibid.

16 *Roper v Simmons*, 543 US 551 (2005): the US Supreme Court held that it is unconstitutional to impose capital punishment for crimes committed while under the age of 18.

17 *Lawrence v Texas*, 539 US 558 (2003): The US Supreme Court held that the Texas statute making it a crime for two persons of the same sex to engage in certain intimate sexual conduct violates the Due Process Clause. See also Justice Kennedy's vote in *Atkins v*

United States would provide the type of judicial relief dictated by global norms," such as those expressed in Article 9(4) of the International Covenant on Civil and Political Rights "imposing a human rights law requirement of habeas-style judicial review," and, Article 9(1) of the Covenant "requiring what seems to be a rough equivalent of due process: "No one shall be deprived of his liberty except on such grounds and in accordance with such procedure as are established by law."[18]

In addition, we draw on the experiences of other countries. To that end, we discuss the relationship between the executive branch and a nation's judiciary in the context of other nations engaged in operational counterterrorism. Our focus is primarily on Israel, where the Israeli Supreme Court, sitting as the High Court of Justice, exercises robust judicial review. As that Court made clear in its seminal decisions regarding Israel's targeted killing policy,[19] the rule of law must take priority over operational counterterrorism policy decisions and applications. That principle, more than any other, must be the guide for how the US develops, applies, and articulates its drone policy.

The broad strokes of our thinking do not obviate the relevance of the age-old adage that "the devil is in the details." Thus, this chapter seeks to address many of the innumerable issues raised by the establishment of a domestic Drone Court. We expect, and welcome, the many questions that our proposal will inevitably generate. Indeed, it is our expectation that criticism will focus on the essence of the proposal from a jurisprudential and constitutional perspective and that many questions will be raised regarding the "nuts and bolts" of its implementation. In fact, creating a dialogue about appropriate institutions and processes to review drone attacks is one of our primary goals in authoring this chapter. We welcome the opportunity to address and discuss various critiques, and some concerns regarding the practicality of the proposed Court are, undoubtedly, valid and legitimate, they should not and must not be perceived to be cause for its rejection nor justify casting it asunder without careful consideration.

And this final introductory note: This chapter incorporates legal and policy discussion alike. In many ways, the two are meshed throughout because of our belief that the issues presented cannot be understood through the lens of law or policy exclusively.

The policy discussion will focus primarily on the question of effectiveness—a difficult to define term whose lack of definition is one of the primary motivations for our proposal.[20] Careful examination of how the Obama administration implements and articulates the necessity for its policy suggests a haphazard and superficial

---

Virginia, 536 US 304 (2002): The US Supreme Court held that executions mentally retarded criminals are "cruel and unusual punishment" prohibited by the Eighth Amendment.

18   Geltzer (n 12) at 770 and footnotes 217, 218.

19   *Public Committee Against Torture in Israel v Government of Israel* (2006) HCJ 769/02.

20   For an exploration into some of the varying aspects of effectiveness when it comes to drone strikes see Chapters 9 to 12 in this volume.

consideration of whether the decision to kill a specific individual is warranted both strategically and tactically. Of course, if the killing of an individual does not lead to tactical or strategic security benefits this raises grave questions of morality (a consequence that will be set aside in this chapter).

Surprisingly missing in the administration's various articulations of the drone policy[21] is a sophisticated discussion that compellingly, consistently, and thoughtfully provides the rationale for the current drone policy. In fact, the DoJ's White Paper, the only public document that currently defines and justifies US drone policy,[22] provides a disturbingly loose application of principles critical to nurturing the rule of law, including its definitions of what constitutes a legitimate target[23] and an imminent threat.[24] The failure to narrowly define these terms results in a drone policy marred by harm to innocent civilians—so-called collateral damage—with all of the negative global ramifications that consistently follow.

A drone policy based on the rule of law would be predicated on criteria-based processes and narrow definitions of self-defense, the legitimacy of a target, and the imminence of the threat. Moreover, any such determinations would be subject to external review. That is what we hope to achieve in the proposal we outline below.

II. Underlying Principles

Our thesis, as noted above, is rooted in the ingenious structure of our constitutional democracy that mandates a separation of powers. The scheme of checks and

---

21 DoJ White Paper (n 9). See also M Zenko, "Talking in Circles: Why Harold Koh's big speech on targeted killings is just more of the same, intentional Obama muddle," Foreign Policy (May 9, 2013) available at <http://www.foreignpolicy.com/articles/2013/05/09/targeted_killings_koh_policy_obama> accessed Dec 2014; A Shapiro, "US Drone Strikes are Justified, Legal Adviser Says," National Public Radio (March 26, 2010) available at <http://www.npr.org/templates/story/story.php?storyId=125206000> accessed Dec 2014; J Swaine, "Barack Obama 'has authority to use drone strike to kill Americans on US soil'" *The Telegraph* (March 6, 2013) available at <http://www.telegraph.co.uk/news/worldnews/barackobama/9913615/Barack-Obama-has-authority-to-use-drone-strikes-to-kill-Americans-on-US-soil.html> accessed Dec 2014.

22 The original OLC memo has now been released in a redacted form by the US Court of Appeals for the Second Circuit, *The New York Times Company et al v United States*, Case No 13–422-cv (April 21, 2014). However, with a redaction of all relevant facts, and an explicit acceptance that all facts would be assumed as valid by the OLC, little is exposed in this memo and thus we continue to analyze the White Paper to best understand the contours of the drone policy: David Barron, Acting Assistant Attorney General, "Re: Applicability of Federal Criminal Laws and the Constitution to Contemplated Lethal Operations Against Shaykh Anwar al-Aulaqi" (July 16, 2010).

23 AN Guiora (n 4).

24 For a historical analysis on the question of "imminence," see Chapter 6 by Steven J. Barela of this volume.

balances seeks to insure fundamental fairness and is a predicate to enforcing the US Constitution's demands in the 5th Amendment that "No person shall ... be deprived of life, liberty or property without due process of law."

The relationship of the separation of powers doctrine to the preservation of individual rights bears special emphasis in today's post-9/11 environment in which proponents of unbridled executive power belittle the importance of individual rights when it is claimed that the security of the nation is at stake. That simplistic logic ignores a fundamental truth about our democracy, which holds that absent appropriate checks and balances individual liberty is likely to be sacrificed. It is critical that each branch of the government have a role in decision-making that affects individual life or liberty, or, conversely, that no branch of the government should be able to make such decisions on its own.

In fact, the United States Supreme Court has spoken eloquently about the relationship of the balance of power among the branches (particularly the judiciary) and the preservation of individual liberty in the context of *both* US citizens *and* foreign nationals. Thus, in 2004, the Court decided *Hamdi v. Rumsfeld*[25] that Hamdi, a US citizen labeled an enemy combatant by the Bush administration, was entitled to judicial review of his detention. In *Hamdi*, a fractured court affirmed the power of the government to detain "enemy combatants," but also mustered eight votes to support the proposition that no US citizen could be held indefinitely without basic due process protections or on the basis of a determination made solely by the executive branch. Justice O'Connor eloquently noted the relationship of the separation of powers to individual liberty:

> [W]e necessarily reject the Government's assertion that separation of powers principles mandate a heavily circumscribed role for the courts in such circumstances. Indeed, the position that the courts must forgo any examination of the individual case and focus exclusively on the legality of the broader detention scheme cannot be mandated by any reasonable view of separation of powers, as this approach serves only to *condense* power into a single branch of government. We have long since made clear that a state of war is not a blank check for the President when it comes to the rights of the Nation's citizens. Whatever power the United States Constitution envisions for the Executive in its exchanges with other nations or with enemy organizations in times of conflict, it most assuredly envisions a role for all three branches when individual liberties are at stake ... It would turn our system of checks and balances on its head to suggest that a citizen could not make his way to court with a challenge to the factual basis for his detention by his government, simply because the Executive opposes making available such a challenge.[26]

---

25   542 US 507 (2004).
26   Ibid at 535, referencing *Youngstown* (n 6) at 587.

Justice Antonin Scalia dissented from the Court's holding, but concurred in rejecting the notion that the executive branch could unitarily determine Hamdi's fate without an independent source of judicial review, noting that for Hamdi to be detained indefinitely would require either a suspension of *habeas corpus* by the Congress or a trial in the nation's criminal courts.[27]

Hamdi involved the rights of a US citizen.[28] The relationship of checks and balances and the separation of powers to individual liberty, however, have been equally eloquently articulated by Justice Kennedy in *Boumediene* for non-US citizens. There, Justice Kennedy, writing for the majority, relied on the separation of powers as the foundation for the Court's holding that the Suspension Clause of the *Writ of Habeas Corpus* could not be denied to Lakhdar Boumediene merely because he had been labeled an enemy combatant *and* was a foreign national hailing from Bosnia-Herzegovina.

Reaching his conclusion, Justice Kennedy engaged in a careful analysis of the history of the *Writ of Habeas Corpus* from the time of the Magna Carta to the present, demonstrating the *Writ*'s purpose to prevent "executive 'imprisonment without any cause'," a history that Kennedy reminds us was "known to the Framers." Kennedy wrote:

> The Framers' inherent distrust of governmental power was the driving force behind the constitutional plan that allocated powers among three independent branches. This design serves not only to make Government accountable but also to secure individual liberty.

Justice Kennedy cited *Youngstown Sheet & Tube Company v. Sawyer* and Justice Jackson's oft-cited concurrence: "[t]he Constitution diffuses power the better to secure liberty."[29] At the same time, Justice Kennedy implied the broad sweep of the separation of powers doctrine, noting its applicability to citizens and non-citizens alike, citing *Yick Wo v. Hopkins*: "The Constitution's separation-of-powers structure, like the substantive guarantees of the Fifth and Fourteenth Amendments, protects persons as well as citizens, foreign nationals who have the privilege of litigating in our courts can seek to enforce separation of powers principles."[30]

Kennedy's conclusion with respect to foreign national Boumediene was soaring:

> Officials charged with daily operational responsibility for our security may consider a judicial discourse on the history of the Habeas Corpus Act of 1679

---

27   Ibid at 555 (Scalia, J, dissenting).

28   At the present time, Hamdi, is a former US citizen, having been freed on the condition that he leave the country and renounce his citizenship.

29   *Youngstown* (n 6) at 635 (Jackson, J, concurring)

30   *Boumediene* (n 10) at 743 (Kennedy, J) citing *Yick Wo v Hopkins,* 118 US 356, at 374 (1886).

and like matters to be far removed from the Nation's present, urgent concerns. Established legal doctrine, however, must be consulted for its teaching. Remote in time it may be; irrelevant to the present it is not. Security depends upon a sophisticated intelligence apparatus and the ability of our Armed Forces to act and to interdict. There are further considerations, however. Security subsists, too, in fidelity to freedom's first principles. Chief among these are freedom from arbitrary and unlawful restraint and the personal liberty that is secured by adherence to the separation of powers. It is from these principles that the judicial authority to consider petitions for habeas corpus relief derives. ... Our opinion does not undermine the Executive's powers as Commander in Chief. On the contrary, the exercise of those powers is vindicated, not eroded, when confirmed by the Judicial Branch.

As stated above, we recognize that Boumediene was detained at Guantánamo, which the Court noted was, for all practical purposes, the equivalent of being detained in the United States. That geographic limitation, however, does not limit Justice Kennedy's sweeping separation of powers language to US citizens or to those on US soil. *Boumediene* stands forthrightly for the proposition that the doctrine of the separation of powers is critical to the preservation of liberty regardless of the country of origin from which a defendant may hale.

At the same time, we also recognize that where the target of a drone attack is a United States citizen, court precedent more strongly supports the proposition that due process rights should be afforded the target. In fact, in *Boumediene*, the Court only held applicable the Suspension Clause of the *Writ of Habeas Corpus*. The *scope* of the applicability of due process rights to foreign nationals remains a contested issue in the courts and in the legal literature.[31] Moreover, Justice Kennedy was clear in his holding that the majority was *not* defining the details of the process due *Boumediene*: "It bears repeating that our opinion does not address the content of the law that governs petitioners' detention. That is a matter yet to be determined."[32]

As we detail below, the precise parameters of the process due either a US or non-US citizen in our proposed Drone Court also are "yet to be determined." A Drone Court presents unique problems that inevitably will entail compromises and require flexibility as it is implemented. Those details are important, but they must not allow us to ignore the fundamental principle that the judiciary must be meaningfully engaged in the decision-making process to preserve our constitutional values. Bluntly stated, the judiciary must be engaged in meaningful

---

31   H Ball, *Bush, Detainees, and the Constitution* (University Press of Kansas 2007); see also Executive Military Order 11/2001; *Hamdi* (n 10); *Rasul* (n 10); *Hamdan* (n 10); *Boumediene* (n 10); DoD Order CSRTS 7/2004 (applied to foreign nationals); Detainee Treatment Act 2005, Public Law No 109–148, Div A, Tit X, 119 Stat 2739; Military Commissions Act 2006, Public Law No 109–366, 120 Stat 2600.

32   *Boumediene* (n 10) at 798.

review of an executive decision to execute a human being. Our proposal simply suggests that a mechanism be established to protect that principle and echoes words in *Boumediene*:

> The laws and Constitution are designed to survive, and remain in force, in extraordinary times. Liberty and security can be reconciled; and in our system they are reconciled within the framework of the law.

We recognize, however, that the application of these principles to drone warfare is controversial and that the courts are divided over whether the executive branch's authority is in fact limited in the context of targeted killings. Thus, as recently as April 2014, the United States District Court for the District of Columbia in *Nasser Al-Aulaqi v. Panetta* held that the family of United States citizen Al-Aulaqi did not have a viable claim under the Fifth Amendment despite the dictates of *Hamdi* and its progeny. Judge Collyer cited the broad discretion of the executive branch to determine matters relating to national security, particularly in times of armed conflict. Specifically, Judge Collyer cited the Fourth Circuit's hold in *Lebron v. Rumsfeld* and turned the separation of powers argument on its head, concluding that the executive branch was shielded from Fifth Amendment claims based on the broad authority of the executive branch when it came to matters involving foreign affairs:

> The reasons for this constitutional structure are apparent. Questions of national security, particularly in times of conflict, do not admit of easy answers, especially not as products of the necessarily limited analysis undertaken in a single case. It is therefore unsurprising that "our Constitution recognizes that core strategic matters of war-making belong in the hands of those who are best positioned and most politically accountable for making them."[33]

While we do not concede the validity of Judge Collyer's reasoning, we recognize the currency of her opinion and the possibility that other courts may follow her lead (in our view, it is a misreading of *Hamdi* and its progeny as well as the underlying rationale for the separation of powers). It is for this reason that we think the best course is for the Congress to pass legislation such as that proposed below to insure that the separation of powers are maintained and that due process principles are upheld in the gravest of all decisions—the taking of one's life after a determination that assassination is warranted to protect the national security.

---

33   *Al-Aulaqi v Panetta,* Civil Action No 12–1192 (RMC) (DDC April 4, 2014) (Collyer, J) quoting *Hamdi,* (n 10) at 531. Here the Court's holding was based on its refusal to imply a *Bivens* remedy in this context. See also *Marbury* (n 11) at 165–66: "the President is invested with certain important political powers, in the exercise of which he is to use his own discretion, and is accountable only to his county in his political character, and to his own conscience."

In proposing, drafting, and enacting such a legislative solution the Congress will be significantly aided by the executive branch which has repeatedly acknowledged that drone strikes implicate due process and that particular factual findings are required before a target may constitutionally be placed on the "kill list." Thus, President Obama stated clearly that "It's very important for the president and the entire culture of our national security team to continually ask tough questions about, are we doing the right thing, are we abiding by the rule of law, are we abiding by due process."[34] He concluded by underlining the necessity of setting "up structures and institutional checks, so that, you know, you avoid any kind of slippery slope into a place where we're not being true to who we are."[35]

Moreover, the Justice Department's White Paper on the use of drones to carry out targeted killings, despite its flaws, details a legal framework under which such a killing would be lawful if a target is "outside a recognized battlefield," specifying particular necessary findings of fact, including: the target be a "senior operational leader of" al-Qaida or an "associated force"; the imminence of the threat; the lack of feasibility of the capture of the target; and, the conclusion that the operation can be conducted in a manner consistent with recognized principles of international and domestic law.

In this sense, the executive branch has correctly identified key determinations that must be made before any order to kill is carried out. What is missing, however, is a mechanism for making sure that those determinations are made consistent with the dictates of the United States Constitution by a reviewing body that comports with the Framer's intent to implement the system of checks and balances that they ingeniously devised. What constitutes imminence, feasibility of capture, and consistency with national and international norms is, as *Marbury v. Madison* taught us centuries ago, the province of the judiciary, not the unfettered discretion of the executive branch. A recent commentary captures the importance of the principle:

> The fundamental rationale behind due process is to check against arbitrary government action. At its core, due process is an amalgamation of what makes the separation of powers a powerful American ideal. The Legislative branch writes the laws—including the ones that dictate charges available against US citizens—that the Executive branch enforces by bringing citizens in violation of the law to be tried before an impartial judicial branch that the Constitution itself or the Legislative branch has established. Times of national crisis will necessarily render some procedures of due process more elastic than times of peace. At the same time, the Court has also recognized that "[w]hatever power the United States Constitution envisions for the Executive in its exchanges

---

34   Jessica Yellin, CNN Chief White House Correspondent, "Drone Program Something You 'Struggle with,' Obama says" CNN Politics (Sept 10, 2012) <http://politicalticker.blogs.cnn.com/2012/09/10/drone-program-something-you-struggle-with-obama-says/> accessed Nov 2014.

35   Ibid.

with other nations or with enemy organizations in times of conflict, it most assuredly envisions a role for all three branches when individual liberties are at stake" (citing *Hamdi*, (n 10) at 536).[36]

The proposed Drone Court seeks to insure that those core principles are observed.

III. Prior Models: Why Replacing the FISA Court Won't Work

In searching for mechanisms to achieve the delicate balance between the protection of constitutional guarantees and national security, it is tempting to rely on models previously adopted and to apply them to the emerging drone warfare landscape. The most often cited model is the Foreign Surveillance Intelligence Act of 1978 and the courts it established to review executive branch requests to engage in foreign intelligence surveillance—the FISC and the Foreign Intelligence Surveillance Court of Review. The temptation to build on the existing FISA structure is understandable. The structure is already in place and the titles of the courts suggest precisely the types of safeguards necessary to achieve rational, fact-based decision-making to ensure that any drone targets warrant the finality that the strike seeks to insure. In fact, many commentators have suggested the wholesale adoption of FISA's judicial structure to drone warfare.[37]

Flaws in FISA's judicial structure that have undermined its intent and made it an ineffective mechanism to protect the rights of United States citizens, however, belie the ease of such a solution. Despite the good intentions of many FISC judges, the FISA court is essentially a "rubber stamp" of the executive branch, which controls the flow of information to a court shrouded in secrecy and non-adversarial in nature, thereby precluding the search for credible and reliable facts on which to base intelligence decisions.

FISA court statistics tell the tale. Since the court's inception 36 years ago, 99.9 percent of executive branch requests to engage in foreign intelligence have been granted. To be precise, in the 33 years from 1979 to 2012, the FISA court granted 33,942 requests for warrants and denied only 11, compiling a denial rate of three

---

36　M Sohn, "Drone Strikes and Due Process: The Role of the Separation of Powers in Lethal Action Against US Citizens Outside Traditional Battlefields," Jolt Digest, Harvard J of Law & Tech (March 6, 2013) available at <http://jolt.law.harvard.edu/digest/digest-note/drone-strikes-and-due-process> accessed Dec 2014.

37　G Greenwald, "The bad joke called 'the FISA court' shows how a 'drone court' would work," *The Guardian* (May 3, 2013) available at <http://www.theguardian.com/commentisfree/2013/may/03/fisa-court-rubber-stamp-drones> accessed Dec 2014; J Harman, "Harman: Drone courts can work," CNN Security Clearance (Feb 19, 2013) available at <http://security.blogs.cnn.com/2013/02/19/harman-drone-courts-can-work/> accessed Dec 2014; S Shane, "Debating a Court to Vet Drone Strikes," *The New York Times* (Feb 8, 2013) available at <http://www.nytimes.com/2013/02/09/world/a-court-to-vet-kill-lists.html?pagewanted=all&_r=0> accessed Dec 2014.

tenths of one percent of the total warrants requested. In the 22 years prior to the September 11, 2001 attacks, 14,036 warrants were issued and *none* were rejected. The 11 denials have come since 2002, but in the 10-year period from 2002 to 2012, the number of warrants granted equaled 19,906, 6,804 more warrants granted than in the 21 years preceding the attacks.[38] FISA's dismal record demands that its flaws be fully understood before proposals to fix the FISA court or, as we suggest here, to establish a Drone Court are proffered.

A. A Brief History of FISA and Its Failures

In a far different political climate than exists today, FISA passed the 95th Congress with overwhelming bipartisan support, garnering only 27 votes in opposition, 26 in the House and one in the Senate.[39] The optimism was palpable, including laudatory statements from both sides of the aisle that unregulated, warrantless foreign intelligence surveillance, which had been assumed legal and engaged in by every administration from Franklin Roosevelt to Jimmy Carter, would finally be reined in by judicial and congressional accountability. FISA's complicated scheme to regulate foreign intelligence, made more so in the post-9/11 era with the passage of the Patriot Act and multiple amendments to FISA, is beyond the purview of this chapter.[40] The general outline of FISA's judicial structure and its vision for the relationship of the branches of the government to each other, however, must be examined to understand FISA's fatal flaws and the need for an entirely different concept to insure the effectiveness of the proposed Drone Court.

The Founding Fathers' conception of the separation of powers was both complex and remarkably simple, taking into account human nature and political reality. That vision is set forth in Federalist Papers 47 and 51, authored by James Madison and Alexander Hamilton. Madison believed that the survival of the nascent Republic depended on the creation of effective checks and balances to insure that no one branch of the government usurped power to the detriment of the nation. "The accumulation of all powers, legislative, executive, and judiciary, in the same hands, whether of one, a few, or many, and whether hereditary, self-appointed, or elective," he wrote, "may justly be pronounced the very definition of tyranny."[41]

---

38   See Foreign Intelligence Surveillance Act Court Orders 1979–2014, Electronic Privacy Information Center (EPIC.org), available at <http://epic.org/privacy/wiretap/stats/fisa_stats.html> accessed Dec 2014; Public Filings—US Foreign Intelligence Surveillance Court, available at <http://www.fisc.uscourts.gov/public-filings?field_case_reference_nid=All&page=8> accessed Dec 2014. See also E Eichelberger, "FISA Court Has Rejected .03 Percent of all Government Surveillance Requests," Mother Jones (June 10, 2013), available at <http://www.motherjones.com/mojo/2013/06/fisa-court-nsa-spying-opinion-reject-request> accessed Dec 2014.

39   124 Cong Rec 10906 (1978) (Nay vote of Senator William Scott).

40   For such an investigation see Brand (n 3).

41   J Madison, "Federalist No. 47," *The Federalist Papers* (1788).

In Federalist No. 51, Madison and Hamilton provided a workable, if not precise, formula to gauge whether a proper separation had been achieved:

> In order to lay a due foundation for that separate and distinct exercise of the different powers of government ... each department should have a will of its own; and consequently should be so constituted that the members of each should have as little agency as possible in the appointment of the members of the others.

The proper balance, concluded Madison and Hamilton, consists in "giving to those who administer each department the necessary constitutional means and personal motives to resist encroachments of the others." Thus, the branches were to be neither separate nor totally distinct; each given the tools to check the other: legislative approval of federal judges, executive veto of legislative decrees, legislative overrides of executive vetoes, judicial review of legislative enactments, and legislative ratification of treaties entered into by the executive branch, among them.

If there was any one branch of the government where independence and separation were particularly critical it was the judicial branch. Hamilton characterized the judiciary as the weakest of the three branches, lacking "influence over either the sword or the purse." From this conclusion flowed an imperative:

> There is no liberty, if the power of judging be not separated from the legislative and executive powers ... [and] [t]he complete independence of the courts of justice is peculiarly essential in a limited Constitution ... which contains certain specified exceptions to the legislative authority; such ... as the prohibitions against bills of attainder, ex post facto laws, and the like. Limitations of this kind can be preserved in practice no other way than through the medium of courts of justice, whose duty it must be to declare all acts contrary to the manifest tenor of the Constitution void. Without this, all the reservations of particular rights or privileges would amount to nothing.[42]

Thomas Jefferson agreed: "The dignity and stability of government in all its branches, the morals of the people and every blessing of society depend so much upon an upright and skillful administration of justice, that the judicial power ought to be distinct from both the legislative and executive and independent upon both, that so it may be a check upon both, as both should be checks upon that."[43]

B. FISA's Fundamentally Flawed Judicial Structure

By any measure, FISA's structure ignored all of these admonitions. If blame can be assigned to any one part of the legislation, the lion's share of it would reside in FISA's prescription for the judiciary. The role of the judges, the manner of

---

42   A Hamilton, "Federalist No. 78," *The Federalist Papers* (1788).
43   Thomas Jefferson to George Wythe, 1776, Papers 1:410.

their selection, and the process and opportunity for review by the FISA court, all coalesced to guarantee that the balance among the branches would go horribly wrong, ensuring that the judiciary's role would be marginal. That the executive branch would exercise unbridled power exacerbated ironically by the imprimatur of judicial approval. Additionally, congressional oversight would prove meaningless at best and corrosive at worst, undermining public confidence in the ability of government to properly balance national security needs against the rights of American citizens—the very goal of FISA from the beginning.

FISA's opponents on the left and the right had predicted such results at the time of its passage. For example, on the right, Yale law professor Robert Bork, who later would be rejected in his bid to sit on the United States Supreme Court, described the judicial role as merely "managerial."[44] On the left, Massachusetts Congressman, Father Robert Drinan who sat on the committee to impeach Richard Nixon, declared "[t]he role of the federal judge in the administration's proposal is almost a degradation of the Federal judiciary, making a mockery and a travesty of the judicial function."[45]

Both noted that the root of the problem stemmed from the limited information supplied to the judge, the *ex parte* nature of the proceedings (including just one party), and the absolute secrecy in which the judge was to operate. Bork wrote:

> How can this be the rule of law if: It would set apart a group of judges who must operate largely in the dark and create rules known only to themselves. Whatever that may be, it debases an important idea to call it the rule of law. It is more like the uninformed, unknown and uncontrolled exercise of discretion.[46]

Drinan noted:

> The judge is not permitted to question the administration's claims in its certification to the FISA court that the information sought is in fact foreign intelligence information. [Even in the case of citizens of the United States,] the judge must accept the certification unless he finds that it is "clearly erroneous" on the basis of the statement submitted with the application [by the executive branch].[47]

Drinan concluded that the FISA judge would be a "rubber stamp" from whom virtually all of the essential background of the requested authorization could be withheld.

---

44   R Bork, "Neutral Principles and Some First Amendment Problems," (1971) 47 Indiana Law J 1.

45   Foreign Intelligence Surveillance Act of 197892 Stat 1783 H5419 1978, Comment of Representative Drinan on June 3, 1977 at H5422.

46   Foreign Intelligence Surveillance Act of 1978, Public Law 92 Stat 1783, E 3601 1978: Statement of Robert Bork, (June 29, 1978) at E3602.

47   123 Cong Rec H5422 (daily ed June 3, 1977) (statement of Rep Drinan).

Four decades later, the court's record proved them both correct. In fact, in 2006, in the wake of Stellar Wind, one of President George W. Bush's illegal surveillance programs,[48] FISA court federal Judge James Robertson resigned in "frustration" over the judicial role afforded by FISA. In July 2013, following disclosures of NSA documents by Edward Snowden, Judge Robertson told CBS News that "anyone who has been a judge will tell you a judge needs to hear both sides of a case." Although Judge Robertson denied that the court acted as a "rubber stamp," he concluded "[t]his process needs an adversary."[49]

Another prediction made by FISA's opponents also came to pass. Wyoming's Republican Senator Malcolm Wallop had worried that "a body of case law is likely to grow without benefit of arguments contrary to the Government."[50] Snowden leaks revealed that that fear, too, had been realized. In 2013, the *Wall Street Journal* reported that the FISA court's definition of "relevance" had allowed the National Security Agency (NSA) to engage in programs such as PRISM under which metadata of billions of phone calls of American citizens are daily collected by the NSA in cooperation with giant third-party internet companies such as Google and Verizon. The FISA court's definition of "relevance" was at odds with a narrower definition decreed by the United States Supreme Court. The newspaper reported that "the FISA court has developed separate precedents, centered on the idea that investigations to prevent national security threats are different from ordinary criminal cases."[51] Precisely how they did it, however, remains a mystery.

Indeed, if one were permitted into FISA's windowless courtroom at 333 Constitution Avenue in Washington D.C. it would appear very different from courtrooms that anchor the system of justice in the United States, operating in near-total secrecy in a non-adversarial setting in which orders are issued by the court on the basis of information provided almost exclusively by the executive branch. Absent are opposition parties, cross-examination, and opposing arguments—the staples of the American justice system.

The same departure from America's constitutional judicial structure is evident in the selection of the FISA judges who are handpicked by the Chief Justice of the

---

48   NSA Inspector General Report on Email and Internet Data Collection under Stellar Wind (Full Document), *The Guardian* (June 27, 2013), available at <http://www.theguardian.com/world/interactive/2013/jun/27/nsa-inspector-general-report-document-data-collection> accessed Dec 2014.

49   "Former Judge Admits Flaws with Secret FISA Court" CBS News (July 9, 2013) available at <http://www.cbsnews.com/news/former-judge-admits-flaws-with-secret-fisa-court/> accessed Dec 2014; See also E Lichtblau, "Judges on Secretive Panel Speak Out on Spy Program," *New York Times* (March 29, 2006) available at <http://www.nytimes.com/2006/03/29/politics/29nsa.html> accessed Dec 2014.

50   S Rep No 95–701, at 94 (1978) reprinted in 1978 USCCAN 3906.

51   J Valentino-Devries, "Secret Court's Redefinition of 'Relevant' Empowered Vast NSA Data-Gathering" *The Wall Street Journal* (July 8, 2013) available at <http://online.wsj.com/news/articles/SB10001424127887323873904578571893758853344> accessed Dec 2014.

United States Supreme Court. While the judges are selected from sitting Article III judges who have been approved by the Senate, their seven-year assignment to the FISA court occurs without the benefit of any congressional input or oversight.

The absence of an independent judiciary in FISA's judicial structure has in turn made effective congressional oversight of the actions of the executive branch impossible. This conclusion is reflected in two phenomena. First, a "rubber stamp" court emboldens the executive branch to act in ways it might not otherwise have acted were true judicial review possible. The dangers inherent in such overreach are particularly ominous "when the request originates with the intelligence community and is surrounded by warnings that the defense or security of the United States would be endangered if the requested authorization for surveillance is not granted."[52]

The Bush administration's handling of its illegal Terrorist Surveillance Program (TSP) demonstrates the dangers inherent in this dynamic. In 2005, the *New York Times* revealed TSP, which authorized warrantless surveillance on a massive scale that clearly violated the dictates of FISA and the Patriot Act. Initially, the Bush administration argued that the program was legal, citing the 9/11 attacks and national security exigencies. Ultimately a simpler path was taken: The administration went to and received *ex post facto* approval from the FISA court. A moment of accountability—the Bush administration's acknowledgment that it had engaged in illegal surveillance—was side-stepped by simply seeking the approval of the court charged with monitoring the illegal activity, thereby making the executive branch unaccountable yet again.[53]

The structure of the FISA court also took its toll on the accountability of the executive branch to the Congress, a consequence that also had been roundly predicted during the FISA debates. Democrats Drinan and California's Senator Eugene Tunney both warned that the reporting mechanism "would not give the Congress adequate information to exercise oversight over the executive branch." Congress, claimed Drinan, "would never know whether abuses—or successes—were occurring under the bill's provisions."[54] Opponents in the House, among them Republican John Ashcroft, who would become President Bush's Attorney General, warned that Congress "could easily be lulled into laziness, feeling that the court was adequately reviewing the situation."[55]

The overall impact of FISA's judicial structure was presciently stated even before the proposed legislation became law. Wyoming's Senator Wallop also contended that the executive branch would not be accountable to Congress because of the impact of a judicial imprimatur of validity:

---

52   Cong Rec H 5422 (daily ed June 3, 1977) (statement of Rep. Drinan).

53   E LaForgia, "US releases documents on NSA surveillance origins" JURIST, Paper Chase (Dec 21, 2013) available at <http://jurist.org/paperchase/2013/12/us-releases-documents-on-nsa-surveillance-origins.php> accessed Dec 2014.

54   S Rep No 94–1035 (1976) at 126.

55   HR Rep No 95–1283, Pt 1 (1978) at 117.

> In a sense the bill succeeds too well. Under it, each and every act of electronic surveillance authorized by the special court would be ipso facto legal ... what could any Congressman or Senator do about any act of surveillance he considered unjust or inappropriate? That act would have been not only requested under congressional standards, but certified as meeting those standards by a Federal judge. For all practical purposes, the Congressman or Senator would face res judicata [a matter finally decided on its merits by a court having competent jurisdiction].[56]

The new Drone Court's structure must avoid all of these problems and restore the separation of powers envisioned by the Framers. Otherwise, one cannot expect the public will have any more confidence that the targeting of US or foreign citizens alike is rooted in credible, rational facts than they have about determinations made by the FISA court.

An additional fundamental proposition bears noting before the proposed structure for the Drone Court is detailed. A government of the people, by the people, and for the people is a core value articulated by Lincoln in his Gettysburg address and known to every schoolchild in America. As basic and obvious as it may seem, its relevance to the protection of fundamental liberties in the context of foreign surveillance and drone strikes cannot be overstated. The simple fact is that secrecy has bred distance between the government and the governed in profound ways. Indeed, we have arrived at a point in our history where government actions related to intelligence gathering and drone strikes are only discernable if an individual commits a criminal act—Edward Snowden being the prime current example. While people disagree about his motives and how he should be viewed and treated, few would argue with the fact that but for Snowden's actions, the microscope under which we currently have placed our surveillance policies would not exist. That fact should tell us something: The government is acting in ways that do not respect Lincoln's words, which we drill into our children as a first lesson in civics.

The response to these assertions is that the threat is so extraordinary and the pace of technology so breathtaking that there is nothing that can or should be done to curb the government's expanding surveillance practices or its increasing use of drones. Yes, the argument goes, transparency in a democracy is important, and yes, the people should participate in the decision-making that affects their lives and rights, particularly the rights to expression and privacy, but those aspirations must give way if we are to protect our democratic values. Absolute secrecy and the pervasive gathering of information, aided by revolutionary technologies that keep us safe, the argument concludes, are necessary imperatives.

In reality, to accept that argument is to end the argument. Such reductionist logic can only result in the unbridled, unchecked authority of the executive branch. It also assumes that there is no middle ground such as FISA sought, but failed to

---

56 S Rep No 95–701 (1978) at 94.

achieve. We reject that logic and that conclusion in our proposal to establish a Drone Court.

IV. How Will the Court Work

To catalogue the FISA court's fundamental flaws is far easier than correcting them. This is precisely because there exists a legitimate question of how much the executive branch's decision-making can be challenged in the area of national security, particularly as it relates to operational counterterrorism. The struggle over the limits of the authority of the president's Article II powers dominated the FISA congressional debates and has already reared its head as the topic of a Drone Court is considered.[57] As noted, we believe that the president's authority to conduct foreign affairs must not trump the rights of those targeted and that an effective means of judicial review as well as congressional oversight must be established. To repeat: "The devil is in the details."

In this section we discuss the inner workings of the Drone Court with the underlying premise that it will function as an inherent part of the nation's judiciary. The Court will display the characteristics that we associate with American justice: a court acting independently and free from any and all external influence, operating as transparently as possible with adversarial processes to insure a serious search for credible and reliable facts. In this sense, there is an existing model—the nation's federal courts established pursuant to Article III of the United States Constitution, courts whose judges the Senate has confirmed and whose processes comport with due process.

Of course, we recognize that the Drone Court cannot replicate ordinary Article III courts and that significant differences will challenge and compromise constitutional due process requirements. The national security interests at stake, the nature of the determinations being made, and the fact that the target has neither been notified of the proceedings nor brought before the court are realities that will require procedures that would not pass constitutional muster in a federal district court criminal proceeding.

There is, however, one principle, that must not be compromised: The Drone Court must be independent from the executive branch. That fundamental proposition must be observed regardless of the sensitive context in which the Court is operating and the difficult circumstances in which it is forced to function. The experience of the FISA court conclusively demonstrates the consequences of doing otherwise.

While we believe that an independent means of judicial review is a *sine qua non* to implement the rule of law, we are under no illusion that the mere existence of a Drone Court independent from the executive branch will insure that the rule of

---

57   N Katyal & R Caplan, "The Surprisingly Stronger Case for the Legality of the NSA Surveillance Program: The FDR Precedent," (2008) 60 Stan Law Rev 1023.

law will be followed in all situations in which determinations about drone strikes are being made. The stakes are simply too high and the consequences too great to assume that fairness and considered decision-making will prevail in all cases. History demonstrates the folly of such a hypothesis.

Nonetheless, the opposite proposition can also be stated with certainty. Absent the exercise of judicial review by an independent court, it will be impossible to impose any check on the executive branch. We believe that the proposed Drone Court can make a significant contribution to the rule of law by serving as a proactive restraint on the executive in the context of operational counterterrorism where its authority has historically been unchallenged. If the principle of an independent judiciary is observed, the Drone Court will reflect the intentions of Madison and Hamilton, and in modern times, Justice Robert Jackson. As previously noted, Justice Jackson decried the "unfettered discretion of the executive branch" in his landmark 1952 Steel Seizure case opinion: "While the Constitution diffuses power the better to secure liberty, it also contemplates that practice will integrate the dispersed powers into a workable government. It enjoins upon its branches separateness but interdependence, autonomy but reciprocity."

Unfortunately, in the area of operational counterterrorism no such interdependence or reciprocity currently exists. The Drone Court can help to incorporate such principles and militate against the powerful and legally tenuous policy aspects of the current drone program, in the short and long term alike.

That said, we are not suggesting that the Court will totally supplant the executive branch's criteria-based, rational decision-making in determining whether a particular drone strike is warranted. Rather, the Court is intended to force the executive branch to proactively present intelligence information that is the basis for a planned drone strike. The Court's fundamental function will be to review and assess that information prior to the implementation of a specific attack in order to determine whether the planned attack comports with judicial standards and criteria.

The challenge posed in the establishment of a Drone Court is to navigate the intersection of executive power and judicial standards. It is at that intersection that the "rubber truly hits the road." Below we offer a roadmap through that controversial terrain predicated on our understanding of constitutional principles, executive power, and independent judicial review. That roadmap details key processes to maintain the delicate balance between national security, the appropriate use of military force, and individual liberty. To find that balance requires asking fundamental questions that include: What court are we talking about, and who will sit as judges? How does the court protect the 6th Amendment's guarantees of the right to confront and cross examine if the accused is not present and has not been notified of the action? What is the nature of the inquiry to determine credible, reliable facts on which to base a decision to kill? What is the standard of proof? How would a decision be appealed?

A. The Structure of The Court, Including its Trial And Appellate Divisions

As noted, we view adoption of the FISA model as a non-starter. The one non-negotiable principle of the Drone Court must be its independence of the executive branch in law and in fact. The FISA court on paper may appear independent of the executive branch, but its processes, including the manner in which judges are selected and its non-adversarial nature, belie such a conclusion.

To answer this first question, we considered and rejected the idea of a new Article III court, consisting of a cadre of newly appointed, Senate-confirmed federal District and Court of Appeals judges whose sole responsibility would be the evaluation of drone strikes. The idea is appealing but the reality of implementation fraught. The country is polarized to the point where even non-controversial federal judicial nominations cannot receive Senate confirmation.[58] That fact, combined with the fear that any such special court with lifetime judicial appointments will become ossified and isolated and captive of the executive branch, lead us to the scheme set forth below.

We propose that 24 current sitting Article III judges—12 from the District Court and 12 from the Court of Appeals be selected to hear cases in an Operational Security Court (OSC) charged with determining whether cause exists to engage in a drone strike to kill a government identified target or targets. The 24 judges will be selected randomly from all sitting federal judges in the United States and serve four-year terms. To ensure geographic diversity the circuits would be divided into four groups and six judges would be drawn from each of the geographical regions (three from the District Court and three from the Court of Appeals). As specific venue rules are adopted for the Court, it might be advisable to rotate the trials in each of the four geographic regions. During its initial phases, judges will be rotated off the Court in groups of six to ensure continuity of expertise as new judges are appointed. Should a judge be selected, his or her regular court duties would be adjusted to accommodate the additional OSC duties. The judge, however, would continue to maintain a caseload and to hear cases not originating in the OSC.

The government would initially file cases in the court's trial division where it will be heard by a panel of 3 of the 12 district court judges. Appeals would be heard in the OSC's appellate division by a panel of 3 of the 12 court of appellate judges. The decision of the appellate division's panel would be final, absent an agreement by 8 of the 12 appeals court judges that the matter should be heard *en banc* by all 12 appellate judges. The final decision of the Court of Appeals would be appealable to the United States Supreme Court.

---

58 S Davis and R Wolf, "U.S. Senate goes nuclear changes filibuster rules," USA Today (Nov 21, 2013) available at <http://www.usatoday.com/story/news/politics/2013/11/21/harry-reid-nuclear-senate/3662445/> accessed Dec 2014: "Senate Majority Leader Harry Reid, D-Nev, pushed through a controversial change to Senate rules that will make it easier to approve President Obama's nominees but threatens to further divide an already polarized Congress."

We suggest this structure for several reasons, including the fact that it eliminates a fundamental FISA flaw. The appointment of FISC judges is solely within the discretion of the Chief Justice of the Supreme Court, leading to a lack of diversity of viewpoint on the FISA courts[59] and charges that the court is biased.

We recognize that random selection of judges and term limits raises the legitimate question of whether a particular judge has sufficient expertise to sit on such a court, understanding that military decisions will require specialized knowledge. We suggest two responses to that concern. First, the OSC will have an office staffed with senior military personnel to advise the court on how to evaluate evidence with regard to any proposed strike.

Second, and perhaps most critically, the only way to truly ensure that people with military expertise make the decision is to put the decision-making responsibility in the hands of the military itself, a move which would eviscerate the fundamental principle of the proposed Drone Court and one which we believe a *sine qua non* for the survival of democratic institutions—an independent judiciary capable of engaging in meaningful judicial review. Democracy, it turns out, has its risks and we believe it critical to consistently err on the side of the equation that can best protect the fundamental values and liberties that bolster our democratic institutions.

B. How Does the Court Protect The 6th Amendment's Guarantees?

While the 6th Amendment guarantees a defendant the right to confront his or her accuser, the protection of that right is, necessarily, not possible in the judicial proceeding we envision. Obviously, the target will not be notified of or present at the proceedings. In the past, our courts have struggled with just how much the 6th Amendment may be compromised in a criminal trial, most notably in the areas of the admissibility of confessions and the validity of exceptions to the hearsay rule.[60] The Drone Court would be very different, completely denying the targeted defendant who has already been sentenced to death by the executive branch with a total denial of access to the forum in which he or she is to be tried. How to handle that situation is clearly one of the biggest hurdles that the Drone Court faces.

To ensure that the intended target is represented in a meaningful adversarial judicial proceeding, we envision that attorneys (who have received the necessary security classification) will represent the target's interests *in absentia* and engage

---

59   The Foreign Intelligence Surveillance Court, WP Politics, *The Washington Post*, available at: <http://www.washingtonpost.com/politics/the-foreign-intelligence-surveillance-court/2013/06/07/4700b382-cfec-11e2–8845-d970ccb04497_graphic.html> accessed Dec 2014; see also R Barnes, "Secrecy of surveillance programs blunt challenges about legality," *The Washington Post* (June 7, 2013) available at <http://www.washingtonpost.com/politics/secrecy-of-surveillance-programs-blunt-challenges-about-legality/2013/06/07/81da327a-cf9d-11e2–8f6b-67f40e176f03_story.html> accessed Dec 2014.

60   *People v Aranda*, 63 Cal.2d 18, 407 P2d 265 (1965); *Bruton v United States*, 391 US 123 (1968).

government witnesses in a rigorous cross-examination—the cornerstone of the search for truth in our adversarial system of justice. The cross-examination will be conditioned on the executive branch sharing with both the Court and the target's attorney the intelligence information in its entirety in a timely manner to allow for adequate preparation by the target's counsel. In addition, counsel for the target would have access to military experts designated by the court who would brief counsel as necessary with regard to the meaning and content of the intelligence reports. We recognize that this solution is not perfect and surely does not comport with 6th Amendment standards. Nonetheless, we believe it a workable solution given the context in which the court's processes are being developed.

C. Burden of Proof

We have considered various burdens of proof that might apply to determine whether the government has made a sufficient showing to justify killing a target, including existing criminal law standards. In developing an appropriate burden of proof to place on the government, we recognize that we are creating a non-traditional judicial paradigm and that any judicial architecture we propose is a treacherous minefield.

Our thoughts about the appropriate burden of proof are also driven by two daunting realities. First, while our proposal incorporates a modicum of representation for the target, such representation falls far short of the standard envisioned by the Confrontation Clause, both in letter and in spirit. Representation is significantly limited because the process will be conducted *in absentia* and the cross-examination focuses on information provided solely by the government. Second, the Court will rule on the legality of the gravest of acts—a person-specific act of operational counterterrorism intended to assassinate. These two realities demand that the government meet a difficult burden of proof consistent with international and domestic law and norms. To *not* impose such a burden is akin to recreating the failed model of the FISA court.

In light of these realities, we proposed a burden of proof of "operational strict scrutiny" predicated on the following principles:

- The *definition of target* must be narrowly defined and applied;
- There must be a *compelling state interest* to kill the target;
- The definition of *imminence must be narrowly defined* and applied;
- That compelling interest must include a showing that there is an *imminent and significant threat to citizens of the United States or its allies*;[61] and,

---

61   See section E below which details a process for unilateral action by the executive branch followed by *post hoc* review to determine whether the perceived threat was in fact "imminent." The exception recognizes that in limited circumstances there may exist a need for such unilateral action prior to a court determination of "imminence." That said, the critical importance of a proper definition of imminence and a finding that imminence exists

- Application of the principle of "alternatives" is essential; that is, the executive must convince the Court that national security will be harmed if the individual is *not* killed which in turn suggests that the executive must demonstrate that the target cannot be captured.

In sum, the standard of operational strict scrutiny is premised on the ability of the government to demonstrate that the information provided by the source is reliable, credible, time relevant and corroborated. Operational strict scrutiny assumes a compelling presentation by the government, subject to cross-examination by appointed counsel and active interjection by the Court, to demonstrate that the intended target is indeed a legitimate target and that the purported threat he or she poses requires and justifies his or her immediate death.

As it's name implies, operational strict scrutiny requires a traditional "strict scrutiny" analysis by the Drone Court, that is, the demonstration that a compelling state interest exists and that the action contemplated is narrowly tailored, that is, that government objectives cannot be achieved by less draconian alternatives (such as capture of the target.) While this burden, undoubtedly, poses significant evidentiary challenges for the executive, it is important to recall that the Court is being asked to authorize the killing of the target. To not impose a strict scrutiny standard on the executive is to continue current administration drone policy that is characterized by unitary executive action, devoid of robust judicial review and transparency, and lacking in narrowly defined criteria and standards.

D. Determining Credible, Reliable Facts on which to Base a Decision To Kill?

The question of the nature of the inquiry to determine credible, reliable facts on which to base a decision to kill requires both a procedural and substantive analysis. With regard to process, the question revolves around evidentiary rules to be applied and special court processes that will need to be devised given the high level security necessarily involved in the inquiry. The trial or hearing would not be open to the public, but during the course of the trial the need for other extraordinary measures might be required, including special *in camera* hearings as well as taking necessary steps to protect the sources. We suggest that those procedures be left to the discretion of the court and developed over time as the Court gains experience with judicial review of targeted killings.

With regard to evidentiary rules, we believe it imperative that the Federal Rules of Evidence (FRE) apply to the proceedings as much as possible. As evidentiary scholars are aware, the Federal Rules, unlike common law rules of evidence and most state evidence codes where the Federal Rules have not been adopted, are intended to be flexible and less rigid than their state counterparts. This is true in

---

cannot be overstated. Imminence is a critical component of operational strict scrutiny only to be excused in limited circumstances subject to a subsequent determination of whether such breach of the requirement was justified.

the areas of relevance, character, and hearsay (see e.g. FRE 807). We believe that the structure of the FRE provides sufficient flexibility for the Court to take into account the security needs of the nation and the interests of the target.

The substantive analysis involved in determining whether the government has met its burden to justify a targeted killing does not lend itself to an exhaustive list of factors to be considered. Every case will turn on its own facts. That said, various areas must be explored in every case to evaluate the substance of the intelligence information provided and the credibility of the source(s) providing the information. In sum, the reliability, viability and relevance of the information along with the credibility of the source must be fully examined through rigorous cross-examination.

The chart below provides examples of lines of questions to be explored with regard to each of these factors.

Table 13.1   Lines of questioning

Test prong	Definition/use
Reliability[1]	Past experiences show the source to be a dependable provider of correct information; requires discerning whether the information is useful and accurate; demands analysis by the case officer whether the source has a personal agenda/grudge with respect to the person identified/targeted.
Viability	Is it possible that an attack could occur in accordance with the source's information? That is, the information provided by the source indicates a terrorist attack that could take place within the realm of the possible and feasible.
Relevance	The information has bearing on upcoming events; consider both the timeliness of the information and whether it is time sensitive imposing the need for an immediate counterterrorism measure.
Corraboration	Another source (who meets the reliability test above) confirms the information in whole or part.

Note: [1] With regard to the reliability and credibility of the sources, it should preliminarily be noted that there are three commonly accepted sources of information relevant to operational counterterrorism: human intelligence (HUMINT); signal intelligence (SIGINT); and open source information. The first two are particularly relevant to the Drone Court. HUMINT, in large part, consists of sources who provide information to the nation state for a variety of reasons whereas SIGINT is the gathering of information through a variety of means including interception of communications and electronic monitoring of individuals.

In addition, the following kinds of questions must be explored, keeping in mind that source protection is essential to continued and effective intelligence gathering.

- Questions regarding the substantive allegations:
  - What is the nature of the suspicious activity?
  - Does the information justify immediate action? Does the information suggest involvement in significant acts of terrorism justifying immediate counterterrorism measures? Or is the information more suggestive than concrete?
  - Is the information sufficient to justify a targeted killing? Would other measures suffice, for example, capture and detention?
  - What are the risks/cost-benefits if the targeted killing is delayed?
  - How time relevant is the source's information?
  - Does the information suggest that the target is not prepared to engage in significant action and that more information should be gathered rather than engaging in an immediate strike? That is, what additional information could be gleaned (premised on the operational feasibility of detention rather than authorizing a targeted killing)?
  - Questions regarding the source:
  - What is the source's background and how does that affect the information provided?
  - Does the source have a grudge or personal "score" to settle?
  - What are the risks to the source by targeting the individual?
  - What are the risks to the source if the intelligence is made public?
  - Does or should the source dictate the appropriate venue for the trial?
  - Does the individual possess information—to varying degrees of specificity—relevant to future acts of terrorism/individuals?

E. "Imminent" Threats: Balancing the Realities

We recognize that our proposal for a Drone Court will inevitably lead to the claim that it is unrealistic and impractical, and, perhaps, even dangerous to the security of the nation. Central to that critique is the notion that the executive branch must be free to act on a moment's notice when the threat is "imminent," uninhibited by any external prior constraint, including judicial approval. The critique continues that even if the executive branch were to involve the judicial branch in the decision to authorize a lethal drone strike, the time pressures involved would not permit a court to make a considered decision.

We do not argue with the notion that an "imminent threat" to national security may require compromises in the juridical process as both a practical and a constitutional matter. We do not suggest that the Drone Court step between the president and their recognized constitutional power to respond immediately to imminent threats to the security of the United States.

We do, however, strenuously contend that the definition of what constitutes an "imminent threat" should be consistent with the history of that phrase and its commonly understood meaning in the context of international law. No president should be free to define "imminent threats" in a subjective manner that stretches the meaning of the phrase beyond any recognizable definition that justifies *any* executive action, including the lethal use of drones, regardless of the actual factual context in which the attack is carried out.

In Chapter 6, Steven J. Barela details the historical development of the phrase "imminent threat," tracing the evolution of "objective," "demonstrable," and "verifiable evidence" as a foundation for claims of self-defense and reminding us that the right to self-defense set forth in Article 51 of the United Nations Charter does not sanction unverifiable, subjective actions by the executive branch. Barela writes:

> As [the legal philosopher H.L.A.] Hart explained, every moral and legal code excludes free, unrestricted violence: "Social life with its rules requiring such forbearances is irksome at times; but it is at any rate less nasty, less brutish, and less short than unrestrained aggression for beings thus approximately equal." If we extrapolate this to the international plane, Article 2(4) of the UN Charter must include a reciprocal limitation. ... It cannot be the case that the exception of self-defense found in Article 51 allows for a wholly subjective and unverifiable trigger to military action.[62]

These concepts, concludes Barela, are "central to the definition of imminence."

The history of the phrase "imminent threat" and its accepted legal foundations have been undermined in their entirety by the executive branch as it seeks to maintain its unilateral, unchecked control of the drone attacks it routinely orders. Indeed, the DoJ White Paper (discussed below), candidly admits to the flexibility it employs to determine whether a threat is "imminent." The White Paper notes that imminence does not "require . . . clear evidence [of] a specific attack . . . in the immediate future."[63] Al-Qaeda's "continually plotting attacks," argues the Justice Department, satisfies any requirement that a threat be "imminent."

The White Paper's standardless definition of an imminent threat is coupled with a remarkably stringent requirement regarding the feasibility of capture which, according to the DoJ, must include a fact-specific analysis—a demonstrably precise and verifiable analysis that the DoJ is willing to forego in its definition of immanency. Such a sieve-like definition renders hollow the DoJ's final conclusion that any lethal drone strike must observe the four law-of-war principles for using force: (1) necessity, (2) distinction, (3) proportionality, and (4) humanity.[64]

---

62   See Chapter 6 at 141 in this volume.
63   DoJ White Paper (n 9) at 7.
64   Ibid at 8.

In light of the White Paper's definition of "imminence," the definition offered by Attorney General Eric Holder in March 2012 (in perfect coincidence with the leaked White Paper) is predictable. Attorney General Holder stated:

> The evaluation of whether an individual presents an "imminent threat" incorporates considerations of the relevant window of opportunity to act, the possible harm that missing the window would cause to civilians, and the likelihood of heading off future disastrous attacks against the United States.[65]

The Attorney General's definition, like the White Paper before it, contradicts the term's historical development and its common meaning in international law. Considerations of the "relevant window of opportunity to act" and "the likelihood of heading off future disastrous attacks" are nowhere defined, leading Barela to rightly conclude that such vague phrases "cannot possibly limit the use of force in the self-administered system" of decision-making that exists in the insulated offices of the executive branch, casting significant doubt on the legitimacy of government claims of self-defense, claims that alternatives other than lethal drone strikes were unavailable, and claims that the use of state power was proportionate to the threat. As we have noted, the administration's opacity with regard to the definition of "imminent threat" opens wide the door for the excessive use of state power.

Thus, while we agree that in the case of "imminent threats" a president may have the power to act unilaterally, we only acknowledge that right if the definition of the phrase is consistent with the modern understanding of the term and accepted international legal norms. Neither pre-condition has been met by the Bush or Obama administrations whose definitions of an "imminent threat" have distorted its meaning beyond recognition. Thus as a first step in dealing with the legitimate question of how a Drone Court should handle "imminent threats," its meaning must be restored to include demonstrable, verifiable, and specific evidence that such an imminent threat in fact exists.

Even if that goal is achieved, we also understand that there will be no perfect solution and that a delicate balance needs to be achieved between the executive branch's constitutional authority to act and the judicial branch's right to review executive action to insure that it is justifiable and not an abuse of power. In section V of this chapter, we propose a scenario that assumes significant time constraints, a matter of hours. Even in that scenario we believe that some accountability can be attained in the expedited hearing format that we describe. Of course, there may be situations where the immanency of the threat—properly defined as detailed above—prevents any manner of review at all. Given the planning that the

---

65 Eric Holder, Attorney General, Speech at Northwestern University School of Law (March 5, 2012) available at <http://www.justice.gov/iso/opa/ag/speeches/2012/ag-speech-1203051.html> accessed Nov 2014.

administration currently claims goes into drone attacks, such instances should be minimal.

However, should such a situation present itself and the executive branch believes that it must act unilaterally, we suggest a two-step process. First, the president would file with the Drone Court an affidavit under seal detailing precisely why he or she considered the threat "imminent" and the evidence on which that belief is based. After the conclusion of the strike, the Drone Court would engage in a *post hoc* review to determine whether the executive branch's determination of its need to act unilaterally to eliminate an "imminent threat" was supported by the demonstrable, credible, and reliable information that the modern history of the phrase and international law demand. Should such a showing be absent from the *post hoc* review, the Court could issue a contempt citation as a written declaration that the president had failed to meet his or her burden with respect to exercising the "imminent threat" exception that we hesitatingly incorporate into our proposal.

Such a procedure would have two salutary effects. First, knowledge by the executive branch that such a *post hoc* review by the Drone Court is required might serve as a check on unlawful unilateral actions in the future. At a minimum, it would create a level of transparency about the information supporting drone attacks which is completely lacking at the present time—a transparency critical to maintaining the checks and balances that serve as the foundation for our democracy.[66] Indeed, it would be a simple matter for the Court to publish summary statistics of affidavits filed by the executive branch and whether the Drone Court upheld the executive branch in its *post hoc* review. To repeat, however, given the meticulous planning that the executive branch currently claims goes into a drone strike, we believe that the need to employ the "imminent threat" exception should be infrequent.

V. Meat to Bones: A Hypothetical Fact Scenario

How will the Drone Court actually work and is it feasible? Explaining by a hypothetical fact situation is, perhaps, the most effective means to bring our proposal to life. We present here a scenario that involves time pressures that put our model to a difficult test. We do so to highlight how it will operate and the limitations that are necessarily involved. Regardless of those limitations, we believe that accountability constrained by time pressures is better than no

---

66   For a general discussion of the White Paper, see Mary Grinman, Laura Fishwick (ed) in Jolt Digest "Department of Justice White Paper Reveals United States Position on Lethal Force Operations Targeting US Citizens" available at <http://jolt.law.harvard.edu/digest/national-security/department-of-justice-white-paper-reveals-united-states-position-on-lethal-force-operations-targeting-u-s-citizens-abroad> accessed Dec 2014.

accountability at all, and, that in most cases, the time pressures will not be as severe as we assume here.

The Scenario

According to intelligence analysts, within the *next five hours,* a target previously under surveillance will be traveling by car. The individual has been under surveillance because of his previous involvement in a terrorist attack and because sources have identified him as involved in this new terrorist attack believed to be in its final planning stages. Given the target's previous involvement and planning for this imminent attack, the executive branch has defined the individual as a legitimate target and ordered the CIA and Department of Defense to make necessary arrangements for a drone attack at the earliest feasible opportunity.

In the context of this scenario, we envision that as soon as the decision to implement a drone attack has been made, the executive branch will notify the Drone Court and request an immediate hearing. The Court would immediately notify a counsel who has been pre-cleared to represent a target. The trial, before a panel of three judges from the Drone Court's trial division, will be scheduled immediately. In the event that the judges are not physically in the same location, the hearing could be conducted electronically (via secured, closed circuit channels).

At the trial, a senior official from the executive branch (presumably from the Department of Defense or other agency designated by the president) will present to the Court *all* relevant intelligence information that is being used to justify the attack. The information would be initially reviewed by the Court and then turned over to defense counsel. The information would be presented to the court by government intelligence analysts and/or senior operational commanders who would be subject to cross-examination by the target's attorney.

The hearing will be conducted in a manner that would allow both sides to fully present their respective cases. While the hearing would be conducted behind closed doors, its rulings would be made public, albeit with sensitive, classified information redacted.

In applying the operational strict scrutiny standard, the Court will assess the following separate, yet related, issues: the reliability of the source; the reliability of the information provided; and, the operational necessity of the proposed attack. That inquiry will include whether corroborating evidence exists to support the determination that the target poses a danger commensurate with the decision to assassinate him or her. Moreover, the Court would apply principles of alternatives, collateral damage, proportionality and military necessity to determine whether the operational decision is justified in accordance with international law. Both domestic and international law would be relevant to the ultimate determination.

As noted above, the burden on the government is deliberately heavy. The strict scrutiny test is intended to ensure narrow application of the drone policy in accordance with the restraints outlined. While the determination of whether a compelling state interest exists necessarily involves considerations of "national security," not all claims of "national security" would justify killing the target; that determination would ultimately be made by the Drone Court in a manner consistent with principles of the rule of law. While "national security" may justify pre-emptive self-defense, whether in accordance with Article 51 of the UN Charter or commonly accepted principles of international law, the definition's legitimacy ultimately depends on its limited application subject to meaningful review.

In the hypothetical presented here, the Court's decision would have to be rendered with speed, and the ability to engage in a meaningful appeal process would be challenging. Nonetheless, if the Drone Court's Appellate Division and United States Supreme Court were put on notice that the proceeding was in progress, the ability to have others reviews the trial court's decision would not necessarily be precluded. It would be necessary of course to provide sufficient time for the government to implement the requested operational plan once judicial approval is obtained.

As noted, this scenario is intentionally extreme and intended to demonstrate how the Drone Court might be forced to act under pressure. The hypothetical clearly demonstrates the limitations that time constraints will necessarily impose on the Court. To those critics who claim that the process is unworkable we offer two thoughts. First, in most instances, as noted, we presume that the time constraints would not be as severe as those posed here. Second, even if the time constraints are severe, our democratic institutions demand an attempt to hold accountable those who make decisions that take the lives of others, particularly when it might involve the loss of innocent lives. The alternative of unfettered, unitary executive power is simply unacceptable if democracy is to survive.

VI. A Concluding Thought

No one wants another 9/11 or any other terrorist act that harms American citizens or its allies. Moreover, few would doubt the ability of the military to glean information about national security threats that are beyond the reach of the civilian population. Those propositions make it difficult to resist the temptation to leave it to the professional intelligence community—the military, the CIA, and the Department of Defense—to make national security judgments about what is in the best interests of the nation, including decisions as to whether an order to kill should be implemented.

It is our belief, however, that, no matter how difficult the task, the distinguishing feature of a vibrant democracy is the ability to resist that temptation and to demand that decisions that might compromise individual rights be subject to meaningful review consistent with the cherished values of our constitutional democracy for which countless lives have been lost over the past two and one half centuries. We recognize that the task is a difficult high-wire act balancing national security and individual rights as well as the proper functions of the branches of the government. Our proposed Drone Court seeks to achieve that proper balance consistent with the Framer's intent, the admonitions of the Supreme Court in cases like *Hamdi* and *Boumediene*, and, most importantly, the values of the Republic that our actions in the name of national security are intended to protect.

Chapter 14

Can UCAVs be Reconciled with Liberal Governance?: The Substantive Law of a Drone Court

Tom Farer and Frédéric Bernard

Whatever one's philosophical or even theological position, a society is not the temple of value-idols that figure on the front of its monuments or in its constitutional scrolls; the value of a society is the value it places upon man's relation to man.

Maurice Merleau-Ponty[1]

I. Introduction: Extraordinary Measures and the Liberal State

The occurrence of a terrorist act inclines governments to initiate preventive measures that grate against human rights. The goal of preventing new terrorist attacks stems partly from the government's responsibility to guarantee its population's safety and partly from the pressure coming from the population to *feel* reassured, a feeling that can at least initially be fostered by violent activity culminating in dead bodies that can plausibly be identified as those of terrorists.[2] At the same time, governments can exploit understandable popular anxiety to tighten their grip on power and, if their democratic vocation has always been notional, suppress criticism and political competition.[3]

In this chapter we address the question of the extent to which it is possible to normalize a particularly problematic special measure, namely targeted killing, without coincidentally subverting the general principles undergirding liberal governance. One of those principles is acceptance of a certain level of societal risk in order to limit the state's capacity to intervene in the quotidian life of the

---

1 Maurice Merleau-Ponty, *Humanism and Terror: The Communist Problem* (John O'Neill trad.) (Transaction Publishers 2000) at xiv.

2 Bruce Ackerman, *Before the Next Attack: Preserving Civil Liberties in an Age of Terrorism* (Yale Univ Press 2006) at 2.

3 David L Altheide, *Terrorism and the Politics of Fear* (Lanham 2006) at 15. To describe this phenomenon, Olivier de Schutter uses the words "windfall effect" ("*effet d'aubaine*"), see Olivier de Schutter, "La Convention européenne des droits de l'homme à l'épreuve de la lutte contre le terrorisme" [2001] Revue universelle des droits de l'homme 185, at 204.

citizenry.[4] Through the accumulation of measures adopted to prevent terrorist acts, a liberal state can undermine itself and over time cease to be liberal;[5] in other words, fundamentally alter what Maurice Merleau-Ponty once called "the value it places upon man's relation to man."

Targeted killing is only one of the array of measures designed to interrupt as early as possible any process that may—or may not—lead to a terrorist act.[6] States' ambition to intervene precociously against individuals and organizations perceived to be hostile and potentially violent is mirrored by aggressive extensions of criminal law to the end of "surrounding" the terrorist act itself by criminalizing related behaviors. To use Walter Laqueur's metaphor, the idea is to "[d]rain the swamp, [so that] the mosquitoes will disappear"[7]. To do so, states have criminalized membership in or financial contributions to nominally philanthropic or social-service organizations deemed supportive of terrorism, as well as speech deemed to incite terrorism.[8] In short, the concept of criminal conspiracy is considerably, one might say dangerously, extended.

In addition governments have increased surveillance of the citizenry, as exemplified in several secret programs (such as PRISM) authorized by the Bush administration and maintained by President Obama to monitor communications.[9] Moreover, with respect to persons suspected of involvement in terrorist conspiracies, some states have loosened, or been accomplices to a loosening of, the prohibition of brutal and degrading interrogation.[10]

---

4   Aharon Barak, *The Judge in a Democracy* (Princeton Univ Press 2006) at 284–285.

5   See the words written in 1943 by the English Catholic thinker Christopher Dawson in *The Judgment of the Nations* (Sheed and Ward 1943) at 8, as quoted by Tom Bingham, *The rule of law* (Penguin 2010) at 159: "As soon as men decide that all means are permitted to fight an evil then their good becomes indistinguishable from the evil that they set out to destroy."

6   David Cole and Jules Lobel, *Less Safe, Less Free: Why America is Losing the War on Terror* (The New Press 2009) at 28: "[V]irtually all of the Bush administration's most controversial initiatives in the war on terror have been defended in preventive terms."

7   Walter Laqueur, *No End to War: Terrorism in the 21st Century* (Continuum 2003) at 11.

8   See the UN Convention, and Security Council Resolution 1373 (2001). See Section 1 *Terrorism Act* (2006): "This section applies to a Statement that is likely to be understood by some or all of the members of the public to whom it is published as a direct or indirect encouragement or other inducement to them to the commission, preparation or instigation of acts of terrorism." Compare Kent Roach, "Must We Trade Rights for Security ? The Choice Between Smart, Harsh, or Proportionate Security Strategies in Canada and Britain" [2005–2006] Cardozo Law Review 2151, at 2176.

9   James Bamford, "They Know Much More Than You Think," *The New York Review of Books* (Aug 15, 2013).

10   David Luban, "Liberalism, Torture and the Ticking Bomb," [2005] *Virginia Law Review* 1425, at 1436–1440. See also Mark Bowden, "The Dark Art of Interrogation," *The Atlantic Monthly*, Oct 2003: "To counter an enemy who relies on stealth and surprise,

Another common preventive measure is the relaxation of rules regarding preventive detention. After 9/11, for instance, the United Kingdom adopted the Anti-Terrorism, Crime and Security Act (2001), Part 4 of which allowed the indefinite detention without charge or trial of immigrants suspected of terrorist connections and who could not be sent back to their country of origin because of the non-refoulement principle.[11] Then, after the July 2005 bombings in London, Prime Minister Tony Blair proposed to extend the length of pre-charge detention from 14 to 90 days, a proposal ultimately rejected by the House of Commons due to an unusual breaking of ranks by members of the majority party.[12] Still another problematical move illustrated by Britain's escalation of preventive tactics is its relaxing of constraints on the use of lethal force by members of the security services. This relaxation in the UK became public knowledge following the killing in the London underground of a Brazilian citizen mistaken for a terrorist bomber.[13]

Popular acceptance of aggressive measures threatening to human rights often stems not only from the trauma inflicted by attacks on civilian targets with their implicit message that no one is safe, but also from the application of exceptional measures primarily against members of an easily identified minority group which shares certain generic characteristics with the perpetrators. A twentieth-century instance in the United States is the sweeping measures taken in 1922 by the US authorities against immigrants (mostly from eastern and southern Europe) after a series of anarchist bombings (the so-called Palmer Raids).[14] After 9/11, the use of terrorist profiles inevitably had a disproportionate effect on Muslims in various Western countries. Obviously, the extensive use of profiles rubs along uneasily with a cardinal value of Western liberal democracies: the principle of non-discrimination on grounds of national origin, religion, or skin color.

Framing the debate as a struggle between "us" and an ethnically, religiously, or racially defined "them" coincidentally tends to produce a discursive characterization of the problem as one of the rights of the population as a whole balanced against the notional ones of a few dangerous individuals. That framing

---

the most valuable tool is information, and often the only source of that information is the enemy itself."

11   Mark Elliott, "United Kingdom: Detention without trial and the war on terror" [2003] International Journal of Constitutional Law 334, at 342–343. This provision has been struck down by the House of Lords in its judgment *A (FC) v Secretary of State for the Home Department* (2004).

12   Matthew Tempest, "Blair defeated on terror bill" *The Guardian* (Nov 9, 2005). Detention length was, however, extended to 28 days. See Stephanie Cooper Blum, "Preventive Detention in the War on Terror: A Comparison of How the United States, Britain, and Israel Detain and Incapacitate Terrorist Suspects" [2008] 3 Homeland Security Affairs 1, at 17.

13   Conor Gearty, *Civil Liberties* (OUP 2007) at 83–84; Tariq Ali, *Rough Music: Blair, Bombs, Baghdad, London, Terror* (Verso 2005).

14   Cass R Sunstein, *Laws of Fear: Beyond the Precautionary Principle* (Harvard Univ Press 2005) at 205.

predictably paves the way for the conclusion that, given the disproportion between these two competing values, the interests of the majority should prevail against what come to be seen as the merely "conditional" rights of individuals or suspect minorities. This reasoning has been used, for instance, to defend the use of torture and other cruel and degrading measures after 9/11.[15] And of course it underlies the rationale for targeted killing.

The argument generally invoked to justify radical measures is that limits on state power—including human rights and law-of-war norms—cannot be applied or at least must shrink in times of crisis, since in Clinton L. Rossiter's words, they constitute "luxury products."[16] Governments explicitly or implicitly take the position that, however regrettable, certain measures are indispensable in practice to protect the population from terrorist attacks.[17]

If one accepts, as we do, Louise Richardson's conclusion that the causes of terrorism lie not in objective conditions of poverty, but in *subjective perceptions of persecution and injustice,*[18] radical preventive measures can function as biographical triggers precisely because those measures bypass norms restraining the unjust exercise of state power. Consequently, they can generate an emotional link between the terrorists and the wider community from which the former seek recruits, funds, and, at worst, non-cooperation with the authorities. The link is a shared sense of victimhood.[19]

------

15   Stefan Sottiaux, *Terrorism and the Limitation of Rights: The ECHR and the US Constitution* (OUP 2008) at 8.

16   Clinton L Rossiter, *Constitutional Dictatorship: Crisis Government in the Modern Democracies* (Princeton Univ Press 1948) at 5: "[T]he complex system of government of the democratic, constitutional State is essentially designed to function under normal, peaceful conditions, and is often unequal to the exigencies of a great national crisis. Civil liberties, free enterprise, constitutionalism, government by debate and compromise—these are strictly luxury products."

17   A complementary argument holds that terrorist acts are "successes" whose energizing effects are likely to produce new terrorist recruits and funds. See Michael Walzer, *Arguing about War* (Yale Univ Press 2005) at 138. The end, however, is identical.

18   Louise Richardson, *What Terrorists Want* (John Murray 2006) at 14. Compare with the notion of "relative deprivation" developed by Ted Robert Gurr, *Why Men Rebel* (Princeton Univ Press 1970) at 24: "Relative deprivation is defined as actors' perception of discrepancy between their value expectations and their value capabilities. Value expectations are the goods and conditions of life to which people believe they are rightfully entitled. Value capabilities are the goods and conditions they think they are capable of getting and keeping."

19   Ibid, Richardson at 251 (about pictures showing evidence of prisoners' abuse in American-controlled prisons such as Abu Ghraib): "We will never know just how many young men were moved to join terrorist groups in anger and outrage at these photographs, nor how many others resolved that they would never lift a finger to help the US in its campaign against terrorism. I believe we can safely assume that the numbers are very large. These photographs, and the failure to repudiate them conclusively by holding the

II. Targeted Killing

We propose restraints on summary execution, whether by drone or other methods, not only in order to defend human rights for their own sake, but also in the belief that they could reduce the terrorist threat. For it seems to us likely that extraordinary measures may actually aggravate the all-to-real danger. As David Kilcullen, a well-known authority on counterterrorism and a former adviser to General David Petraeus, has warned, there is

> a "ladder of extremism" that shows the progress of a jihadist. At the bottom is the vast population of mainstream Muslims, who are potential allies against radical Islamism as well as potential targets of subversion, and whose grievances can be addressed by political reform. The next tier up is a smaller number of "alienated Muslims," who have given up on reform. Some of these join radical groups, like the young Muslims in North London who spend afternoons at the local community center watching jihadist videos. They require "ideological conversion"—that is, counter-subversion, which Kilcullen compares to helping young men leave gangs ... A small number of these individuals, already steeped in the atmosphere of radical mosques and extremist discussions, end up joining local and regional insurgent cells, usually as the result of a "biographical trigger—they will lose a friend in Iraq, or see something that shocks them on television." With these insurgents, the full range of counterinsurgency tools has to be used, including violence and persuasion. The very small number of fighters who are recruited to the top tier of Al-Qaeda and its affiliated terrorist groups are beyond persuasion or conversion. "They're so committed you've got to destroy them," Kilcullen said. "But you've got to do it in such a way that you don't create new terrorists."[20]

A. On the Realism of International Law

Whatever else it may be, terrorism is violence that transgresses moral limits.[21] Since a governmental response emphasizing the summary application of lethal force risks transgressing moral limits, it correspondingly threatens to blur the distinction between legitimate and illegitimate violence.

That potential for blurring the distinction between authorized and private violence should incentivize governments to bind themselves publicly and in practice to employ lethal means strictly within the normative constraints which

---

most senior people responsible, has made the crucial task of driving a wedge between the terrorists and the communities that produce them immeasurably more difficult."

20   George Packer, "Knowing the enemy: Can social scientists redefine the 'war on terror'?" *The New Yorker* (Dec 18, 2006).

21   Alex P Schmid, *Political Terrorism: A Research Guide to Concepts, Theories, Data Bases, and Literature* (North-Holland Publishing Company 1984) at 109.

apply universally: the human rights and humanitarian law conventions, the principles of customary international law underlying those conventions, and the United Nations Charter. All are relevant because respect for international law extends both to a state's relationship with other states (the UN Charter) and to a state's relationship with not only its own nationals but with humanity as a whole, the bearer in our time of indissoluble rights. A persuasive declaration of intent to operate within the confines of global law also requires demonstrations of respect for interpretations of that law emanating from the courts and other bodies authorized to interpret it, such as the International Court of Justice, the Human Rights Committee established by the International Covenant on Civil and Political Rights (ICCPR), and the various regional human rights courts and commissions.

Persons eager to confront transnational terrorism by every means which strikes them as tactically efficient often disparage international law and especially human rights and humanitarian law and the UN Charter restraints on the use of force as rigid obstacles to an efficient counterterror strategy, obstacles built on Utopian premises in less dangerous times. Their arguments grossly understate the law's tolerance of exceptional albeit not unlimited measures for protecting the security of the citizen.

Take the example of secret surveillance. By its nature, secret surveillance reduces the potential scope of the right to privacy. What a liberal interpreter of the right must do is identify the optimal point for reconciling individual rights and community interests and devise or reinforce means for restraining excessive zeal. Restraint requires a balancing test applied by judicial or at least quasi-judicial bodies rather than the surveillance bureaucrats who face punishment for failure to detect real conspiracies but not for excessive zeal in interpreting their mandate.[22] Parallel to the assessment process in cases of collateral damage from artillery and aerial bombardment, agents of the state must weigh the degree of proposed encroachment on the right to privacy against the importance of the information sought, the probability that the proposed incursion on privacy rights will produce that information, and the difficulty of obtaining it by less intrusive means. And the burden of persuasion must rest on the state since it, after all, has far superior means for meeting it than an appointed or self-appointed representative of the public or the necessarily unknowing target.

Other rights, for instance to association and expression, also are subject to this balancing test. The test is written into the very statement of many rights in that

---

22   See, for instance, Article 8 § 2 European Convention of Human Rights (ECHR): "There shall be no interference by a public authority with the exercise of this right except such as is in accordance with the law and *is necessary in a democratic society in the interests of national security, public safety or the economic wellbeing of the country, for the prevention of disorder or crime, for the protection of health or morals, or for the protection of the rights and freedoms of others*" (our emphasis).

the ICCPR declares them subject to limits required for public health and security.[23] Moreover, most rights are derogable in times of emergency.[24] It is therefore apparent that international law provides states with considerable license to employ measures intended to prevent and/or deter. Only a few rights—principally, of course, the right to life, the right to protection from torture and cruel, inhuman and degrading treatment and the right to due process—are unqualified.[25] Not only the ICCPR, but also most regional human rights agreements, contain derogation clauses.[26]

They are there precisely to prevent emergency conditions from creating legal black holes; they are there, in other words, to inhibit abusive suspension or denial of rights.[27] Pursuant to these emergency clauses, when a state faces a public emergency or a war, it is allowed to suspend rights to the extent strictly necessary under the circumstances. In the context of Northern Ireland in the 1990s, for example, the United Kingdom derogated from the ICCPR and from the European Convention of Human Rights (ECHR) regarding the length of the period between the arrest of "terrorist" suspects and the judicial control of their detention.[28]

Nevertheless, some advocates argue that awful dangers stemming from new technologies make virtually any but the most elastic limits on executive discretion untenable.[29] It is true that individuals can cause far greater damage today than they ever could before. However, technological change is not one-sided and governments also enjoy today far more powerful technologies to identify and track dangerous persons, to effect their capture, and, where absolutely necessary, to eliminate them, the use of drones (UCAVs) being a paramount example. Thus it is not demonstrable that the march of technology requires governments to transcend individual rights in order to protect collective interests.

B. Needed: A Warrant to Kill

Although there exists no agreed upon definition of targeted killings, they may be usefully described as "the intentional, premeditated and deliberate use of lethal force, by States or their agents acting under color of law, or by an organized armed group in armed conflict, against a specific individual who is not in the physical custody

---

23   See Articles 19 (freedom of expression) and 22 (freedom of association) of the International Covenant on Civil and Political Rights (ICCPR).

24   See Article 4 ICCPR (state of emergency).

25   Human Rights Committee, *General Comment No 29*. See also the *Belmarsh* case (*A and others v Secretary of State for the Home Department* [2004] UKHL 56).

26   See, for instance, Article 15 ECHR and Article 27 American Convention on Human Rights (ACHR).

27   Joan Fitzpatrick, *Human Rights in Crisis: The International System for Protecting Rights During States of Emergency* (Univ of Penn Press 1994).

28   See *Brannigan and Mc Bride v UK*, App no 14553/89 (May 26, 1993).

29   See, for instance, Walter Laqueur, *The New Terrorism: Fanaticism and the Arms of Mass Destruction* (OUP 2000).

of the perpetrator."[30] To this point, it appears that US- and UK-controlled drones have carried out targeted killings in Afghanistan, Libya, Iraq, Pakistan, Yemen, and Somalia, while Israeli drones have conducted similar missions in Palestinian territories.[31] In the view of the current UN Report of the Special Rapporteur on the Promotion and Protection of Human Rights and Fundamental Freedoms while Countering Terrorism, the majority of these drone strikes have taken place in the context of armed conflicts as defined by international humanitarian law.[32] We will assess and consider the implications of his conclusion below.

The resort to drones is highly tempting for governments, firstly because they allow the killing of potential terrorists without any risk to their armed forces.[33] Secondly, targeted killing, if highly discriminating, appears less likely than periodic incursions by special forces to generate among the general population in the country where targets are located a feeling of being the objects of foreign aggression.

Whatever their operational virtues, killer drones are a weapon not a strategy. When employed outside the context of a large-scale and sustained armed conflict, in a literal sense they summarily execute persons believed to be engaged in criminal conspiracies.

Killing other than in large-scale combat between conventional forces or in defense of the self or of innocent third parties from imminent lethal threats is not easily reconciled with legal or, for that matter, moral norms. Targeted killing as defined above obviously negates the target's presumption of innocence as well as all other guarantees of a fair trial. In addition to this, targeted killings often cause the death of bystanders.[34] In doing so, targeted killings cross a crucial moral line, the one that bars the killing of innocent people.[35] Being *prima facie* "transgressive," as we suggested above they are probably one of the means—along with torture and limitless confinement without charge and trial—most likely to play the role of David Kilcullen's biological triggers.

Sharing our concerns about the legal character of drone killing, Professors Amos Guiora and Jeffrey Brand propose in the previous chapter of this book

---

30   Philip Alston, "The CIA and Targeted Killings Beyond Borders" [2011] *Harvard National Security Journal* 283, at 298.

31   Anthony Dworkin, "Drones and targeted killing: Defining a European position," in European University Institute, *RSCAS Policy Papers* PP 2013/77, Oct 2013, 47, at 47.

32   Ben Emmerson, Report of the Special Rapporteur on the Promotion and Protection of Human Rights and Fundamental Freedoms while Countering Terrorism, *Report*, UN Doc A/68/389, (Sept 18, 2013) at 22–23.

33   Kenneth Anderson, "Targeted Killing in U.S. Counterterrorism Strategy and Law" Brookings (May 11, 2009) at 8.

34   See the statistics provided by Emmerson (n 32). See also Human Rights Watch, "Between a Drone and Al-Qaeda: The Civilian Cost of US Targeted Killings in Yemen" (Oct 2013).

35   See Ronald Dworkin, *Is Democracy Possible Here?* (Princeton Univ Press 2006) at 27.

the creation of a US Drone Court, "to ensure that any drone policy is conducted in accordance with the rule of law." To that end, they argue, the procedure for choosing targets and executing strikes must be transparent and assure to the fullest extent possible that innocents are not targeted and that collateral damage, if any, is not disproportionate to the imperative of preventing a terrorist attack.

Structurally, the proposed Drone Court's framework is a creative effort, undertaken in the face of a severe challenge to the security of the citizen and the state, to preserve long-established individual rights by means of new procedures. The Court, which would have primary jurisdiction to authorize the killing of citizens and non-citizens alike, would be independent from the executive branch and composed of current Article III sitting judges. Adversarial proceedings would be, if not fully respected, at least roughly approximated by the appointment of security-cleared attorneys representing the targets *in abstentia* and having access to the entire intelligence information gathered by the executive branch. The Court's determination would be subject to appeal.

Whatever process of authorization and *post facto* review may evolve at both the national and international levels, even if it is no more than a stream of opinions from what the late Oscar Schachter, referring to the global community of international law scholars, judges and practitioners, called the "invisible college,"[36] it will be in need of a body of substantive law. Our purpose is to illuminate that body as best we can.

III. The Substantive Tests of Legality: A Holistic Approach

A holistic approach demands procedures, which draw on all the relevant parts of the international legal system, to contain the use of UCAVs. In the words of the previous Special Rapporteur, Philip Alston:

> When a State conducts a targeted killing in the territory of another State with which it is not in armed conflict, whether the first State violates the sovereignty of the second is determined by the law applicable to the use of inter-State force, while the question of whether the specific killing of the particular individual(s) is legal is governed by IHL [international humanitarian law] and/or human rights law.[37]

---

36 Oscar Schachter, "The Invisible College of International Lawyers" (1977) 72 NW Univ Law Rev 217.

37 Philip Alston, Special Rapporteur on extrajudicial, summary or arbitrary executions, *Report*, UN Doc A/HRC/14/24 Add.6 (May 28, 2010) at 11. See also Nils Melzer, "Keeping the Balance Between Military Necessity and Humanity: A Response" (2010) 42 NYU J Int'l L and Politics 829, at 897.

The right to life is thus surrounded by a double layer of legal protection. The outer layer consists of restraint on the use of force by states. Human rights norms and, in the event of an armed conflict, humanitarian legal norms constitute the inner layer. It follows that the legitimacy of any case of targeted killing is a function of its compliance with norms in both layers.

Some authors have suggested banning drones altogether on the grounds that they can reasonably be characterized as intrinsically "cruel" weapons and therefore violative of humanitarian law. Not surprisingly, this proposal finds very little support among legal scholars, much less practitioners.[38] Drones, after all, carry the same type of weapons as those found on piloted planes. If "cruel" is construed to include weapons that tend to aggravate the risk and harshness of collateral damage, then drones, because of their susceptibility to precise targeting, are potentially less cruel.[39]

A. Outer Layer: The Presumption against Armed Intervention

We join scholars from a variety of jurisprudential schools in treating Articles 2(4) and 51 of the UN Charter as the point of departure for assessing the legality of any projection of force by a state across national frontiers.[40] Like them, we conclude that there is a presumption against the legality of transnational force projection for purposes other than self-defense against an actual or imminent armed attack or with the permission of the government of the state where cross-border force is employed or under the authorization of the UN Security Council.

There is abundant disagreement about the elasticity of the Charter language. Disagreement ranges over a number of issues. The first and probably the most important for our purposes is what constitutes "imminence," a word that does not appear in the Charter.[41] In the early years after the Charter's adoption, it

---

38   In 2009, former senior law lord, Lord Bingham declared in an interview: "From time to time in the history of international law various weapons have been thought to be so cruel as to be beyond the pale of human tolerance. I think cluster bombs and landmines are the most recent examples. It may be—I'm not expressing a view—that unmanned drones that fall on a house full of civilians is a weapon the international community should decide should not be used." Quoted in Murray Wardrop, "Unmanned drones could be banned, says senior judge" *The Telegraph* (July 6, 2009).

39   See Michael N Schmitt, "Precision attack and international humanitarian law" 87 [2005] IRRC 445.

40   See Christine Gray, "The Charter Limitations on the Use of Force : Theory and Practice," in Vaughan Lowe, Adam Roberts, Jennifer Welsh, Dominik Zaum (eds), *The United Nations Security Council and War: The Evolution of Thought and Practice since 1945* (OUP 2010) at 86.

41   See Article 51 of the UN Charter: "Nothing in the present Charter shall impair the inherent right of individual or collective self-defence if an armed attack occurs against a Member of the United Nations, until the Security Council has taken measures necessary to maintain international peace and security. Measures taken by Members in the exercise of

was possible to argue that the right of self-defense could be exercised only after an attack had occurred, the interpretation actually proposed by Hans Kelsen, a leading international legal scholar of that era.[42] Other scholars countered by noting that the Charter referred to self-defense as an "inherent right" and argued that the word "inherent" should be construed in light of the occasions when the right to self-defense had previously been invoked by states. Since the claim of self-defense had not in the past been limited to instances of armed attack, much less imminent armed attack, and since waiting could put the defending state at a grave disadvantage, Kelsen's literal reading almost certainly did not reflect the intentions of the drafters.[43] They were, after all, unlikely to have adopted a suicide pact. The failure of a subsequent Soviet proposal to define aggression as being first to launch force across a border seemed to confirm the evolving view of scholars, reflecting the practice of states, that preemption is not illegitimate.[44]

A second area of disagreement, also important for assessing compliance with the outer layer of normative restraint on targeted killing, is whether episodic armed incursions which stop well short of seriously threatening the political independence and military capabilities of a state and do relatively little damage to the state's economic assets before the intruding forces withdraw (or are liquidated) can be aggregated into a continuing armed attack justifying a massive response by the targeted state at a time or times of its choosing, a response designed to destroy the other state's capacity for future incursions.[45] Can even a single incursion together with threatened future ones be treated as the trigger of an international armed conflict? If it can, then the victimized state's kinetic responses against the other state's combatants would not be preemptive or preventive but simply tactical moves in an ongoing "war."

Many scholars and, it would appear, the majority of judges of the International Court of Justice insist that legal preemption must satisfy the so-called Caroline principle.[46] Under that principle, as US Secretary of State Daniel Webster wrote in 1842, action can only be justified if there is "a necessity of self-defense, instant,

---

this right of self-defence shall be immediately reported to the Security Council and shall not in any way affect the authority and responsibility of the Security Council under the present Charter to take at any time such action as it deems necessary in order to maintain or restore international peace and security."

42   Hans Kelsen, *The Law of the United Nations: A Critical Analysis of its Fundamental Problems* (Frederick A Praeger 1950) at 797–798.

43   Myres S McDougal, "The Soviet-Cuban Quarantine and Self-Defense," 57 [1963] Am J Int'l Law 546; Richard Falk, "The Beirut Airport Raid," (1969) 63 Am J Int'l Law 415.

44   Benjamin B Ferencz, "A Proposed Definition of Aggression: By Compromise and Consensus" [1973] 22 Int'l and Comp Law Q 407.

45   Mary Ellen O'Connell & Mirakmal Niyazmatov, "What is Aggression? Comparing the *Jus ad Bellum* and the ICC Statute," (2012) 10 J of Int'l Criminal Justice 189.

46   RY Jennings, "The *Caroline* and *McLeod* cases," [1938] 32 Am J Int'l Law 82.

overwhelming and leaving no choice of means, and no moment of deliberation."[47] In addition, the preemptive measures must be no more than necessary to disable the poised attackers; in other words, once the poised force is disabled, the target state cannot, without Security Council authorization, use the occasion to install a more pliant regime or shatter the aggressor state's capacity to renew its threat in the foreseeable future.[48] Webster did not, however, face the issue of periodic attacks, which, in their aggregate, can reasonably be deemed to create a state of war. Today the United States apparently embraces the aggregation approach and in that way elides the charge that in its counterterror moves it has gone beyond preemption to the preventive use of force, a step that would have little support among legal scholars or governments.[49]

A third area of disagreement is whether force can be used in self-defense against non-state actors. Although the International Court of Justice ruled, in the *Wall* case, that the Charter does not allow the use of force in self-defense against a non-state actor other than one whose actions are imputable to the host state,[50] most commentators believe that a state should be able, within the meaning of Article 51 of the UN Charter, to defend itself against a non-state armed attack, at least where there is large-scale transnational violence.[51]

Despite the important uncertainties we have just described, we favor caution in stretching the elasticity of "imminence" to the point where the line between preemption and preventive war begins to disappear. And we would correspondingly limit the aggregation claim to cases where a state is attacking an organized armed group marked by a hierarchy of command which is conducting or indisputably conspiring to conduct a sustained campaign against the nationals and/or institutions of the state invoking its right to self-defense. We favor caution out of concern for the general good and also concern for the interests of all those states—the great majority—with a stake in a relatively peaceful and stable international system. Liberal capitalist democracies are foremost among them.

Another dangerous blurring of normative lines arises from the fact that some states show signs of beginning to conflate the right to make war—*jus ad bellum*—and the law applicable during a war—*jus in bello*. A state manifests this conflation when it acts as if it enjoyed a broad discretion to employ whatever

47   Ibid at 89.

48   Vaughan Lowe, *International Law* (OUP 2007) at 279.

49   Office of Legal Counsel, US Department of Justice (DoJ White Paper) *Lawfulness of a Lethal Operation Directed Against a US Citizen Who is a Senior Operational Leader of Al-Qa'ida or An Associated Force*, (Nov 8, 2011) at 7. See also Eric Holder, Attorney General, Speech at NW Univ School of Law (March 5, 2012) available at <http://www.justice.gov/iso/opa/ag/speeches/2012/ag-speech-1203051.html> accessed Nov 2014.

50   International Court of Justice (ICJ), *Legal Consequences of the Construction of a Wall in the Occupied Palestinian Territory* (July 9, 2004) at para 139.

51   Claus Kreß, "Some Reflections on the International Legal Framework Governing Transnational Armed Conflicts" (2010) 15 J of Conflict and Sec Law 245; Christian J Tams, "The Use of Force Against Terrorists" [2009] 20 Eur J Int'l Law 359.

means it chooses to "defeat" terrorism, even to the point where it must employ a strained if not risible interpretation of humanitarian law. Restraints on means are as integral to world order as restraints on recourse to force. Defenders of the main elements of the current order—countries with the most to lose from the collapse of normative constraint on the use of force—must struggle to maintain the *jus in bello*, the accumulated product of carefully negotiated norms reflecting the traumatic experience of mass slaughter.[52]

B. Inner Layer: Individual Protection

As explained above, the holistic approach holds that every layer of international law must be respected by states engaged in targeted killing. Compliance with restraints on force projection is one obligation. Respect for human rights and humanitarian law is the other.

1. Human rights law

As drone attacks usually take place in foreign countries, the first question to address is whether human rights norms apply to a state's conduct beyond its borders. Although human rights instruments were initially conceived as duties states assumed in relation to persons within their respective territories,[53] international monitoring bodies and a significant group of scholars have rejected that limiting interpretation of the relevant instruments.[54] In light of judicial decisions and the commentary of scholars, one can now say with confidence that human rights norms apply beyond national frontiers to all territories a state effectively controls. With respect to territory beyond their zones of continuous control, state responsibility is contested.[55] However, there is an emerging body of influential opinion finding responsibility under certain circumstances.[56]

Recently, in the matter of *Mohammad Munaf v. Romania*,[57] the Human Rights Committee (the body responsible for monitoring compliance with the ICCPR) addressed this question. An Iraqi-American national claimed that Romania had violated his rights under the Covenant because its Embassy in Iraq had handed him over to the United States Army, which then submitted him to torture. In order

---

52   Jasmin Moussa, "Can *Jus Ad Bellum* Override *Jus in Bello*? Reaffirming the Separation of the Two Bodies of Law" [2008] 90 IRRC 963.

53   See, for instance, Ed Bates, *The Evolution of the European Convention on Human Rights: From its Inception to the Creation of a Permanent Court of Human Rights* (OUP 2010) at 110–111.

54   See Human Rights Committee, *General Comment No 31: The Nature of the General Legal Obligation Imposed on States Parties to the Covenant*, (March 29, 2004).

55   Christof Heyns, Special Rapporteur on extrajudicial, summary or arbitrary executions, *Report*, UN Doc A/68/382 (Sept 13, 2013) at 10.

56   See Theodor Meron, "Extraterritoriality of Human Rights Treaties," [1995] 89 Am J Int'l Law 78, at 81.

57   UN Doc CCPR/C/96/D/1539/2006 (Aug 21, 2009), communication No 1539/2006.

to determine whether the Covenant was applicable, the Committee affirmed its jurisprudence "that a State party may be responsible for extraterritorial violations of the Covenant, *if it is a link in the causal chain that would make possible violations in another jurisdiction.* Thus, an extraterritorial violation must be a necessary, and for the state concerned, foreseeable consequence of its actions."[58]

The European Court of Human Rights adopted a similar approach in the case of *Issa and others v. Turkey* (2004),[59] which concerned applications made by six women living in northern Iraq, near the Turkish border, against Turkey following the (alleged) forced disappearance of their sons and husbands at the hand of the Turkish army. In determining whether the applicants' relatives came within the jurisdiction of Turkey, the Court held that:

> a State may also be held accountable for violation of the Convention rights and freedoms of persons who are in the territory of another State but who are found to be under the former State's authority and control through its agents operating— whether lawfully or unlawfully—in the latter State. ... Accountability in such situations stems from the fact that Article 1 of the Convention cannot be interpreted so as to allow a State party to perpetrate violations of the Convention on the territory of another State, which it could not perpetrate on its own territory.[60]

In this context, it should also be noted that, while a territorial state's consent to an attack may "heal" possible problems regarding the outer layer of protection, this consent cannot legitimate violations of human rights on the state's territory.[61]

In cases of targeted killing by drones, the attacking state is obviously "a link in the causal chain" that leads to the death of the targeted individual and is acting in a way it could not replicate legally on its own territory other than in the exceptional case where lethal force is necessary to defend the life of the arresting

---

58   Ibid at para 14.2 (emphasis added). Based on that reasoning, the Committee held Mr Munaf's claim admissible, while holding at the same time that Romania had not breached his rights, since it didn't know at the time what his future treatment would be.

59   Application No 31821/96 (Nov 16, 2004).

60   Ibid at para 71 (reference omitted). See also the Inter-American Commission of Human Rights, *Coard et al v the United States*, Case no 10.951, Report No 109/99 (Sept 29, 1999) at § 37: "Given that individual rights inhere simply by virtue of a person's humanity, each American State is obliged to uphold the protected rights of any person subject to its jurisdiction. While this most commonly refers to persons within a state's territory, it may, under given circumstances, refer to conduct with an extraterritorial locus where the person concerned is present in the territory of one state, but subject to the control of another state— usually through the acts of the latter's agents abroad."

61   See, for instance, ECtHR Grand Conseil, *El-Masri v "The Former Yugoslav Republic of Macedonia,"* App No 39630/09 (Dec 13, 2012) at § 206: "[T]he respondent State must be regarded as responsible under the Convention for acts performed by foreign officials on its territory with the acquiescence or connivance of its authorities."

officers or innocent third parties. There are thus solid grounds in the decisions of the Committee and the European Court for arguing that human rights in general apply extraterritorially. These grounds are even more solid regarding the right to life since Article 6 of the ICCPR calls it "inherent" ("Every human being has the inherent right to life").

The United States chooses to disagree, most recently in a statement before the Human Rights Committee.[62] This claim, if unchallenged, would enable the US to frame the debate in purely national terms of respect for domestic law (in particular the Constitution).[63] Hence debate in the US about the legality of drone killing has been largely conducted in terms of the respective constitutional powers of the

---

62   Charlie Savage, "U.S., Rebuffing U.N., Maintains Stance That Rights Treaty Does Not Apply Abroad," *The New York Times* (March 13, 2014). We should note, however, that there is no unanimity within the United States, since Legal Adviser Harold Koh advocated in a long and detailed memo that the US should change its official position as regards the obligation to *respect* human rights: see Office of the Legal Adviser, *Memorandum Opinion on the Geographic Scope of the International Covenant on Civil and Political Rights* (Oct 19, 2010).

63   Gary D Solis, *The Law of Armed Conflict: International Humanitarian Law in War* (Cambridge Univ Press 2010) at 542. The same debate takes place in the United States about the extraterritorial application of the UN 1984 Convention Against Torture. In its response to the Questions Asked by the Committee Against Torture (the body responsible for overseeing the implementation of the Convention) on May 5, 2006, United States Legal Adviser John B Bellinger III declared: "In response to Question 44, and the Committee's question about the geographic scope of Article 16, as ratified by the United States, I would like to emphasize that by its terms, Article 16 of the Convention obliges States Parties 'to prevent in any territory under its jurisdiction other acts of cruel, inhuman or degrading treatment or punishment which do not amount to torture.'" (our emphasis). Clearly this legal obligation does not apply to activities undertaken outside of the "territory under [the] jurisdiction" of the United States. The United States does not accept the concept that "de facto control" equates to territory under its jurisdiction. There is nothing in the text or the *travaux* of the Convention indicating that the two are equivalent." Here again, Legal Adviser Harold Koh issued a memorandum supporting the view that the geographic scope of the Convention against torture is not limited to the US national territory: see Office of the Legal Adviser, *Memorandum Opinion on the Geographic Scope of the Convention Against Torture and Its Application in Situations of Armed Conflict,* (Jan 21, 2013). See also Beth Van Schaack, "The United States' Position on the Extraterritorial Application of Human Rights Obligations: Now is the Time for Change," (2014) 90 Int'l Law Studies 20. Very recently, the United States adopted the latter position before the UN Committee Against Torture. See Mary E McLeod, Opening Statement (Nov 12–13, 2014): "There should be no doubt, the United States affirms that torture and cruel, inhuman, and degrading treatment and punishment are prohibited at all times in all places, and we remain resolute in our adherence to these prohibitions." Available at <https://geneva.usmission.gov/2014/11/12/ acting-legal-adviser-mcleod-u-s-affirms-torture-is-prohibited-at-all-times-in-all-places/#. VGS_-S5wT0g.facebook> accessed Dec 2014. See also Harold Koh's blog comment, "America's 'Unequivocal Yes' to the Torture Ban," available at <http://justsecurity. org/17551/americas-unequivocal-yes-torture-ban/>: "this is the first time in more than two

executive and the Congress in matters relating to the use of force and the possible role of the Federal Courts in protecting individual rights and adjudicating conflicts between the other two branches.[64]

Clearly, the right to life operates within national territories to bar any government agency from setting out to kill someone because they are thought to be very dangerous.[65] The police are entitled to use deadly force only in self-defense or in the belief that killing is necessary to stop a suspect from killing or grievously wounding a third party. Deadly force may also be employed as a last resort to prevent the escape of a person convicted of the most serious crimes or a person whom the authorities believe, on the basis of highly persuasive evidence, has committed a crime punishable by the most severe sentences its law allows.[66]

The great weight of legal opinion supports the conclusion that the initiation of armed conflict, even a conflict of an international character, does not automatically suspend the duties a state has assumed by becoming a party to a human rights treaty.[67] As noted above, it simply allows the concerned state or states to declare a state of exception and suspend non-derogable rights to the extent necessary for the effective conduct of hostilities.

Foremost among the non-derogable rights, of course, is the right not to be *arbitrarily* deprived of life. The word "arbitrarily" is the nexus that allows the right to life to be adapted to the exigencies of war. Humanitarian law clarifies the content of the right to life—as well as the detention regime of combatants[68]—as

---

decades that the United States moved away from a strict territorial reading of a human rights treaty."

64   See, for instance, David Cole, "The Drone Memo: Secrecy Made It Worse," *The New York Review of Books Blog* (June 24, 2014).

65   Inter-American Commission on Human Rights, *Report on Terrorism and Human Rights*, OEA/Ser.L/V/II.116 (Oct 22, 2002) at § 87; see also ECtHR Grand Conseil, *McCann and others v UK*, App No 18984/91 (Sept 27, 1995); and Human Rights Committee, *Husband of Maria Fanny Suárez de Guerrero v Columbia*, communication No. R.11/45 (Jan 20, 1978).

66   Regarding Osama bin Laden's death, see "Osama bin Laden: Statement by the UN Special Rapporteurs on summary executions and on human rights and count terrorism" (May 6, 2011): "In respect of the recent use of deadly force against Osama bin Laden, the United States of America should disclose the supporting facts to allow an assessment in terms of international human rights law standards. For instance it will be particularly important to know if the planning of the mission allowed an effort to capture Bin Laden." See also the UN General Assembly Res 34/169, *Code of Conduct for Law Enforcement Officials* (Dec 17, 1979), as well as the *Basic Principles on the Use of Force and Firearms by Law Enforcement Officials* adopted by the Eighth UN Congress on Prevention of Crime and Treatment of Offenders (Sept 7, 1990).

67   See HRC, *General Comment No 31* (n 54).

68   Marco Sassòli and Laura M. Olson, "The relationship between international humanitarian and human rights law where it matters: admissible killing and internment of fighters in non-international armed conflicts" [2008] 90 IRRC 599.

a *lex specialis* defining what constitutes an "arbitrary" deprivation of life in a situation of armed conflict.[69]

As the International Court of Justice put it in its *Legality of the Threat or Use of Nuclear Weapons* advisory opinion,

> the protection of the International Covenant on Civil and Political Rights does not cease in times of war, except by operation of Article 4 of the Covenant whereby certain provisions may be derogated from in a time of national emergency. Respect for the right to life is not, however, such a provision. In principle, the right not arbitrarily to be deprived of one's life applies also in hostilities. The test of what is an arbitrary deprivation of life, however, then falls to be determined by the applicable *lex specialis*, namely, the law applicable in armed conflict which is designed to regulate the conduct of hostilities.[70]

In these conditions, a lethal attack can be qualified as non-arbitrary under human rights law only if it takes places in a manner consistent with humanitarian law, which is why Philip Alston once stated that while "extrajudicial executions," "summary executions," and "assassinations" are by definition illegal, targeted killings may be legal in the circumstance of armed conflict.[71] Conversely, human rights law forbids targeted killings outside the framework of an armed conflict unless it can be shown that the killing was the only means to prevent an imminent terrorist attack.[72]

2. Humanitarian law

The place of drones within humanitarian law turns on questions of the law's scope and its substance. On the question of scope, humanitarian law applies to both international (state vs state) and non-international (state vs non-state actors) armed conflicts. Although with respect to the latter its substance is more limited, both clearly outlaw summary execution of civilians and both allow the killing of combatants.

---

69   Emmerson (n 32) at 18.

70   ICJ (n 50) at § 25. See also Marko Milanovic, "Lessons for human rights and humanitarian law in the war on terror: comparing *Hamdan* and the *Israeli Targeted Killings* case" [2007] 89 IRRC 373, at 391; see also William Abresch, "A Human Rights Law of Internal Armed Conflict: The European Court of Human Rights in Chechnya" [2005] 16 Eur J of Int'l Law 741.

71   Alston (n 37) at 5. The European Convention on Human Rights adopts a slightly different approach, but arrives at the same result. Thus, Article 2 of the ECHR, which protects the right to life, goes further linguistically than Article 6 ICCPR in that it forbids any intentional killing (not only arbitrary ones). However, Article 15 of the ECHR allows derogation from Article 2 "in respect of deaths resulting from lawful acts of war." Thus, the main difference between the two instruments is that states need formally to derogate from the ECHR in order to comply with it.

72   Gary D Solis, "Targeted Killing and the Law of Armed Conflict," [2007] 60 Naval War College Rev 127, at 134–135: "Without an ongoing armed conflict the targeted killing of a civilian, terrorist or not, would be assassination—a homicide and a domestic crime."

Legal scholars, practitioners and the guardian of humanitarian law, the International Committee of the Red Cross (ICRC), have all agreed in the past that certain cumulative objective conditions must be met before a struggle reaches the threshold of a non-international armed conflict:[73] They are a certain intensity of violence and a sufficient degree of organization of the parties to the conflict.

The US deems the threshold criteria to have been satisfied in all of the cases to date where it has engaged in targeted killing. Therefore in all cases it was entitled to attack the targeted persons anywhere on earth, at least where it had the permission of the government of the territory where it struck or that government was unable or unwilling to arrest the combatant and turn him over to US authorities.[74] Presumably it will deploy this claim in defense of future killings.

One problem the US faces in maintaining its legal position is the apparent opposition of the ICRC.[75] The latter has rejected the notion that the violence perpetrated by loosely associated terrorist groups operating within different states can be merged for purposes of reaching the scale-and-intensity threshold. They have instead argued that the situation within each state—for example, Yemen, Somalia, Algeria, Nigeria, Mali, Libya—must be considered separately.[76] If and to the extent humanitarian law applies—as it certainly does to the crescendoing conflict with the Islamic Caliphate in the Levant (ISIL)—it requires a state to comply with the three core principles of humanitarian law: necessity, distinction, and proportionality. Although we cannot describe them in detail here, we can briefly draw attention to the following points. First, the principle of necessity requires that the killing must advance a tactical or strategic end. Secondly, the principle of distinction draws a sharp line between full-time combatants and "civilians" who may episodically participate directly in or consciously facilitate combat

---

73   See International Tribunal for the former Yugoslavia, *Prosecutor v Dusko Tadic, Decision on the Defence Motion for Interlocutory Appeal on Jurisdiction*, case no IT-94-I (Oct 2, 1995) at § 70; International Criminal Tribunal for Rwanda, *The Prosecutor v Alfred Musema* (Jan 27, 2000) at § 248; International Criminal Court, *Prosecutor v Thomas Lubanga Dyilo* (March 14, 2012) at §§ 536–538. See also Mary Ellen O'Connell, "The Choice of Law Against Terrorism" [2010] 4 J of Nat'l Sec Law & Policy 343.

74   Harold Hongju Koh, Legal Advisor to the State Department, "Speech: The Obama Administration and International Law," Annual Meeting of the American Society of International Law (March 25, 2010), available at <http://www.State.gov/s/l/releases/remarks/139119. htm> accessed June 2014: "As I have explained, as a matter of international law, the United States is in an armed conflict with al-Qaeda, as well as the Taliban and associated forces, in response to the horrific 9/11 attacks, and may use force consistent with its inherent right to self-defense under international law." See also Vaughan Lowe, "Security Concerns and National Sovereignty in the Age of World-Wide Terrorism," in Ronald St John Macdonald / Douglas M. Johnston (eds), *Towards World Constitutionalism* (Brill 2005) at 663.

75   ICRC, *International humanitarian law and the challenges of contemporary armed conflicts*, 31IC/11/5.1.2 (2011) at 10–11.

76   See Noam Lubell and Nathan Derejko, "A global battlefield? Drones and the geographical scope of armed conflict" (2013) J of Int'l Criminal Justice 78.

operations. The latter can be targeted only when they are participating or taking steps to participate in combat. Thirdly, the principle of proportionality requires that reasonably foreseeable collateral damage must not be disproportionately large in relation to the importance of the military objective and the difficulty of achieving that objective by means less threatening to the civilian population.

3. Toward consistent application of the relevant law

Reconciling the practice of drone killing with liberal values requires more than a declaration of intent to comply with the UN Charter and human rights and humanitarian law. Good intentions are a necessary point of departure. Translating good intentions into quotidian reality demands an effective decision-making and review process. More specifically, effective translation requires a process that is relatively transparent and holds practitioners accountable for their actions. It is not by coincidence that transparency and accountability are cornerstones of human rights law.[77]

Transparency requires that the criteria for targeting and the authority approving drone strikes be clearly identified and that, to the extent that disclosure does not compromise key sources, the authorizing authority can publicly justify its decisions and confirm that they were made on the basis of a factually compelling record and through the reasoned application of the relevant law.[78]

The release of the redacted 2010 Justice Department memorandum concerning the targeting of Anwar al-Awlaki in Yemen is a first step in this direction—an action required by court ruling.[79] So may be the indication by President Obama that control of lethal counterterrorism operations conducted outside areas of active hostilities would be transferred from the Central Intelligence Agency (CIA) to the Department of Defense,[80] although it is not absolutely clear that, when it comes to irregular warfare, the Pentagon is less secretive than the CIA.

Secondly, transparency, along with accountability, must mark not only the initial authorization of a lethal strike but also its *post facto* evaluation,[81] particularly where

---

77   See HRC, *General Comment No 31* (n 54).

78   Alston, "The CIA ... " (n 30). Article 51 of the UN Charter also requires states to report their self-defense measures to the Security Council, which entails a transparency requirement.

79   US Court of Appeals for the Second Circuit, *The New York Times Company et al v United States*, Case No 13–422-cv (April 21, 2014): David Barron, Acting Assistant Attorney General, "Re: Applicability of Federal Criminal Laws and the Constitution to Contemplated Lethal Operations Against Shaykh Anwar al-Aulaqi" (July 16, 2010).

80   President Barack Obama, Prepared Remarks at the National Defense University on the Administration's counter-terrorism policy, Washington DC (May 23, 2013) available at   <http://www.whitehouse.gov/the-press-office/2013/05/23/remarks-president-national-defense-university> accessed Dec 2014.

81   See Israel High Court of Justice, *The Public Committee Against Torture et al v The Government of Israel*, HCJ 769/02 (Dec 14, 2006) at paras 39, 40 and 60: "After

there appears to be collateral damage.[82] The Drone Court proposed by Professors Guiora and Brand in the previous chapter is the most promising instrument for institutionalizing the substantive standards and process requirements we have tried to identify.

However, at its inception, the Drone Court would have to address some difficult questions concerning the applicable law and its substance. For instance:

- Who would determine the applicable law? In particular, would the Drone Court (and the court reviewing its decisions) be expected to follow the executive's position or would it be independent in its assessment—thus being allowed to disagree with the US official position? This question is not merely theoretical, as illustrated by the contemporary discussion on the extraterritorial scope of major UN treaties such as the International Covenant on Civil and Political Rights and the Covenant against Torture.[83]
- How would the work of human rights bodies such as the Human Rights Committee or the Committee against Torture be received in the Drone Court's jurisprudence? Would they help shape the Drone Court's views or would they merely be discarded as foreign law?
- Would there be a mechanism allowing the Drone Court to synchronize its views with those of other judicial bodies, thereby preventing the creation of a "unilateral" US view on drones?

The answers given to these questions by the Drone Court would help to predict how the proposed court would balance legitimate national security concerns with individual rights and the integrity of the international legal system.

In the absence of national courts with the requisite independence and will, the enforcement of governing norms will depend on the International Criminal Court, which will have jurisdiction where killings are carried out in the territory of state parties, and on the de-legitimating power of Professor Schacter's "invisible college." In all events, human rights norms should be construed to require compensation for death and injury to non-combatants.

IV. Recapitulation

As previously noted, targeted killing by drones or other means must comply with UN Charter law on the use of force and with human rights and humanitarian law, the two levels of legal protection for individual rights. At this point, the US

---

each targeted killing, there must be a retroactive and independent investigation of the identification of the target and the circumstances of the attack."

82 See Michael Schmitt, "Investigating Violations of International Law in Armed Conflict" [2011] 2 Harvard Nat'l Sec J 31.

83 See (n 63–64) and accompanying text.

government appears to regard itself as justified in eliminating not only persons playing significant roles in substantial, hierarchically organized groups like ISIL, Al-Qaeda in the Maghreb, the Haqqani network, and the Shabab in Somalia, but any person associated with such groups even where they pose no immediate threat to US nationals or the nationals of allied states.[84]

We believe this position goes beyond the applicable legal restraints on the use of force, which we would summarize as follows:

1. Despite the uncertainties and controversies surrounding the resort to force in international relations (the "outer layer"), drone attacks outside of a state's territory must rely either on the consent of the concerned state or on a plausible claim of self-defense against an armed attack. The attacker must, however, be a hierarchically organized group able to sustain operations over time and which has attacked or is manifestly preparing to attack the drone-deploying state.

2. In order to justify a targeted killing under the inner layer of normative restraint, a state has to demonstrate that the conditions of a non-international armed conflict are met. It must establish that there is: (i) a non-state armed group identifiable as such based on criteria that are objective and verifiable (for instance, a minimal level of organization and engagement of the group in collective, armed, anti-government action); and (ii) a minimal threshold of intensity and duration. In addition it must demonstrate that it has no viable law-enforcement alternative.[85] Under traditional views the group had to have some territorial base, but given the contemporary ease of moving constantly across borders, many commentators argue that this criterion has become obsolete.

3. Persons who apparently are not integral members of an armed group but associate with it informally and episodically can be targeted only where it is established that they are actively engaged in or moving into place in order to participate in an armed attack.

4. Group members who are non-combatants such as fund raisers and contributors or media advisors and producers cannot be targeted (of course they are subject to arrest and prosecution).

5. In all cases, targeted killings must respect the principles of necessity and proportionality. Collateral damage should raise a rebuttable presumption of disrespect for the proportionality principle.

6. Outside the context of an armed conflict within the meaning of the Geneva Conventions as construed by the ICRC, persons suspected of planning or executing terrorist acts can be killed only when state officials can

---

84   See Helen Duffy, *The "War on Terror" and the Framework of International Law* (Cambridge Univ Press 2005) at 448–449.

85   Kenneth Roth, "The Law of War in the War on Terror" (Jan/Feb 2004) Foreign Affairs.

demonstrate by clear and convincing evidence that killing was the only means available to prevent an imminent act of terrorism.

V. Hypotheticals: ISIL in Iraq and Syria and the Afghani Taliban in Pakistan

Perhaps for purposes of clarification it would be useful to show how these generalizations would apply to the conflict with ISIL and the Afghan Taliban. The struggle between ISIL and the government of Iraq is an internal armed conflict in which the US participates at the invitation of the Iraqi government. Because of that invitation, its combat activities in Iraq are consistent with the UN Charter. Under the laws of war it can target any ISIL combatant anywhere in the country whether they be anonymous grunts or Abu Bakr Al-Baghdati himself, who appears to be the military as well as the political leader of ISIL.

Admittedly the case would be somewhat more complicated if Al-Baghdati established a rebel government with a formal division between civilian and military leaders, declared himself the head of state, and appointed a colleague as general and chief of staff. When the United States bombed Muammar Qaddafi's palace in 1994, it claimed it was not targeting the head of state as such, but rather smashing a center of command and control. The "shock-and-awe" bombardment of Baghdad by US-led coalition forces in 2003 targeted various of Saddam Hussein's palaces. And while never conceding that it sought to kill him in this initial onslaught, the extent, severity, and targeting of the bombardment at least implied the coalition governments had confidence in a license to kill the state's operational head. The same could be said of the NATO forces that bombed Qaddafi's home-office complex in Tripoli during the 2011 intervention. It is very doubtful that if such a license, or what one might call "tolerance," exists today in the global system of order it extends to nominal heads of state: presidents and monarchs whose role is essentially ceremonial and where operational power rests in the hands of a prime minister and her cabinet.

As for US sorties in Syrian air space, the Syrian government has stated that it has not authorized US cross-border operations—no doubt in hope of extracting from the US recognition of the Assad regime's legitimacy. The legality of the use of force in Syria is therefore harder to justify under the Charter, although an argument may be made for collective self-defense against an attacking force that moves at will across the Iraq-Syrian border because of the Syrian government's inability to control a substantial part of its territory.

The same UN Charter-based argument is available to justify drone strikes in Pakistan against senior figures in the Afghan Taliban and the Haqqani network with undifferentiated political/military roles operating out of bases in Pakistan. But according to US government officials, such strikes have not been limited to senior figures. There have also been "signature strikes" against houses or compounds believed to be owned or controlled by the organizations. This type of strike has apparently been designed in the hope of killing senior figures, but in any event thinning out the ranks of combatants and deterring recruitment. However, raid planners are reported to have operated on the premise that civilians of fighting age who are round and about the premises can be deemed to be combatants.

Since the Director of the CIA, the agency which has had operational responsibility for the strikes, concedes that the agency has not had the capacity to measure the military efficacy of such strikes, in particular whether they foster recruitment of new combatants in numbers exceeding those killed or otherwise contribute positively to Taliban strength, their military utility is uncertain. If military utility is uncertain and all persons of fighting age killed or maimed are presumed to be combatants, the proportionality test imposed by humanitarian law, and through it by human rights law, is effectively nullified except in instances where it can be established that women, children, and old men were also among the casualties. And even then, against what is their death or crippling to be balanced? Simply against number of presumed combatants killed or crippled as a goal in itself? However dubious the moral answer and uncertain the strategic one, the legal answer may be "yes."

In an armed conflict, killing enemy combatants has historically been deemed a legitimate military purpose however dubious its strategic consequences. But what is legally and morally intolerable is simply presuming that persons of military age found in the area where Taliban combatants live are themselves combatants. The attacking government must offer much more persuasive evidence than mere proximity, even if it is recurrent, if, for purposes of claiming proportionality, the attacker wants to aggregate proven fighters and those persons merely presumed to have that status. In this connection it is necessary to recall that, in the view of the ICRC, the semi-official steward of humanitarian law, even persons who occasionally join a conflict retain civilian status when they are not actually fighting or moving into position to strike a blow.

VI. Conclusion

Overall, we believe that the enumerated criteria will prevent targeted killing from descending into a system of state-sanctioned murder that would not merely blur but actually eliminate the line between transgressive and legitimate violence. That being said, we recognize that an exceptional case could conceivably appear, like a black swan, which does not satisfy these criteria but which on moral grounds demands immediate and decisive measures.

Imagine, for instance, the following scenario. The President of the United States is suddenly presented with information from an utterly credible source that a scientist in a small laboratory in a remote part of Belarus, having succeeded in so manipulating the Ebola virus' genetic structure as to make it as contagious as the common cold, has agreed to sell the new virus to an unknown country or group and the transfer is scheduled to occur with an hour. The United States happens to have an East-European-based drone with hellfire missiles that can reach the laboratory in approximately 30 minutes.

If in this hypothetical case the President orders a drone killing, there would be a *prima facie* violation of international law. However, we believe that in such an improbable scenario where preemptive action serves the deep values that lie behind human rights and humanitarian law, the President could properly invoke in her defense the principle of "necessity."[86] As former US State Department Legal Advisor Harold Koh recently wrote:

> [D]o courts tell ambulance-drivers who ran red lights to prevent deaths that their actions are illegal because their actions might encourage ambulance-chasers to do the same thing? . . . Or do we define the contours [of] a narrow "affirmative defense" that would render lawful otherwise illegal behavior? [87]

We believe that the application by independent judicial bodies of the rules we have outlined will enable liberal democratic societies to defend themselves without in the process eroding those constraints on executive power that are essential for the survival of liberal democratic societies. The prospect is considerably bleaker if we must rely for their application only on international public opinion and the contingency of a prudent President appropriately sensitive to the risk that waging drone warfare beyond the bounds set by legal norms will (a) generate more terrorists than it kills, (b) create for violent militants in target countries a more supportive environment than they could achieve on their own, and (c) erode the painfully accumulated moral and legal restraints on assassination within and among nations.

---

86   This reasoning enjoys kinship with Ian Brownlie's analogy of humanitarian intervention and euthanasia. We outlaw euthanasia, Brownlie argued, because we fear the slippery slope and assume that in those unusual cases where euthanasia seems morally justified, then prosecutors will not indict or juries will not convict. In the case of humanitarian intervention, he concluded, we should treat it as *prima facie* illegal and thus place a heavy burden of persuasion on intervenors to convince the "invisible college" that all other means to avert a terrible human catastrophe had been exhausted, that their motive was in fact humanitarian, and that they reasonably believed the intervention would do much more good than harm. See Ian Brownlie, "Humanitarian Intervention," in John Morton Moore (ed), *Law and Civil War in the Modern World* (John Hopkins Press 1974) at 217.

87   Harold Hongju Koh, Syria and the Law of Humanitarian Intervention, Part II: International Law and the Way Forward, EJIL: TALK! (Oct 4, 2013), available at <http://www.ejiltalk.org/syria-and-the-law-of-humanitarian-intervention-part-ii-international-law-and-the-way-forward> accessed Dec 2014.

Conclusion
Defending Legitimacy

Steven J. Barela

At the initiation of this project a specific conclusion was not envisioned. The idea was to investigate the three proposed realms of legitimacy as they relate to drone killings across international borders, a design that does not necessarily point in the direction of policy solutions. By peering through the lenses of *legality*, *morality*, and *efficacy* a series of worthwhile views into the legitimacy of drones could be offered to enhance the public understanding of an issue that is only growing in importance. This indebted editor believes that the contributing authors have done precisely this. Unexpectedly, this book project also advanced in a manner that has produced a keen policy proposal meant to integrate these three basic tenants of legitimacy.

As has been seen, this starting point has simultaneously produced descriptive and normative contributions. That is, the authors have had to explore both what is and what ought to be,[1] to be able to express a complete view on each of their research questions. The current applicable law, our moral understandings of lethal force and war, along with judgments of their effectiveness as a policy tool, cannot be currently presented as definitive since unmanned combat air vehicles have so drastically changed the practical and political calculations for exercising force across borders; both views are required.

First, many contributions are descriptive in the sense that they aim at discerning what exactly the rules governing unmanned targeted killings are and what they say. On crucial points today—to name but a few, the legality of a state's resort to force in response to multiple low-level attacks, its use of force following attacks attributable to non-state entities, and the extraterritorial effects of human rights treaties—there is no fixed or agreed upon jurisprudence (or even doctrine). For this reason, the authors are at times unable to provide the reader with a definitive answer as to the black letter of the law, preferring—wisely—to present the positions expressed in the debate by diverse participants (states, International Committee of the Red Cross, various bodies of the United Nations, etc.).

Many chapters also contain normative passages in which the authors describe and defend the rules (or their interpretations) that, according to them, are more likely to lead to a use of drones that is legal, moral, or effective. Given the room

---

1 For an early illumination of the "is vs ought" problem see the Scottish philosopher David Hume, *A Treatise of Human Nature*, Book III, Part I, Section I (BiblioBazaar 2006) at 257.

for disagreement created by debates occurring on major aspects of the issue, this normative aspect of the book is not at all surprising. In fact, I would venture to say that the thoughtful and lucid views expressed by these learned scholars are extremely useful at this particular moment. As we are experiencing a time of great tension brought on by the emerging technology of unmanned combat aerial vehicles (UCAVs), and perhaps a resettling of the applicable norms, it is during such a time of adjustment that careful opinion and advocacy are necessary. Thus, this normative aspect of the invited work is most welcome.

Additionally, in order to provide further context it is also useful to briefly elaborate on the distinction between offensive and defensive measures presented in the Introduction to this volume. If "offensive" measures are used to describe actions directly taken to prevent an individual or a group from committing future harm (mainly through killing, interrogation, and detention), "defensive" measures may be defined as including any state action designed to preserve or even advance the principles and values on which the state is founded. Therefore defending its legitimacy as well as its *raison d'être*, even (if not especially) when faced with a terrorist threat, can be understood as representing a strategic imperative for the state. Offensive measures are simply possible tactics that should never trump defensive strategy.

This is not to say that a state's response to terrorism can be either wholly offensive or wholly defensive. In fact, sound counterterrorist operations will be composed of both. However, the defensive aspect of a state response to terrorism is of particular relevance when determining the scope of the offensive measures. For the sake of legitimacy, these aggressive measures must be as narrow as possible and always remain openly justifiable politically, an objective that necessitates full respect for the applicable body of laws and norms.

Considered as a whole,[2] a robust offensive approach, with its very strong preventive content, is at odds with the traditional liberal vision. As this world view is built on the acceptance of a certain level of societal risk in order to limit the state's capacity to intervene in the quotidian life of the citizenry,[3] acting in advance of a violation requires supposition of future acts and engenders the government with the power to act (violently) on such speculation. In other words, the liberal state implicitly accepts that some persons will abuse the broad freedoms— its distinguishing feature—in order to protect against an overreaching by the

---

2   Arthur Chaskalson, "Preserving Civil Liberties in an Age of Global Terror: International Perspectives" [2007] Cardozo Law Rev 11, at 13: "It is also important that particular measures should not be looked at in isolation; one must look at the cumulative effect of all the measures that are taken to combat terrorism. It is the cumulative effect of the erosion of established rights and the power of courts over a period of time, which changes society. And therein lies the danger."

3   Aharon Barak, *The Judge in a Democracy* (Princeton Univ Press 2006) at 284–285.

government.[4] Risk is no doubt distressing, but it is also an inherent part of our human condition.

We should bear in mind that, historically, the occurrence of a terrorist act almost inevitably prompts governments to initiate forceful offensive measures—often radical and violent—aimed at preventing any recurrence. The objective of preventing new terrorist attacks comes partly from the government's responsibility to guarantee its population's safety, and partly from the pressure coming from the population to *feel* reassured.[5]

This concept of defending legitimacy was persuasively presented (albeit in slightly different terms) by Michael Ignatieff as the offensive tactics employed in the US "war on terror" were becoming more widely known: "It is the paramount duty of political leaders in a democracy under attack to keep the forces of order intently focused on the political requirement of maintaining *legitimacy*."[6] Ignatieff identified the attackers of September 11th as persons who were willing to slaughter innocent civilians, thus totally rejecting established laws and institutions, and warned that those who confront them must not do the same in order to protect the very legitimacy of their society under attack. Ignatieff forcefully counseled:

> The only cure for nihilism is for liberal democratic societies—their electorates, their judiciary, and their political leadership—to insist that force is legitimate only to the degree that it serves defensible political goals. This implies a constant exercise of due diligence: strict observance of rules of engagement regarding the use of deadly force and the avoidance of collateral damage.[7]

Concluding with a Policy Proposal

Nevertheless, despite the benefits envisioned for exploring the use of armed drones through the varied lenses of legitimacy, it had not been originally supposed that presenting these views would lead to a policy for defending the legitimacy of the government under attack by a terrorist group(s). The fact that a policy proposal

---

4   Frédéric Bernard, *L'Etat de droit face au terrorisme* (Schulthess 2010): "[Risk] is indeed inevitable and represents the price to be paid to secure these [fundamental] rights collectively" at 289 (my translation); see also former Justice Stephen Breyer, "Symposium on Terrorism, Globalization and the Rule of law: an Introduction" (2006) 27 Cardozo Law Rev 1981: "well functioning constitutional democracies are characterized by their willingness to take risks to secure liberty" at 1985.

5   Bruce Ackerman, *Before the Next Attack: Preserving Civil Liberties in an Age of Terrorism* (Yale Univ Press 2006) at 2. See also my Chapter 11 exploring the "opinion" of security.

6   Michael Ignatieff, *The Lesser Evil* (Princeton Univ Press 2004) at 143 (my emphasis).

7   Ibid at 143–44.

has materialized to do so is, in fact, an unforeseen and acutely welcome product of the project itself. To explain this development it is necessary to revisit the genesis of our endeavor.

Through my counterterrorism studies I had come to know Amos Guiora as both an experienced practitioner and an accomplished academic author. Since I knew that he had long expressed the need for counterterrorism policies to be legal, moral, and effective, he was contacted straightaway upon the launching of this undertaking. When Guiora told me that he was working on a policy proposal for creating a Drone Court, I was somewhat hesitant as I did not yet see where such a chapter would exactly fit in this volume. Quickly setting aside any doubts, I immediately assured him this book was the correct place for such an intriguing contribution. Shortly afterwards, Guiora brought the skilled and experienced scholar Jeffery Brand onboard as a coauthor. The two of them proceeded to make their case for a Drone Court through their persuasive writing, in-depth research, and a continued exchange of views through articles and reports dealing with armed drones specifically and counterterrorism generally. Through all of this work together over the life of the project it became clear that their policy proposal was indeed the proper conclusion to this volume as a genuine attempt to integrate the elements of legitimacy.

As has been seen in their chapter, Guiora and Brand suggest the need for a court to be established by the legislature, administered by the judiciary, in order to oversee the lethal authority exercised by the executive. In other words, their proposal rests on the constitutional cornerstone of the separation of powers principle—a reasoning of political philosophy found in the landmark 2008 US Supreme Court *Boumediene* decision.[8] Not only is it important to understand the tension created between the divided powers of government when a democratic society comes under attack by way of terrorist methods,[9] one must also recognize that this principle plays an important role in evaluations of legitimacy within the US constitutional structure.

To add to this closing proposal, the transatlantic team of Tom Farer and Frédéric Bernard were brought together to address the general principles undergirding liberal governance: an idea that once again represents a foundational tenet of legitimacy in the United States. As discussed above, terrorist violence poses a real danger to liberal societies because it can tempt those who govern to turn their back on long-accepted codes of limited government in pursuit of perfect security. Hence Farer and Bernard's exploration of a holistic approach to the substantive international law a Drone Court would need to navigate and apply in the complicated circumstances created is certainly a valuable contribution to the assessments of legitimacy.

---

8 *Boumediene v Bush,* 553 US 723 (2008).

9 See generally Andrea Bianchi and Alexis Keller (eds), *Counterterrorism: Democracy's Challenge* (Hart Publishing 2008).

While there is little doubt that the establishment of a court would at least offer the opportunity for legality to play the role of a check against executive power in its use of deadly force, questions might still remain as to whether morality and efficacy also are to be included in such a forum. In light of conventional views this would appear to be a valid concern on its face, yet the history of legal philosophy and current judicial practice illuminates why this disquiet is unwarranted.

As addressed in this volume, the belief in a distinct cleavage between legality and morality overlooks what the fathers of legal positivism—John Austin and Jeremy Bentham—have themselves acknowledged. H.L.A. Hart notably wrote about these two pivotal legal minds who founded this theory that now dominates. Hart found that although they indeed separated with the utmost of clarity "the law as it is from law as it ought to be,"[10] they still found "frequent coincidence of positive law and morality," as Austin once put it.[11] Hart identified the overlap between the two realms as laws prohibiting murder, violence, and theft, explaining "such rules are so fundamental that if a legal system did not have them there would be no point in having any other rules at all."[12] Since deadly force is the most grave and disturbing facet of the use of UCAVs and terrorism alike, a Drone Court would be dealing with an act that must be constrained in some way to be considered moral. As a result, creating a real check or limitation on the use of force with combat drones most certainly means taking a step in the direction of morality.

As for the concern over efficacy, it should not be overlooked that courts have at times found a way for this critical element to also become a part of legal calculation—in particular on questions of counterterrorism. One of the clearest examples of the use of efficacy for the rendering of a provision of a law outside the bounds of legality comes from the German Federal Constitutional Court in 2006.[13] In the wake of 9/11 the parliament was convinced that the shooting down of an aircraft presumed to be hijacked was a feasible means to address a specific security alarm. However, the Court ruled that the statute "must pursue a legitimate end," and this law simply "could never achieve what it pretended to pursue"[14] because it could never be known definitively on the ground that the aircraft was to be turned into a projectile weapon. The Court determined that the policy was not legal or legitimate because it was considered to be inefficacious.[15]

---

10   HLA Hart, "Positivism and the Separation of Law and Morals" (1958) 71 Harvard Law Review 593, at 594.

11   John Austin, *The Province of Jurisprudence Determined* and *The Uses of the Study of Jurisprudence* (Weidenfeld and Nicolson 1954) at 162.

12   Hart (n 10) at 621.

13   For discussion of the decision in English see, O Lepsius, "Human Dignity and the Downing of Aircraft: The German Federal Constitutional Court Strikes Down a Prominent Anti-terrorism Provision in the New Air-Transport Security Act" (2006) 7 German Law Journal 761.

14   Ibid at 774.

15   In addition, the court also applied a Kantian moral logic to strike down this legislation because it would reduce the individual onboard to no more than means for

For an additional example, one can look to the 1999 ruling handed down by the Israeli Supreme Court on the General Security Service's methods of interrogation.[16] In essence, it was found that an interrogation is required to be "reasonable"—an important judicial tool for the former Court President Aharon Barak who authored this decision.[17] While there are various passages that highlight the inclusion of efficacy into this legal context,[18] there is one point where the Court explicitly expressed in its ruling against specific techniques because they could not be considered as a part of a "fair and *effective* interrogation."[19]

In sum, it is indeed possible that morality and efficacy would come to play a significant role in the proposed Drone Court. To what extent this occurs certainly remains an open question, just as the degree and manner in which the various bodies of international law are applied in such a court cannot be foretold (in this regard, there are valid and important questions raised by Farer and Bernard

---

furthering security interests of the state and others. "The state must respect human beings' constitutional right to dignity. The Court concluded that § 14 of the Air-transport Security Act showed no respect for the well-being of those on board the target airplane. They are simply treated as part of the aircraft, which has to be destroyed since it has become a weapon. The Court explained that, under these circumstances, the state has given up the passengers' lives for the purpose of the presumptive prevention of a severe danger. The Court found that such a legal treatment neglects the constitutional status of the individual as a subject with dignity and inalienable rights. When the law takes their death into account as unavoidable damage for the benefit of other objectives, the Court explained, it transforms persons into things and delegalizes them (verdinglicht und zugleich entrechtlicht). The state denies the protection of the law to those who, as passengers-turned-victims in the aircraft, ought to be protected. Doing this, the Court reasoned, the law denies to those on board the value the constitution attributes to every human being," at 767.

16   *Public Committee Against Torture in Israel v The State of Israel*, Israeli High Court of Justice, HCJ5100/94 (Sept 6, 1999).

17   For a description of this tool in an extrajudicial writing see Barak (n3): "[t]he key test here is reasonableness. Put simply, the executive must act reasonably, for an unreasonable act is an unlawful act. In many cases the test of reasonableness allows for only one possibility, which the executive *must* choose. Sometime, however, the reasonableness test allows for several possibilities, thereby creating a 'zone of reasonableness.' The executive has freedom of choice within this range. The principle of separation of powers requires the executive, rather than the judiciary, to choose one possibility within this zone. But the principle of separation of powers requires the court, rather than the executive, to determine the limits of the zone of reasonableness" at 248.

18   Israeli High Court of Justice (n13): "Of course, it is possible to conduct an *effective* investigation without resorting to violence. Within the confines of the law, it is permitted to resort to various sophisticated techniques. Such techniques—accepted in the most progressive of societies—can be *effective* in achieving their goals. In the end result, the legality of an investigation is deduced from the propriety of its purpose and from its methods" at 23–4 (my emphasis).

19   Ibid at 27 (my emphasis).

in their chapter).[20] The part that all three would come to play in a Drone Court would depend entirely on the legislation that would create it, and the judges' interpretation of that law and their constitutional mandate. However, creating a forum in which *legality, morality,* and *efficacy* can be openly discussed certainly represents a positive—if only preliminary—step toward defending legitimacy in the world's current foremost user of UCAVs for cross-border counterterrorism. Genuine movement in this direction is needed, even if it is imperfect.

Cases in Point: Signature Strikes and the Feasibility of Capture

In April 2015 there were a pair of developments that disturbingly demonstrate the pressing need for a policy solution. There was the shocking disclosure that the Obama administration had never abandoned the tenuous (at best) practice of using armed drones to kill persons whose behavior indicates they are likely members of a terrorist group or militants, even if their identity is unknown. In addition, at the beginning of that month a man was taken into custody and arraigned before a federal court in Brooklyn on terrorism charges; yet two years earlier key parts of the US government argued that he should be placed on the so-called drone kill list because his capture was deemed infeasible. While each of the revelations requires a renewed extended analysis, the opportunity will be taken here to simply explain how they serve to demonstrate the potency and practicality of the policy proposal put forward.

Again, the development of this book project is telling on the first question. When the chapter outline was being designed in the fall of 2013 the practice of killing individuals based on their pattern of life ("signature" strikes)—rather than known identity and guilt ("personality" strikes)[21] —was taken by many to have been markedly reduced, if not entirely discontinued. The well-publicized Presidential Policy Guidance (PPG) signed in May of that year[22] was thought to

---

20   See the end of Section III, Chapter 14. As the US has been aptly described as using its constitutional structure of tripartite government to avoid full application of international treaties this is of real concern (see eg Andrea Bianchi, "International Law and US Courts: The Myth of Lohengrin Revisited" (2004) 15, 4 Eur J of Int'l Law 751). To see this recently played out one can look to the historic US Supreme Court *Hamdan* decision which timidly applied the Geneva Conventions through the Uniform Code of Military Justice and deftly avoided the question treated by the lower courts of whether the treaty is "self-executing" or "judicially enforceable" (*Hamdan v Rumsfeld*, 548 US 557 (2006) at 627–628).

21   First exposing "signature strikes" vs "personality strikes", see A Entous, S Gorman and J Barnes, "U.S. Tightens Drone Rules" *Wall Street Journal* (Nov 4, 2011). See also the oft cited J Becker and S Shane, "Secret 'Kill List' Proves a Test of Obama's Principles and Will" *New York Times* (May 29, 2012).

22   President Barack Obama, "U.S. Policy Standards and Procedures for the Use of Force in Counterterrorism Operations Outside the United States and Areas of Active

significantly circumscribe the practice, even if the language in the document was opaque.[23]

At that point in time an entire chapter on signature strikes was under consideration due to the immense elasticity inherent to such a practice as it raises quite troublesome questions of law, morals and effectiveness. To wit, striking individuals whose identity is unknown cannot possibly be the *targeted* killing so often described as the Obama administration's policy; and it clearly cannot fit within the pronounced standard of "[n]ear certainty that the terrorist target is present".[24] Nevertheless, due to the presumed abandonment of this policy, it was decided that dedicating a whole chapter to this topic could make this book out of date for discussions of current policy.

All the same, the critical question of the legitimacy of signature strikes was surely not left out of this volume as it is earnestly addressed in several different chapters: Chapter 2 by P. Grzebyk; Chapter 10 by M. Madej; and Chapter 12 by R. Kolb.[25] Additionally, the policy proposal of a Drone Court offers a constructive avenue for tackling this imprudent practice as demonstrated by the *New York Times* debate led by Guiora and Brand in the wake of the news of the accidental killing of two hostages with an armed drone in Pakistan.[26] As it turns out, the analysis and proposed solution found here are indeed timely and required because the supposed halt to signature strikes never occurred.

President Obama himself made the deadly error known in an expression of condolences to the families of the hostages,[27] and with it came the revelation that

---

Hostilities" known as the "Presidential Policy Guidance" (PPG hereinafter), The White House (May 22, 2013) at 2.

23   T Zakaria and M Hosenball, "U.S. drone guidelines could reduce 'signature strikes'" *Reuters* (May 23, 2013) available at <http://www.reuters.com/article/2013/05/24/us-usa-obama-speech-guidelines-idUSBRE94N03520130524>; G Dyer, "Obama rewrites rules for drone strikes in terror war" Financial Times (May 23, 2013) available at <http://www.ft.com/intl/cms/s/0/5d6c44c4-c3cb-11e2-8c30-00144feab7de.html#axzz3e5hBSly1>; Clearly demonstrating the extent to which the belief of cessation had taken hold, former counterterrorism czar Richard Clarke affirmed on ABC's *This Week* that "President Obama said he was going to stop [signature strikes], and clearly he didn't." For a discussion of the factual basis for this statement see L Carroll, "Did Obama pledge to stop using 'signature strikes'?" Pundit Fact, *Tampa Bay Times* (April 28, 2015) available at <http://www.politifact.com/punditfact/statements/2015/apr/28/richard-clarke/did-obama-pledge-stop-using-signature-strikes/> all accessed June 2015.

24   PPG (n 22) at 2. While one might argue that a militant structure can be a "terrorist target", this statement clearly refers to a specific person as the adjective "present" can only refer to a moving/moveable object.

25   At 62-66, 262-264, and 313-315 in this volume.

26   A Guiora, J Brand, M Hakimi, M O'Connell, S Vladeck, and N Rao, "Room for Debate: Should a Court Approve all Drone Strikes?" *New York Times* (April 24, 2015).

27   A Entous, D Paletta and F Schwartz, "American, Italian Hostages Killed in CIA Drone Strike in January" *Wall Street Journal* (April 23, 2015); P Lewis, S Ackerman and J

the CIA did not know who it was striking.[28] Of keen interest, as this information was released the name and a claim of high rank—Ahmed Farouq, "deputy emir of al Qaeda in the Indian Subcontinent"—was pinned on one militant killed.[29] Nonetheless, it is suggested that independent analysts and the public should be wary of such specifics. Details of identity and guilt bolster the rationale for targeting specific individuals, but this was simply not the case.[30]

Shots fired, presumably at the most suspicions characters or their assumed hideouts, just might kill the suspected terrorist actor that intelligence services had been warning about. However, the fact that such an action happens to neutralize a specific hidden target does not mean that this translates into legal, moral and effective (past or future) policy. The use of deadly force by a government should be constrained by valid limiting principles beforehand, not justified with cherry picked results of shots that auspiciously hit a previously unknown target.[31] The legitimacy of lethal action ultimately turns on the criteria in place and the evidence used. Yet, this incident reveals ever shifting standards and "the government categorically refuses to reveal the evidence on which specific signature strikes are based."[32]

Boone, "Obama regrets drone strike that killed hostages but hails US for transparency" *The Guardian* (London, April 23, 2015).

28   M Mazzetti and E Schmitt, "First Evidence of a Blunder: 2 Extra Bodies" *New York Times* (April 23, 2015); S Ackerman, S Siddiqui and P Lewis, "White House admits: we didn't know who drone strike was aiming to kill" *The Guardian* (London, April 23, 2015).

29   G Botelho and J Sciutto, "Ahmed Farouq: Leader of al Qaeda's Indian branch killed by US" CNN.com (April 23, 2015) available at <http://edition.cnn.com/2015/04/23/world/ahmed-farouq-al-qaeda/> accessed June 2015; see also Lewis, Ackerman and Boone (n 27) and ibid, Ackerman, Siddiqui and Lewis. At the time of this disclosure another name and claim of high level import was also released of another US citizen killed in a different drone strike in January—Adam Gadahn, or Azzam the American who was "long seen as an important Qaeda propagandist who, as a member of the terror network's media arm, As Sahab, played instrumental roles including translator, video producer and cultural interpreter": E Schmitt, "Adam Gadahn Was Propagandist for Al Qaeda Who Sold Terror in English" *New York Times* (April 23, 2015); see also all the above references in this note that mention Adam Gadahn. However, neither of these US citizens were actually targeted for drone killing.

30   The attack was said to target a compound based on "near-certain" intelligence that suggested it was being frequented by at least one Al-Qaeda leader: Lewis, Ackerman and Boone (n 27). While an Al-Qaeda compound might be a legal target, the evidence must support such a classification. The frequency of a visitor, no matter how certain, would not automatically render a structure a legal target—it would need to be "permanently dedicated to that purpose": K J Heller, "'One Hell of a Killing Machine': Signature Strikes and International Law" (2013) 11 J Int'l Criminal Justice 89, at 96. See also (n 24).

31   Also on this exact point, see the article detailing the fluky killing of a top leader in Yemen: G Miller, "CIA didn't know strike would hit al-Qaeda leader" *Washington Post* (June 17, 2015).

32   Heller (n 30) at 104.

It cannot be overlooked that there are absolutely no available figures on the number of signature strikes that have taken place. However, alongside this fundamental dearth of data there are a couple of general indicators. We have the reporting in the initial disclosure of this practice in 2011: "The bulk of CIA's drone strikes are signature strikes."[33] In addition, multiple non-government reports citing unnamed officials have led one scholar to conclude: "The vast majority of drone strikes conducted by the CIA have been signature strikes." Apart from this, next to nothing is known about the frequency or quantity of signature strikes.[34] Yet despite the announced policy shift, we have learned that the PPG of 2013 never actually applied to Pakistan,[35] and that Yemen has fallen into a similar category.[36] Considering that the overwhelming majority of armed drone operations outside the territory of an armed conflict have been carried out in these two countries,[37] it is quite problematic that the controls on deadly force continue to be misrepresented and are subject to change where it matters most.

Additionally, the "capture is not feasible" standard, often stated and explicitly reiterated in the same 2013 PPG document, has also been starkly brought into question with the indictment of US citizen Mohanad Mahmoud al Farekh on April 2, 2015. According to unnamed sources within the government, both the Pentagon and the CIA argued in 2013 that al Farekh met all the criteria for a drone killing, clearly indicating that his capture was not believed to be possible. The Justice Department was not convinced and fought to keep him off the list. This arguably and appropriately saved his life since his capture was indeed proven to be achievable two years later, along with the notable fact that he was indicted for a crime that would bring a maximum 15 year prison sentence.[38]

There are a host of important questions raised by this case. For example, what were the parameters used by the Pentagon and CIA to arrive at the conclusion that al Farekh could not be captured and why did it differ from the Justice Department? Did the existence of a new technology to kill from a distance influence their assessment? On what grounds did the Justice Department argue that a US citizenship required further assessment—i.e. what and why is there a difference in standards between citizens and non-citizens? As it was reported that members of

---

33   Entous, Gorman and Barnes (n 21).

34   See Heller (n 30): at 90.

35   A Entous "Obama Kept Looser Rules for Drones in Pakistan: Waived requirement to show proposed targets pose imminent threat to the US" *Wall Street Journal* (April 26, 2015) at A1.

36   For a discussion of the shelving of higher standards in Yemen see Miller (n 31).

37   See Tables 10.1 and 10.2.

38   M Mazzetti and E Schmitt, "Terrorism Case Renews Debate Over Drone Hits" *New York Times* (April 12, 2015) at A1.

Congress took part, what is the legal authority for US lawmakers to be involved in discussions over the execution/killing of individuals?[39]

As important as these questions are, the essential point for us here is that the pivotal feasibility of capture standard is by no means settled or jointly understood. In essence, both of these cases underscore the fact that the standards for drone killing are shuffling with the circumstances and subject to different interpretations. This is a real problem for the legitimate exercise of cross-border lethal force that demands a solution.

Not coincidentally, the proposal found in Chapter 13 of this volume aims to restrain this precise type of instability when it comes to legal and moral standards, not to mention producing data that can be measured for efficacy. Of fundamental importance here, the months and years of debate and surveillance clarify the fact that there is undoubtedly sufficient time for a court to intervene and interpret the applicable law—in this case, there is ample opportunity to set fixed and sound precedents on signature strikes and the feasibility of capture.

Undermining Legitimacy Abroad

However, bolstering legitimacy at home certainly does not automatically translate into the same outcome abroad. The conspicuous reality is that the governments ostensibly in charge where these remotely controlled aircraft are operating have had enormous trouble explaining their position on the drone strikes publicly, and even maintaining their status as the legitimate governing authority. Specifically, the governments in Pakistan and Yemen have had obvious, and sometimes dreadful, difficulties in this regard.

During the ramping up of the Obama administration's drone program,[40] the government of Pakistan made public denials of their consent for the US to exercise this lethal force in its territory.[41] However, there have also been credible reports that Pakistan has given secret consent for these UCAV strikes within its territory,[42]

---

39   See David Cole "Targeted Killing: The New Questions" *The New York Review of Books Blog* (April 14, 2015) available at <http://www.nybooks.com/blogs/nyrblog/2015/apr/14/targeted-killing-new-questions/> accessed June 2015.

40   See Table 10.1.

41   See eg Ben Emmerson "Statement of the Special Rapporteur Following Meetings in Pakistan," United Nations Human Rights, Office for the High Commissioner of Human Rights, Islamabad (March 14, 2013) available at <http://www.ohchr.org/en/NewsEvents/Pages/DisplayNews.aspx?NewsID=13148&LangID=E> accessed Dec 2014: "The position of the Government of Pakistan is quite clear. It does not consent to the use of drones by the United States on its territory and it considers this to be a violation of Pakistan's sovereignty and territorial integrity."

42   See eg Greg Miller and Bob Woodward, "Secret Memos Reveal Explicit Nature of US, Pakistan Agreement on Drones" Washington Post (Oct 24, 2013) available at <http://www.washingtonpost.com/world/national-security/top-pakistani-leaders-secretly-backed-

and at times in exchange for having enemies of the Pakistani government added to the list of those targeted by US drones.[43] To be sure, if proper permission exists (even in secret), the international legality of the use of force flips from heads to tails; what was thought to be illegal, is legal.[44] Hence, this is not a point of lawfulness. Nonetheless, there are other extremely important consequences that can be found.

As presented in the Introduction to this volume, German sociologist and political economist Max Weber put forward a definition of the state that has become pivotal to Western political thought. Weber presented the defining concept of the state as, "a human community that (successfully) claims the *monopoly of the legitimate use of physical force* within a given territory."[45] While secret consent to allow lethal force can give assurances that the ruling government will not retaliate for such force, it does not secure any guarantees from the population that they will continue to deem the ruling government as legitimate. Relinquishing the monopoly on force would certainly imply, according to Weber's definition, that the regime in office is not fulfilling its primary duty; viz it is outsourcing justice.

One indication of this problem can be found in the protests that shut down a NATO supply route through Pakistan into landlocked Afghanistan from late 2013 into 2014.[46] Thousands of people were mobilized by Pakistani politician and former cricket star Imran Khan who tapped into the deep resentment of some of the local population who see the armed drone strikes as a counterproductive strategy and a violation of sovereignty. The three-month blockade was brought to an end after a court ruled it to be illegal, even if Khan claimed victory for

---

cia-dronecampaign-secret-documents-show/2013/10/23/15e6b0d8–3beb-11e3-b6a9-da62c264f40e_story.html> accessed March 2015.

43   Mark Mazzetti, *The Way of the Knife: The CIA, a Secret Army, and a War at the Ends of the Earth* (Penguin Press 2013) at 103–113.

44   In this volume see Chapter 1 by Christian J. Tams and James G. Devaney, and Chapter 12 by Robert Kolb. Discussions over this question of consent with these specialists of international law revealed genuine concern on this issue. As consent can be offered secretly or tacitly, and establishing who is the proper authority for providing it, is not only extremely difficult but also of direct consequence, these authors suggest that this question requires substantial research in the future.

45   Max Weber, "Politics as a Vocation" in HH Gerth and C Wright Mills (eds. and trans.) *From Max Weber: Essays on Sociology* (OUP 1946) at 78 (original emphasis).

46   Tim Craig, "Imran Khan, a Pakistani provincial leader, complicates NATO plans for Afghanistan" Washington Post (Jan 4, 2014) available at <http://www.washingtonpost.com/world/asia_pacific/imran-khan-a-pakistani-provincial-leader-complicates-nato-plans-for-afghanistan/2014/01/04/fef921a4–68da-11e3-8b5b-a77187b716a3_story.html> accessed March 2015; Al Jazeera America, Wire Services, "Drone protesters in Pakistan block NATO supply route" Al Jazeera (Nov 23, 2013) available at <http://america.aljazeera.com/articles/2013/11/23/thousands-of-droneprotestersinpakistanblocknatosupplyroute.html> accessed March 2015.

having brought about change in the drone policy.[47] The essential point here is not about the disruption of an important supply route—even if this is of consequence. Instead, it is of concern to see a politician gain significant ground by vehemently opposing the exercise of foreign lethal force on Pakistani territory because it goes directly to the question of consent. If we are to take Weber's definition of the state seriously, extending consent to a foreign power to exercise lethal force at its own discretion inside of one's territory touches the essence of the state itself and brings the legitimacy of the commanding authority that does so into question.

Moreover, as these words are written in March of 2015, the internationally recognized president of Yemen, Abd-Rabbo Mansour Hadi, suffered an evaporation of his status as the legitimate authority in that territory when key military bases and commanders were captured by Houthis rebels and he was forced to flee to Saudi Arabia.[48] Particular events that signaled the significance of this breakdown was the arrest of his defense minister along with the capture of an airfield outside of the southern city of Aden on March 25, 2015[49]—one that was critical to US special forces' counterterrorism campaign as well as drone operations in that country.[50]

While the degree of legitimacy President Hadi possessed before these developments was already up for considerable debate, the territory of Yemen has now devolved into a space where a number of external actors try to exert their own influence (most notably Iran and Saudi Arabia). Hadi was a vocal supporter of the US drone strikes and had emphasized his own personal involvement.[51]

---

47   Salaman Masood, "Political Party Ends Blockade of NATO Supply Route in Pakistan" *New York Times* (Feb 27, 2014) available at <http://www.nytimes.com/2014/02/28/world/asia/political-party-ends-blockade-of-nato-supply-route-in-pakistan.html?_r=0> accessed March 2015. This is not the first time that the supply route had been closed; in 2011–12 the route was closed for seven months in reaction to a US air raid that mistakenly killed 24 Pakistani soldiers.

48   Hakim Almasmari, "Yemen's President Emerges in Saudi Arabia After Fleeing Country" *The Wall Street Journal* (March 26, 2015) available at <http://www.wsj.com/articles/five-gulf-states-and-egypt-team-up-in-yemen-action-1427361004?tesla=y> accessed March 2015.

49   Frank Gardner, BBC Security correspondent, "Yemen Crisis: President Hadi Flees as Houthi Rebels Advance" *BBC News.com* (March 25, 2015) available at <http://www.bbc.com/news/world-middle-east-32048604> accessed March 2015.

50   See Eric Schmitt, "Out of Yemen, US Is Hobbled in Terror Fight" *New York Times* (March 22, 2015) available at http://www.nytimes.com/2015/03/23/us/politics/out-of-yemen-us-is-hobbled-in-terror-fight.html?_r=0> accessed March 2015; Jackie Northam, "US Confirms it is Supporting Saudi Military Operations in Yemen" *NPR news—The Two-Way, Breaking News* (March 25, 2015) available at <http://www.npr.org/blogs/thetwo-way/2015/03/25/395294977/yemens-president-flees-palace-as-rebels-continue-advance> accessed March 2015.

51   See Aiden Warren and Ingvild Bode, *Governing the Use-of-Force in International Relations: The Post 9/11 US Challenge on International Law* (Palgrave Macmillan 2014) at 123.

Before him, Ali Abdullah Saleh, the ruling president for 33 years, attempted to deny US responsibility for the UCAV airstrikes and even claimed that some detonations were in fact conducted by the Yemeni military or other local actors.[52] Yet Saleh was forced to step down by popular pressure in 2011 and we can consequently say that this represents the downfall of two ruling authorities—a collapse of legitimacy—in the territory of Yemen in just four years during the drone operations.[53]

Granted, this does pose a chicken or egg problem; these strikes often take place in countries where the government is already weak and it does not maintain full control over its territory.[54] Nevertheless, exercising deadly force within such territory can certainly not help this problem. It is suggested that we keep an eye on the strength of those governing powers in the places where drone strikes are taking place as one way to gauge soundness of counterterrorism operations with armed drones. There are veritable challenges presented to the legitimacy of local governments when lethal force is exercised within their territory without any public process explaining the justification, or at times without even any recognition of who is doing the killing. Hence, this is a recommended place where our gaze can be affixed to assess a less often noted implication of cross-border drone violence—an undermining of legitimacy. Stable and legitimate governments might not always directly contribute to increasing security by monitoring threats from violent individuals and groups, but they surely are more reliable than the instability of multiple actors vying for control of the territory.

I thus close with a few questions we might keep in mind as debates over armed drones develop in the coming years:

- How do we judge the legitimacy of the governments in Pakistan, Yemen, and Somalia where we know that drone strikes have regularly occurred?
- Has the uncoerced pull toward compliance with the ruling government been strengthened or weakened by drone strikes?
- What does it mean to security interests to have a ruling government collapse and leave no monopoly of power within a territory?

The existence of such difficult questions and many gray areas concerning the codified law and recognized norms makes this volume and its contributions especially timely. Our conversations, both international and domestic, must be informed and intelligent as drone technology rapidly proliferates and further strikes from untold countries risk being justified on the arguments thus far put

---

52   Ibid.

53   See Table 10.2.

54   For an investigation of this problem see eg Ashley Deeks, "'Unwilling or Unable': Toward a Normative Framework for Extraterritorial Self-Defense" (2012) 52(3) Virginia J of Int'l Law 483.

forward. If we accept the claim that UCAVs "are here to stay,"[55] then fleshing out their legal, moral, and effective use is an imperative for maintaining legitimacy in the societies that deploy them and within the territories where they strike.

---

55 Christof Heyns, Special Rapporteur on Extrajudicial, Summary or Arbitrary Executions, *ReportA/68/382* (Sept 13, 2013) at 4.

Index

References in **bold** are for tables and figures are shown in *italics*.